Practical Time Series Analysis
Prediction with Statistics and Machine Learning

Aileen Nielsen

Beijing · Boston · Farnham · Sebastopol · Tokyo

Practical Time Series Analysis

by Aileen Nielsen

Copyright © 2020 Aileen Nielsen. All rights reserved.

Published by O'Reilly Media, Inc., 1005 Gravenstein Highway North, Sebastopol, CA 95472.

O'Reilly books may be purchased for educational, business, or sales promotional use. Online editions are also available for most titles (*http://oreilly.com*). For more information, contact our corporate/institutional sales department: 800-998-9938 or *corporate@oreilly.com*.

Development Editor: Jeff Bleiel

Acquisitions Editor: Jonathan Hassell

Production Editor: Katherine Tozer

Copyeditor: Rachel Monaghan

Proofreader: Sonia Saruba

Indexer: WordCo Indexing Services, Inc.

Interior Designer: David Futato

Cover Designer: Karen Montgomery

Illustrator: Rebecca Demarest

October 2019:   First Edition

Revision History for the First Edition

2019-09-20:   First Release
2019-12-13:   Second Release

See *http://oreilly.com/catalog/errata.csp?isbn=9781492041658* for release details.

978-1-492-04165-8

[LSI]

Table of Contents

Preface

Weather, stock markets, and heartbeats. They all form time series. If you're interested in diverse data and forecasting the future, you're interested in time series analysis.

Welcome to *Practical Time Series Analysis*! If you picked this book up, you've probably already noticed that time series data is everywhere. Time series data grows increasingly ubiquitous and important as the big data ecosystem expands. For better or worse, sensors and tracking mechanisms are everywhere, and as a result there are unprecedented amounts of high-quality time series data available. Time series are uniquely interesting because they can address questions of causality, trends, and the likelihood of future outcomes. This book will take you through the major techniques that are most commonly applied to time series to address such questions.

Time series data spans a wide range of disciplines and use cases. It can be anything from customer purchase histories to conductance measurements of a nano-electronic system to digital recordings of human language. One point we discuss throughout the book is that time series analysis applies to a surprisingly diverse set of data. Any data that has an ordered axis can be analyzed with time series methods, even if that ordered axis is not time per se. Traditional time series data, such as stock data and weather patterns, can be analyzed with time series methods, but so can quirky data sets such as spectrographs of wine, where the "time" axis is actually an axis of frequency. Time series are everywhere.

Who Should Read This Book

There are two kinds of intended readers for this book. The first and larger category of reader is that of a data scientist who has rarely worked with time series data. This person could be an industry veteran or a junior analyst. The more experienced data analyst can skim the introductory conceptual areas of each chapter but will still benefit from this book's discussions about best practices as well as pitfalls of working with time series data. A newer data analyst might consider working through the book in its entirety, although I have tried to keep each topic as self-contained as possible.

The second category of reader is someone supervising analytics at an organization with an extensive in-house data collection. If you are a member of this group, you will still need some technical background, but it's not necessary that you be currently coding in your professional life. For such a reader, this book is useful to point out opportunities for your organization to use time series analysis even if it is not currently practiced in-house. This book will point you to new kinds of questions and analyses your organization can address with your existing data resources.

Expected Background

With respect to coding, you should have some familiarity with R and Python, especially with certain fundamental packages (in Python: NumPy, Pandas, and scikit-learn; and in R: data.table). The code samples should be readable even without all the background, but in that case you may need to take a short detour to familiarize yourself with these packages. This is most likely the case with respect to R's data.table, an underused but highly performant data frame package that has fantastic time functionality.

In all cases, I have provided brief overviews of the related packages, some example code, and descriptions of what the code does. I also point the reader toward more complete overviews of the most used packages.

With respect to statistics and machine learning, you should have some familiarity with:

Introductory statistics
    Ideas such as variance, correlation, and probability distributions

Machine learning
    Clustering and decision trees

Neural networks
    What they are and how they are trained

For these cases, I provide a brief overview of such concepts within the text, but the uninitiated should read more deeply about them before continuing with some chapters. For most topics, I provide links to recommended free online resources for brief tutorials on the fundamentals of a given topic or technique.

Why I Wrote This Book

I wrote this book for three reasons.

First, time series is an important aspect of data analysis but one that is not found in the standard data science toolkit. This is unfortunate both because time series data is increasingly available and also because it answers questions that cross-sectional data cannot. An analyst who does not know fundamental time series analysis is not making the most of their data. I hoped that this book could fill an existing and important void.

Second, when I started writing this book, I knew of no centralized overview of the most important aspects of time series analysis from a modern data science perspective. There are many excellent resources available for traditional time series analysis, most notably in the form of classic textbooks on statistical time series analysis. There are also many excellent individual blog posts on both traditional statistical methods and on machine learning or neural network approaches to time series. However, I could not identify a single centralized resource to outline all these topics and relate them to one another. The goal of this book is to provide that resource: a broad, modern, and practical overview of time series analysis covering the full pipeline for time series data and modeling. Again, I hoped that this book could fill an existing and important void.

Third, time series is an interesting topic with quirky data concerns. Problems associated with data leakage, lookahead, and causality are particularly fun from a time series perspective, as are many techniques that apply uniquely to data ordered along some kind of time axis. Surveying these topics broadly and finding a way to catalog them was another motivation to write this book.

Navigating This Book

This book is organized roughly as follows:

History
> Chapter 1 presents a history of time series forecasting, all the way from the ancient Greeks to modern times. This puts our learning in context so we can understand what a traditional and storied discipline we study in this book.

All about the data
> Chapters 2, 3, 4, and 5 tackle problems relating to obtaining, cleaning, simulating, and storing time series data. These chapters are tied together by their concern for all that comes before you can actually perform time series analysis. Such topics are rarely discussed in existing resources but are important in most data pipelines. Data identification and cleaning represent a large portion of the work done by most time series analysts.

Models, models, models
> Chapters 6, 7, 8, 9, and 10 cover a variety of modeling techniques that can be used for time series analysis. We start with two chapters on statistical methods, covering standard statistical models, such as ARIMA and Bayesian state space models. We then apply more recently developed methods, such as machine learning and neural network, to time series data, highlighting the challenges of

data processing and data layout when time series data is used for fitting models that are not inherently time aware, such as decision trees.

Post-modeling considerations

Chapters 11 and 12 cover accuracy metrics and performance considerations, respectively, to provide some guidance as to what you should consider after you have performed your first iterations of time series modeling.

Real-world use cases

Chapters 13, 14, and 15 provide case studies from healthcare, finance, and government data, respectively.

Comments on recent happenings

Chapters 16 and 17 briefly cover recent time series developments and predictions for the future. Chapter 16 is an overview of a variety of automated time series packages, some open source and developed as an academic effort, and some coming from large tech companies. These tools are under active development as efforts ramp up to improve time series forecasting at scale via automated processes. Chapter 17 discusses some predictions for the future of time series analysis as the big data ecosystem grows and as we learn more about how big data can assist time series analysis.

In general, I recommend reading through a chapter before attempting to work with the code. There are usually a few new concepts introduced in each chapter, and it can be helpful to give these some attention before turning to your keyboard. Also, in most cases the code to execute certain models is relatively simple so that the conceptual understanding will be your main skill gained, with knowledge of the APIs of important packages being a secondary benefit that will come much more easily if you pay attention to the concepts.

Also, the book is written so that it is most sensibly read from beginning to end (with later chapters referring to concepts covered earlier in the book), but, again, I purposely kept the chapters as self-contained as possible to enable more experienced readers to jump around as convenient.

Online Resources

The GitHub repository for this book (*https://oreil.ly/time-series-repo*) includes much of the code discussed in this book in a form where you can run it on data sets, which are also provided in the repository. In some cases, variable names, formatting, and such will not be identical to that produced in the book, but it should be easily linked (for example, in some cases variable names were shortened in the book due to formatting constraints).

Additionally, should you enjoy video presentations, I have two online tutorials covering some of the content of this book with a Python focus. If you want to complement this book with a video presentation, consider the following resources:

- Time Series Analysis (*https://youtu.be/JNfxr4BQrLk*) (SciPy 2016)
- Modern Time Series Analysis (*http://bit.ly/32YnPht*) (SciPy 2019)

Conventions Used in This Book

The following typographical conventions are used in this book:

Italic
: Indicates new terms, URLs, email addresses, filenames, and file extensions.

`Constant width`
: Used for program listings, as well as within paragraphs to refer to program elements such as variable or function names, databases, data types, environment variables, statements, and keywords.

`Constant width bold`
: Shows commands or other text that should be typed literally by the user.

`Constant width italic`
: Shows text that should be replaced with user-supplied values or by values determined by context.

 This element signifies a tip or suggestion.

 This element signifies a general note.

 This element indicates a warning or caution.

Using Code Examples

Supplemental material (code examples, exercises, etc.) is available for download at *https://github.com/PracticalTimeSeriesAnalysis/BookRepo*.

This book is here to help you get your job done. In general, if example code is offered with this book, you may use it in your programs and documentation. You do not need to contact us for permission unless you're reproducing a significant portion of the code. For example, writing a program that uses several chunks of code from this book does not require permission. Selling or distributing a CD-ROM of examples from O'Reilly books does require permission. Answering a question by citing this book and quoting example code does not require permission. Incorporating a significant amount of example code from this book into your product's documentation does require permission.

We appreciate, but do not require, attribution. An attribution usually includes the title, author, publisher, and ISBN. For example: "*Practical Time Series Analysis* by Aileen Nielsen (O'Reilly). Copyright 2020 Aileen Nielsen, 978-1-492-04165-8."

If you feel your use of code examples falls outside fair use or the permission given above, feel free to contact us at *permissions@oreilly.com*.

O'Reilly Online Learning

 For almost 40 years, *O'Reilly Media* has provided technology and business training, knowledge, and insight to help companies succeed.

Our unique network of experts and innovators share their knowledge and expertise through books, articles, conferences, and our online learning platform. O'Reilly's online learning platform gives you on-demand access to live training courses, in-depth learning paths, interactive coding environments, and a vast collection of text and video from O'Reilly and 200+ other publishers. For more information, please visit *http://oreilly.com*.

How to Contact Us

Please address comments and questions concerning this book to the publisher:

O'Reilly Media, Inc.
1005 Gravenstein Highway North
Sebastopol, CA 95472
800-998-9938 (in the United States or Canada)

707-829-0515 (international or local)
707-829-0104 (fax)

We have a web page for this book, where we list errata, examples, and any additional information. You can access this page at *https://oreil.ly/practical-time-series-analysis*.

Email *bookquestions@oreilly.com* to comment or ask technical questions about this book.

For more information about our books, courses, conferences, and news, see our website at *http://www.oreilly.com*.

Find us on Facebook: *http://facebook.com/oreilly*

Follow us on Twitter: *http://twitter.com/oreillymedia*

Watch us on YouTube: *http://www.youtube.com/oreillymedia*

Acknowledgments

Thank you to the two technical reviewers for this book, Professors Rob Hyndman and David Stoffer. Both were exceptionally gracious in allotting time to review this book and providing copious and helpful feedback and new ideas for what to cover. The book is substantially better than it would have been without their input. I particularly thank Rob for pointing out missed opportunities in the original draft to highlight alternative methodologies and many interesting sources of time series data. I particularly thank David for offering skepticism regarding overly automated approaches to analysis and for pointing out when I was unduly optimistic about certain tools. Thank you for all your help, knowledge, and deeply experienced perspective on time series analysis.

I am grateful to my editor at O'Reilly, Jeff Bleiel, who has been most encouraging, supportive, and helpful in reviewing these many pages over the past year. I also owe many thanks to my production editor, Katie Tozer, who has been so patient with the process of cleaning up this book and guiding me through technical glitches in production. Thank you to Rachel Monaghan for careful and excellent copyediting. Thank you to Rebecca Demarest, who created many of the illustrations in this book and helped me clean up many untidy graphics. Thanks also go to Jonathan Hassell, who took this project and ran with it and somehow convinced O'Reilly to publish it. Finally, thank you to Nicole Tache, who first reached out to me a few years ago to work with O'Reilly. To everyone at O'Reilly, thank you for letting me undertake this project and for all your support throughout the process.

I am grateful to many readers who gave feedback or proofreading assistance along the way, including Wenfei Tong, Richard Krajunus, Zach Bogart, Gabe Fernando, Laura

Kennedy, Steven Finkelstein, Liana Hitts, and Jason Greenberg. Having their input was exceptionally helpful. Thank you for reading and for offering feedback.

I thank my mother (and role-model), Elizabeth, for lifelong love, support, and discipline. I thank my father, John, and Aunt Claire for all their love and support for my education over many years. Most importantly, I thank my husband, Ivan, and my son, Edmund Hillary, for their patience and enthusiasm in supporting the book and excusing the time I spent away from family while writing.

Any mistakes or bugs you find in this book are mine and mine alone. I welcome feedback at *aileen.a.nielsen@gmail.com*.

Time Series: An Overview and a Quick History

Time series data and its analysis are increasingly important due to the massive production of such data through, for example, the internet of things, the digitalization of healthcare, and the rise of smart cities. In the coming years we can expect the quantity, quality, and importance of time series data to grow rapidly.

As continuous monitoring and data collection become more common, the need for competent time series analysis with both statistical and machine learning techniques will increase. Indeed, the most promising new models combine both of these methodologies. For this reason, we will discuss each at length. We will study and use a broad range of time series techniques useful for analyzing and predicting human behavior, scientific phenomena, and private sector data, as all these areas offer rich arrays of time series data.

Let's start with a definition. *Time series analysis* is the endeavor of extracting meaningful summary and statistical information from points arranged in chronological order. It is done to diagnose past behavior as well as to predict future behavior. In this book we will use a variety of approaches, ranging from hundred-year-old statistical models to newly developed neural network architectures.

None of the techniques has developed in a vacuum or out of purely theoretical interest. Innovations in time series analysis result from new ways of collecting, recording, and visualizing data. Next we briefly discuss the emergence of time series analysis in a variety of applications.

The History of Time Series in Diverse Applications

Time series analysis often comes down to the question of causality: how did the past influence the future? At times, such questions (and their answers) are treated strictly within their discipline rather than as part of the general discipline of time series analysis. As a result, a variety of disciplines have contributed novel ways of thinking about time series data sets.

In this section we will survey a few historical examples of time series data and analysis in these disciplines:

- Medicine
- Weather
- Economics
- Astronomy

As we will see, the pace of development in these disciplines and the contributions originating in each field were strongly tied to the nature of the contemporaneous time series data available.

Medicine as a Time Series Problem

Medicine is a data-driven field that has contributed interesting time series analysis to human knowledge for a few centuries. Now, let's study a few examples of time series data sources in medicine and how they emerged over time.

Medicine got a surprisingly slow start to thinking about the mathematics of predicting the future, despite the fact that prognoses are an essential part of medical practice. This was the case for many reasons. Statistics and a probabilistic way of thinking about the world are recent phenomena, and these disciplines were not available for many centuries even as the practice of medicine developed. Also, most doctors practiced in isolation, without easy professional communication and without a formal recordkeeping infrastructure for patient or population health. Hence, even if physicians in earlier times had been trained as statistical thinkers, they likely wouldn't have had reasonable data from which to draw conclusions.

This is not at all to criticize early physicians but to explain why it is not too surprising that one of the early time series innovations in population health came from a seller of hats rather than from a physician. When you think about it, this makes sense: in earlier centuries, an urban hat seller would likely have had more practice in recordkeeping and the art of spotting trends than would a physician.

The innovator was John Graunt, a 17th-century London haberdasher. Graunt undertook a study of the death records that had been kept in London parishes since the early 1500s. In doing so, he originated the discipline of demography. In 1662, he published *Natural and Political Observations . . . Made upon the Bills of Mortality* (See Figure 1-1).

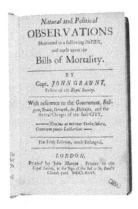

Figure 1-1. John Graunt's actuarial tables were one of the first results of time-series-style thinking applied to medical questions. Source: Wikipedia (https://perma.cc/ 2FHJ-67SB).

In this book, Graunt presented the first *life tables*, which you may know as *actuarial tables*. These tables show the probability that a person of a given age will die before their next birthday. Graunt, as the first person known to have formulated and published life tables, was also the first documented statistician of human health. His life tables looked something like Table 1-1, which is taken from some Rice University statistics course notes (*https://perma.cc/HU6A-9W22*).

Table 1-1. A sample of John Graunt's life tables

Age	Proportion of deaths in the interval	Proportion surviving until start of interval
0–6	0.36	1.0
7–16	0.24	0.64
17–26	0.15	0.40
27–36	0.09	0.25

Unfortunately, Graunt's way of thinking mathematically about human survival did not take. A more connected and data-driven world began to form—complete with nation states, accreditation, professional societies, scientific journals, and, much later, government-mandated health recordkeeping—but medicine continued to focus on physiology rather than statistics.

There were understandable reasons for this. First, the study of anatomy and physiology in small numbers of subjects had provided the major advances in medicine for centuries, and most humans (even scientists) hew to what works for them as long as possible. While a focus on physiology was so successful, there was no reason to look further afield. Second, there was very little reporting infrastructure in place for physicians to tabulate and share information on the scale that would make statistical methods superior to clinical observations.

Time series analysis has been even slower to come into mainstream medicine than other branches of statistics and data analysis, likely because time series analysis is more demanding of recordkeeping systems. Records must be linked together over time, and preferably collected at regular intervals. For this reason, time series as an epidemiological practice has only emerged very recently and incrementally, once sufficient governmental and scientific infrastructure was in place to ensure reasonably good and lengthy temporal records.

Likewise, individualized healthcare using time series analysis remains a young and challenging field because it can be quite difficult to create data sets that are consistent over time. Even for small case-study-based research, maintaining both contact with and participation from a group of individuals is excruciatingly difficult and expensive. When such studies are conducted for long periods of time, they tend to become canonical in their fields—and repeatedly, or even excessively researched—because their data can address important questions despite the challenges of funding and management.[1]

Medical instruments

Time series analysis for individual patients has a far earlier and more successful history than that of population-level health studies. Time series analysis made its way into medicine when the first practical electrocardiograms (ECGs), which can diagnose cardiac conditions by recording the electrical signals passing through the heart, were invented in 1901 (see Figure 1-2). Another time series machine, the electroencephalogram (EEG), which noninvasively measures electrical impulses in the brain, was introduced into medicine in 1924, creating more opportunities for medical practitioners to apply time series analysis to medical diagnosis (see Figure 1-3).

Both of these time series machines were part of a larger trend of enhancing medicine with repurposed ideas and technologies coming out of the second Industrial Revolution.

---

1 Examples include the British Doctors Study and the Nurses' Health Study.

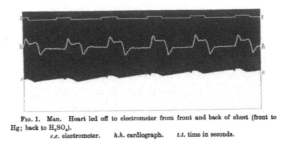

FIG. 1. Man. Heart led off to electrometer from front and back of chest (front to Hg; back to H₂SO₄).
e.e. electrometer.   h.h. cardiograph.   t.t. time in seconds.

Figure 1-2. An early ECG recording from the original 1877 paper by Augustus D. Waller, M.D, "A Demonstration on Man of Electromotive Changes Accompanying the Heart's Beat" (https://perma.cc/ZGB8-3C95). The earliest ECGs were difficult to construct and use, so it was a few more decades before they became a practical tool for physicians.

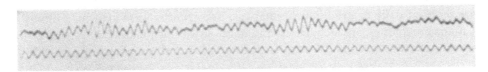

Figure 1-3. The first human EEG recording, from 1924. Source: Wikipedia (https://oreil.ly/P_M4U).

ECG and EEG time series classification tools remain active areas of research for very practical purposes, such as estimating the risk of a sudden cardiac crisis or a seizure. These measurements are rich sources of data, but one "problem" with such data is that it tends to be available only for patients with specific ailments. These machines do not generate long-range time series that can tell us more broadly about human health and behavior, as their measurements are seldom applied for long periods of time or before a disease has emerged in a patient.

Luckily, from a data analysis point of view, we are moving past the era where ECGs and the like are the dominant medical time series available. With the advent of wearable sensors and "smart" electronic medical devices, many healthy humans take routine measurements automatically or with minimal manual input, leading to the ongoing collection of good longitudinal data about both sick and healthy people. This is in stark contrast to the last century's medical time series data, which was almost exclusively measured on sick people and which was very limited in access.

As recent news coverage has shown, a variety of nontraditional players are entering the medical field, ranging from enormous social media companies to financial institutions to retail giants.[2] They likely all plan to use large data sets to streamline healthcare. There aren't just new players in the healthcare field—there are also new techniques. The personalized DNA-driven medicine means that time series *data* is increasingly measured and valued. Thanks to burgeoning modern healthcare data sets, both healthcare and time series analysis will likely evolve in the coming years, particularly in response to the lucrative data sets of the healthcare sector. Hopefully this will happen in such a way that time series can benefit everyone.

Forecasting Weather

For obvious reasons, predicting the weather has long been a preoccupation to many. The ancient Greek philosopher Aristotle delved into weather with an entire treatise (*Meteorology*), and his ideas about the causes and sequencing of the weather remained dominant until the Renaissance. At that time, scientists began to collect weather-related data with the help of newly invented instruments, such as the barometer, to measure the state of the atmosphere. They used these instruments to record time series at daily or even hourly intervals. The records were kept in a variety of locations, including private diaries and local town logbooks. For centuries this remained the only way that Western civilization tracked the weather.

Greater formalization and infrastructure for weather recording arrived in the 1850s when Robert FitzRoy was appointed the head of a new British government department to record and publish weather-related data for sailors.[3] FitzRoy coined the term *weather forecast*. At the time, he was criticized for the quality of his forecasts, but he is now regarded to have been well ahead of his time in the science he used to develop them. He established the custom of printing weather forecasts in the newspaper; they were the first forecasts printed in *The Times* of London. FitzRoy is now celebrated as the "father of forecasting."

In the late 19th century—hundreds of years after many atmospheric measurements had come into use—the telegraph allowed for fast compilations of atmospheric conditions in time series from many different locations. This practice became standard in many parts of the world by the 1870s and led to the creation of the first meaningful data sets for predicting local weather based on what was happening in other geographic locations.

---

2 See, for example, Darrell Etherington, Amazon, JPMorgan and Berkshire Hathaway to Build Their Own Healthcare Company," TechCrunch, January 30, 2018, https://perma.cc/S789-EQGW; Christina Farr, Facebook Sent a Doctor on a Secret Mission to Ask Hospitals to Share Patient Data," CNBC, April 5, 2018, https://perma.cc/65GF-M2SJ.

3 This same Robert FitzRoy was captain of the HMS *Beagle* during the voyage that took Charles Darwin around the world. This voyage was instrumental in providing evidence to Darwin for the theory of evolution by natural selection.

By the turn of the 20th century, the idea of forecasting the weather with computational methods was vigorously pursued with the help of these compiled data sets. Early endeavors at computing the weather required a spectacular amount of effort but gave poor results. While physicists and chemists had well-proven ideas about the relevant natural laws, there were too many natural laws to apply all at once. The resulting system of equations was so complex that it was a notable scientific breakthrough the first time someone even attempted to do the calculations.

Several decades of research followed to simplify the physical equations in a way that increased accuracy and computational efficiency. These tricks of the trade have been handed down even to current weather prediction models, which operate on a mix of known physical principles and proven heuristics.

Nowadays many governments make highly granular weather measurements from hundreds or even thousands of weather stations around the world, and these predictions are grounded in data with precise information about weather station locations and equipment. The roots of these efforts trace back to the coordinated data sets of the 1870s and even earlier to the Renaissance practice of keeping local weather diaries.

Unfortunately, weather forecasting is an example of the increasing attacks on science that reach even into the domain of time series forecasting. Not only have time series debates about global temperatures been politicized, but so have more mundane time series forecasting tasks, such as predicting the path of a hurricane (*https://perma.cc/ D9GG-FND2*).

Forecasting Economic Growth

Indicators of production and efficiency in markets have long provided interesting data to study from a time series analysis. Most interesting and urgent has been the question of forecasting future economic states based on the past. Such forecasts aren't merely useful for making money—they also help promote prosperity and avert social catastrophes. Let's discuss some important developments in the history of economic forecasting.

Economic forecasting grew out of the anxiety triggered by episodic banking crises in the United States and Europe in the late 19th and early 20th centuries. At that time, entrepreneurs and researchers alike drew inspiration from the idea that the economy could be likened to a cyclical system, just as the weather was thought to behave. With the right measurements, it was thought, predictions could be made and crashes averted.

Even the language of early economic forecasting mirrored the language of weather forecasting. This was unintentionally apt. In the early 20th century, economic and weather forecasting were indeed alike: both were pretty terrible. But economists' aspirations created an environment in which progress could at least be hoped for, and so a variety of public and private institutions were formed for tracking economic data. Early economic forecasting efforts led to the creation of economic indicators and tabulated, publicly available histories of those indicators that are still in use today. We will even use some of these in this book.

Nowadays, the United States and most other nations have thousands of government researchers and recordkeepers whose jobs are to record data as accurately as possible and make it available to the public (see Figure 1-4). This practice has proven invaluable to economic growth and the avoidance of economic catastrophe and painful boom and bust cycles. What's more, businesses benefit from a data-rich atmosphere, as these public data sets permit transportation providers, manufacturers, small business owners, and even farmers to anticipate likely future market conditions. This all grew out of the attempt to identify "business cycles" that were thought to be the causes of cyclical banking failures, an early form of time series analysis in economics.

BUSINESS CYCLE REFERENCE DATES		DURATION IN MONTHS		
Peak	Trough	Contraction	Expansion	Cycle
Quarterly dates are in parentheses		Peak to Trough	Previous trough to this peak	Trough from Previous Trough / Peak from Previous Peak
	December 1854 (IV)	--	--	--	--
June 1857(II)	December 1858 (IV)	18	30	48	--
October 1860(III)	June 1861 (III)	8	22	30	40
April 1865(I)	December 1867 (I)	32	46	78	54
June 1869(II)	December 1870 (IV)	18	18	36	50
October 1873(III)	March 1879 (I)	65	34	99	52
March 1882(I)	May 1885 (II)	38	36	74	101
March 1887(II)	April 1888 (I)	13	22	35	60
July 1890(III)	May 1891 (II)	10	27	37	40
January 1893(I)	June 1894 (II)	17	20	37	30
December 1895(IV)	June 1897 (II)	18	18	36	35
June 1899(III)	December 1900 (IV)	18	24	42	42
September 1902(IV)	August 1904 (III)	23	21	44	39

Figure 1-4. The US federal government funds many government agencies and related nonprofits that record vital statistics as well as formulate economic indicators. Source: National Bureau of Economic Research (https://www.nber.org/cycles/cyclesmain.html).

Much of the economic data collected by the government, particularly the most newsworthy, tends to be a proxy for the population's overall economic well-being. One example of such vital information comes from the number of people requesting unemployment benefits. Examples include the government's estimates of the gross domestic product and of the total tax returns received in a given year.

Thanks to this desire for economic forecasting, the government has become a curator of data as well as a collector of taxes. The collection of this data enabled modern economics, the modern finance industry, and data science generally to blossom. Thanks to time series analysis growing out of economic questions, we now safely avert many more banking and financial crises than any government could have in past centuries. Also, hundreds of time series textbooks have been written in the form of economics textbooks devoted to understanding the rhythms of these financial indicators.

Trading markets

Let's get back to the historical side of things. As government efforts at data collection met with great success, private organizations began to copy government recordkeeping. Over time, commodities and stock exchanges became increasingly technical. Financial almanacs became popular, too. This happened both because market participants became more sophisticated and because emerging technologies enabled greater automation and new ways of competing and thinking about prices.

All this minute recordkeeping gave rise to the pursuit of making money off the markets via math rather than intuition, in a way driven entirely by statistics (and, more recently, by machine learning). Early pioneers did this mathematical work by hand, whereas current "quants" do this by very complicated and proprietary time series analytic methods.

One of the pioneers of *mechanical trading*, or time series forecasting via algorithm, was Richard Dennis. Dennis was a self-made millionaire who famously turned ordinary people, called the Turtles, into star traders by teaching them a few select rules about how and when to trade. These rules were developed in the 1970s and 1980s and mirrored the "AI" thinking of the 1980s, in which heuristics still strongly ruled the paradigm of how to build intelligent machines to work in the real world.

Since then many "mechanical" traders have adapted these rules, which as a result have become less profitable in a crowded automated market. Mechanical traders continue to grow in number and wealth, they are continually in search of the next best thing because there is so much competition.

Astronomy

Astronomy has always relied heavily on plotting objects, trajectories, and measurements over time. For this reason, astronomers are masters of time series, both for calibrating instruments and for studying their objects of interest. As an example of the long history of time series data, consider that sunspot time series were recorded in ancient China as early as 800 BC, making sunspot data collection one of the most well-recorded natural phenomena ever.

Some of the most exciting astronomy of the past century relates to time series analysis. The discovery of variable stars (which can be used to deduce galactic distances) and the observation of transitory events such as supernovae (which enhance our understanding of how the universe changes over time) are the result of monitoring live streams of time series data based on the wavelengths and intensities of light. Time series have had a fundamental impact on what we can know and measure about the universe.

Incidentally, this monitoring of astronomical images has even allowed astronomers to catch events as they are happening (*https://perma.cc/2TNK-2TFW*) (or rather as we are able to observe them, which may take millions of years).

In the last few decades, the availability of explicitly timestamped data, as formal time series, has exploded in astronomy with a wide array of new kinds of telescopes collecting all sorts of celestial data. Some astronomers have even referred to a time series "data deluge."

Time Series Analysis Takes Off

George Box, a pioneering statistician who helped develop a popular time series model, was a great pragmatist. He famously said, "All models are wrong, but some are useful."

Box made this statement in response to a common attitude that proper time series modeling was a matter of finding the best model to fit the data. As he explained, the idea that any model can describe the real world is very unlikely. Box made this pronouncement in 1978, which seems bizarrely late into the history of a field as important as time series analysis, but in fact the formal discipline was surprisingly young.

For example, one of the achievements that made George Box famous, the Box-Jenkins method—considered a fundamental contribution to time series analysis—appeared only in 1970.[4] Interestingly, this method first appeared not in an academic journal but rather in a statistics textbook, *Time Series Analysis: Forecasting and Control* (Wiley). Incidentally this textbook remains popular and is now in its fifth edition.

The original Box-Jenkins model was applied to a data set of carbon dioxide levels emitted from a gas furnace. While there is nothing quaint about a gas furnace, the 300-point data set that was used to demonstrate the method does feel somewhat outmoded. Certainly, larger data sets were available in the 1970s, but remember that they were exceptionally difficult to work with then. This was a time that predated

---

4 The Box-Jenkins method has become a canonical technique for choosing the best parameters for an ARMA or ARIMA model to model a time series. More on this in Chapter 6.

conveniences such as R, Python, and even C++. Researchers had good reasons to focus on small data sets and methods that minimized computing resources.

Time series analysis and forecasting developed as computers did, with larger data sets and easier coding tools paving the way for more experimentation and the ability to answer more interesting questions. Professor Rob Hyndman's history of forecasting competitions (*https://perma.cc/32LJ-RFJW*) provides apt examples of how time series forecasting competitions developed at a rate parallel to that of computers.

Professor Hyndman places the "earliest non-trivial study of time series forecast accuracy" as occurring in a 1969 doctoral dissertation at the University of Nottingham, just a year before the publication of the Box-Jenkins method. That first effort was soon followed by organized time series forecasting competitions, the earliest ones featuring around 100 data sets in the early 1970s.[5] Not bad, but surely something that could be done by hand if absolutely necessary.

By the end of the 1970s, researchers had put together a competition with around 1,000 data sets, an impressive scaling up. Incidentally, this era was also marked by the first commercial microprocessor, the development of floppy disks, Apple's early personal computers, and the computer language Pascal. It's likely some of these innovations were helpful. A time series forecasting competition of the late 1990s included 3,000 data sets. While these collections of data sets were substantial and no doubt reflected tremendous amounts of work and ingenuity to collect and curate, they are dwarfed by the amount of data now available. Time series data is everywhere, and soon everything will be a time series.

This rapid growth in the size and quality of data sets owes its origins to the tremendous advances that have been made in computing in the past few decades. Hardware engineers succeeded in continuing the trend described by Moore's Law—a prediction of exponential growth in computing capacity—during this time. As hardware became smaller, more powerful, and more efficient, it was easy to have much more of it, affordably—to create everything from miniature portable computers with attached sensors to massive data centers powering the modern internet in its data-hungry form. Most recently, wearables, machine learning techniques, and GPUs have revolutionized the quantity and quality of data available for study.[6]

Time series will no doubt benefit as computing power increases because many aspects of time series data are computationally demanding. With ramped-up

---

5 That is, 100 separate data sets in different domains of various time series of different lengths.

6 Given the array of gadgets humans carry around with them as well as the timestamps they create as they shop for groceries, log in to a computer portal at work, browse the internet, check a health indicator, make a phone call, or navigate traffic with GPS, we can safely say that an average American likely produces thousands of time series data points every year of their life.

computational and data resources, time series analysis can be expected to continue its rapid pace of development.

The Origins of Statistical Time Series Analysis

Statistics is a very young science. Progress in statistics, data analysis, and time series has always depended strongly on when, where, and how data was available and in what quantity. The emergence of time series analysis as a discipline is linked not only to developments in probability theory but equally to the development of stable nation states, where recordkeeping first became a realizable and interesting goal. We covered this earlier with respect to a variety of disciplines. Now we'll think about time series itself as a discipline.

One benchmark for the beginning of time series analysis as a discipline is the application of autoregressive models to real data. This didn't happen until the 1920s. Udny Yule, an experimental physicist turned statistical lecturer at Cambridge University, applied an autoregressive model to sunspot data, offering a novel way to think about the data in contrast to methods designed to fit the frequency of an oscillation. Yule pointed out that an autoregressive model did not begin with a model that assumed periodicity:

> When periodogram analysis is applied to data respecting any physical phenomenon in the expectation of eliciting one or more true periodicities, there is usually, as it seems to me, a tendency to start from the initial hypothesis that the periodicity or periodicities are masked solely by such more or less random superposed fluctuations—fluctuations which do not in any way disturb the steady course of the underlying periodic function or functions…there seems no reason for assuming it to be the hypothesis most likely a priori.

Yule's thinking was his own, but it's likely that some historical influences led him to notice that the traditional model presupposed its own outcome. As a former experimental physicist who had worked abroad in Germany (the epicenter for the burgeoning theory of quantum mechanics), Yule would certainly have been aware of the recent developments that highlighted the probabilistic nature of quantum mechanics. He also would have recognized the dangers of narrowing one's thinking to a model that presupposes too much, as classical physicists had done before the discovery of quantum mechanics.

As the world became a more orderly, recorded, and predictable place, particularly after World War II, early problems in practical time series analysis were presented by the business sector. Business-oriented time series problems were important and not overly theoretical in their origins. These included forecasting demand, estimating future raw materials prices, and hedging on manufacturing costs. In these industrial use cases, techniques were adopted when they worked and rejected when they didn't. It probably helped that industrial workers had access to larger data sets than were available to academics at the time (as continues to be the case now). This meant that sometimes practical but theoretically underexplored techniques came into widespread use before they were well understood.

The Origins of Machine Learning Time Series Analysis

Early machine learning in time series analysis dates back many decades. An oft-cited paper from 1969, "The Combination of Forecasts," analyzed the idea of combining forecasts rather than choosing a "best one" as a way to improve forecast performance. This idea was, at first, abhorrent to traditional statisticians, but ensemble methods have come to be the gold standard in many forecasting problems. Ensembling rejects the idea of a perfect or even significantly superior forecasting model relative to all possible models.

More recently, practical uses for time series analysis and machine learning emerged as early as the 1980s, and included a wide variety of scenarios:

- Computer security specialists proposed anomaly detection as a method of identifying hackers/intrusions.
- Dynamic time warping, one of the dominant methods for "measuring" the similarity of time series, came into use because the computing power would finally allow reasonably fast computation of "distances," say between different audio recordings.
- Recursive neural networks were invented and shown to be useful for extracting patterns from corrupted data.

Time series analysis and forecasting have yet to reach their golden period, and, to date, time series analysis remains dominated by traditional statistical methods as well as simpler machine learning techniques, such as ensembles of trees and linear fits. We are still waiting for a great leap forward for predicting the future.

More Resources

- On the history of time series analysis and forecasting:

    Kenneth F. Wallis, "Revisiting Francis Galton's Forecasting Competition," Statistical Science 29, no. 3 (2014): 420–24, https://perma.cc/FJ6V-8HUY.
        This is a historical and statistical discussion of a very early paper on forecasting the weight of a butchered ox while the animal was still alive at a county fair.

    G. Udny Yule, "On a Method of Investigating Periodicities in Disturbed Series, with Special Reference to Wolfer's Sunspot Numbers," Philosophical Transactions of the Royal Society of London. Series A, Containing Papers of a Mathematical or Physical Character 226 (1927): 267–98, https://perma.cc/D6SL-7UZS.

Udny Yule's seminal paper, one of the first applications of autoregressive moving average analysis to real data, illustrates a way to remove the assumption of periodicity from analysis of a putatively periodic phenomenon.

J.M. Bates and C. W. J. Granger, "The Combination of Forecasts," Organizational Research Quarterly 20, No. 4 (1969): 451–68, https://perma.cc/9AEE-QZ2J.

This seminal paper describes the use of ensembling for time series forecasting. The idea that averaging models was better for forecasting than looking for a perfect model was both new and controversial to many traditional statisticians.

Jan De Gooijer and Rob Hyndman, "25 Years of Time Series Forecasting," International Journal of Forecasting 22, no. 3 (2006): 443–73, https://perma.cc/84RG-58BU.

This is a thorough statistical summary of time series forecasting research in the 20th century.

Rob Hyndman, "A Brief History of Time Series Forecasting Competitions," Hyndsight blog, April 11, 2018, https://perma.cc/32LJ-RFJW.

This shorter and more specific history gives specific numbers, locations, and authors of prominent time series forecasting competitions in the last 50 years.

- On domain-specific time series histories and commentary:

NASA, "Weather Forecasting Through the Ages," Nasa.gov, February 22, 2002, https://perma.cc/8GK5-JAVT.

NASA gives a history of how weather forecasting came to be, with emphasis on specific research challenges and successes in the 20th century.

Richard C. Cornes, "Early Meteorological Data from London and Paris: Extending the North Atlantic Oscillation Series," PhD diss., School of Environmental Sciences, University of East Anglia, Norwich, UK, May 2010, https://perma.cc/NJ33-WVXH.

This doctoral thesis offers a fascinating account of the kinds of weather information available for two of Europe's most important cities, complete with extensive listings of the locations and nature of historic weather in time series format.

Dan Mayer, "A Brief History of Medicine and Statistics," in Essential Evidence-Based Medicine (Cambridge, UK: Cambridge University Press, 2004), https://perma.cc/WKU3-9SUX.

This chapter of Mayer's book highlights how the relationship between medicine and statistics depended greatly on social and political factors that made data and statistical training available for medical practitioners.

Simon Vaughan, "Random Time Series in Astronomy", Philosophical Transactions of the Royal Society A: Mathematical, Physical and Engineering Sciences 371, no. 1984 (2013): 1–28, https://perma.cc/J3VS-6JYB.

Vaughan summarizes the many ways time series analysis is relevant to astronomy and warns about the danger of astronomers rediscovering time series principles or missing out on extremely promising collaborations with statisticians.

## CHAPTER 2

Finding and Wrangling Time Series Data

In this chapter we discuss problems that might arise while you are preprocessing time series data. Some of these problems will be familiar to experienced data analysts, but there are specific difficulties posed by timestamps. As with any data analysis task, cleaning and properly processing data is often the most important step of a timestamp pipeline. Fancy techniques can't fix messy data.

Most data analysts will need to find, align, scrub, and smooth their own data either to learn time series analysis or to do meaningful work in their organizations. As you prepare data, you'll need to do a variety of tasks, from joining disparate columns to resampling irregular or missing data to aligning time series with different time axes. This chapter helps you along the path to an interesting and properly prepared time series data set.

We discuss the following skills useful for finding and cleaning up time series data:

- Finding time series data from online repositories
- Discovering and preparing time series data from sources not originally intended for time series
- Addressing common conundrums you will encounter with time series data, especially the difficulties that arise from timestamps

After reading this chapter, you will have the skills needed to identify and prepare interesting sources of time series data for downstream analysis.

Where to Find Time Series Data

If you are interested in where to find time series data and how to clean it, the best resource for you in this chapter depends on which of these is your main goal:

- Finding an appropriate data set for learning or experimentation purposes
- Creating a time series data set out of existing data that is not stored in an explicitly time-oriented form

In the first case, you should find existing data sets with known benchmarks so you can see whether you are doing your analysis correctly. These are most often found as contest data sets (such as Kaggle) or repository data sets. In these cases, you will likely need to clean your data for a specific purpose even if some preliminary work has been done for you.

In the second case, you should think about effective ways to identify interesting time-stamped data, turn it into a series, clean it, and align it with other timestamped data to make interesting time series data. I will refer to these found in the wild data sets as *found time series* (this is my own term and not technical language).

We discuss both prepared data sets and found time series next.

Prepared Data Sets

The best way to learn an analytical or modeling technique is to run through it on a variety of data sets and see both how to apply it and whether it helps you reach a concrete goal. In such cases it is helpful to have prepared options.

While time series data is everywhere, it is not always easy to find the kind of data you want when you want it. If you often find yourself in this position, you will want to get familiar with some commonly used time series data repositories, so we'll discuss a few options you should consider.

The UCI Machine Learning Repository

The UCI Machine Learning Repository (*https://perma.cc/M3XC-M9HU*) (see Figure 2-1) contains around 80 time series data sets, ranging from hourly air quality samples in an Italian city to Amazon file access logs to diabetes patients' records of activity, food, and blood glucose information. These are very different kinds of data, and a look at the files shows they reflect distinct ways of tracking information across time, yet each is a time series.

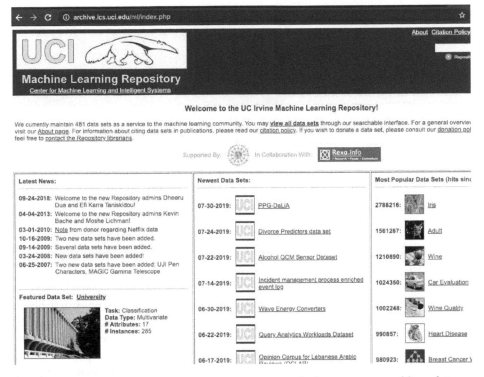

Figure 2-1. The UCI Machine Learning Repository includes an annotated list of time series data sets.

Consider the very first data set (*https://perma.cc/8E7D-ESGM*) listed under the Time Series section in the UCI repository, which is a data set about absenteeism at work (see Figure 2-2).

A quick look at the data reveals that the time columns are limited to "Month of absence," "Day of the week," and "Seasons," with no column for year. There are duplicate time indices, but there is also a column indicating employee identity so that we can differentiate between these duplicate time points. Finally, there are various employee attribute columns.

Machine Learning Repository
Center for Machine Learning and Intelligent Systems

Absenteeism at work Data Set
Download: Data Folder, Data Set Description

Abstract: The database was created with records of absenteeism at work from July 2007 to July 2010 at a courier company in Brazil.

Data Set Characteristics:	Multivariate, Time-Series	Number of Instances:	740	Area:	Business
Attribute Characteristics:	Integer, Real	Number of Attributes:	21	Date Donated	2018-04-05
Associated Tasks:	Classification, Clustering	Missing Values?	N/A	Number of Web Hits:	92522

Figure 2-2. The absenteeism at work data set is the first in the list of time series data sets in the UCI Machine Learning Repository.

This data set could be quite challenging to process, because you would first need to determine whether the data was all from one year or whether the cycling of months from 1 to 12 several times through the row progressions indicates that the year is changing. You would also need to decide whether to look at the problem in the aggregate, from a total absenteeism per unit time perspective, or to look at absenteeism per ID for those reported in the data set (see Figure 2-3). In the first case you would have a single time series, whereas in the latter case you would have multiple time series with overlapping timestamps. How you looked at the data would depend on what question you wanted to answer.

Figure 2-3. The first few lines of one file in the Australian sign language data set. As you can see, this is a wide file format.

Contrast the absenteeism data set with another data set early on in the list, the Australian Sign Language signs data set (*https://perma.cc/TC5E-Z6H4*), which includes recordings from a Nintendo PowerGlove as subjects signed in Australian Sign Language. The data is formatted as wide CSV files, each within a folder indicating the individual measured and with a filename indicating the sign.

In this data set, the columns are unlabeled and do not have a timestamp. This is nonetheless time series data; the time axis counts the time steps forward, regardless of when the actual events happened. Notice that for the purpose of thinking about signs as time series, it doesn't matter what the unit of time is; the point is the sequencing rather than the exact time. In that case, all you would care about is the ordering of the event, and whether you could assume or confirm from reading the data description that the measurements were taken at regular intervals.

As we can see from inspecting these two data sets, you will run into all sorts of data munging challenges. Some of the problems we've already noticed are:

- Incomplete timestamps
- Time axes can be horizontal or vertical in your data
- Varying notions of time

The UEA and UCR Time Series Classification Repository

The UEA and UCR Time Series Classification Repository (*https://perma.cc/56Q5-YPNT*) is a newer effort providing a common set of time series data available for experimentation and research in time series classification tasks. It also shows a very diverse set of data. We can see this by looking at two data sets.

One data set is a yoga movement classification task. The classification task (*https://perma.cc/U6MU-2SCZ*) involves distinguishing between two individual actors who performed a series of transitions between yoga poses while images were recorded. The images were converted to a one-dimensional series. The data is stored in a CSV file, with the label as the leftmost column and the remaining columns representing time steps. Time passes from left to right across the columns, rather than from top to bottom across the rows.

Plots of two sample time series per gender are shown in Figure 2-4.

Univariate Versus Multivariate Time Series

The data sets we have looked at so far are *univariate* time series; that is, they have just one variable measured against time.

Multivariate time series are series with multiple variables measured at each timestamp. They are particularly rich for analysis because often the measured variables are interrelated and show temporal dependencies between one another. We will encounter multivariate time series data later.

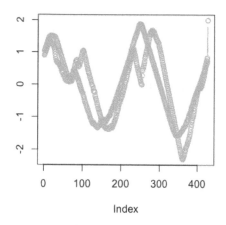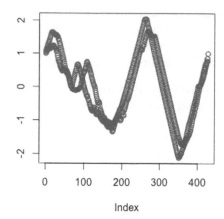

Figure 2-4. Plots of a male and a female actor performing a yoga move repeatedly. We plot two sample time series per actor. There are no explicit time labels on the x-axis. Rather than the unit of time, what is important is whether the x-axis data points are evenly spaced, as they are presented here.

Also consider the wine data set (*https://perma.cc/CJ7A-SXFD*), in which wines were classified by region according to the shapes of their spectra. So what makes this relevant to time series analysis? A *spectrum* is a plot of light wavelength versus intensity. Here we see a time series classification task where there is no passage of time at all. Time series analysis applies, however, because there is a unique and meaningful ordering of the x-axis, with a concrete meaning of distance along that axis. Time series analysis distinguishes itself from cross-sectional analysis by using the additional information provided by ordering on the x-axis, whether it is time or wavelength or something else. We can see a plot of such a "time" series in Figure 2-5. There is no temporal element, but we are nonetheless looking at an ordered series of data, so the usual ideas of time series apply.[1]

---

1 To learn more about this kind of data analysis, also see references on

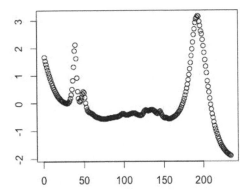

Figure 2-5. A sample spectrum of the wine sampled in the UCI wine data set. Peaks in the curve indicate wavelength regions that have particularly high rates of absorption. The wavelengths are uniformly spaced along the x-axis, whereas the y-axis indicates the rate of absorption, also on a linear scale. We can use time series analysis to compare curves, such as the one above, to one another.

Government time series data sets

The US government has been a reliable provider of time series data for decades, or even centuries. For example, the NOAA National Centers for Environmental Information (*https://perma.cc/EA5R-TP5L*) publishes a variety of time series data relating to temperatures and precipitation at granularities as fine as every 15 minutes for all weather stations across the country. The Bureau of Labor Statistics (*https://www.bls.gov/*) publishes a monthly index of the national unemployment rate. The Centers for Disease Control and Prevention (*https://perma.cc/Y6KG-T948*) publishes weekly flu case counts during the flu season. The Federal Reserve Bank of St. Louis (*https://fred.stlouisfed.org/*) offers a particularly generous and helpful set of economic time series data.

For initial forays into time series analysis, I recommend that you access these real-world government data sets only for exploratory analysis and visualization. It can be difficult to learn on these data sets because they present extremely complicated problems. For example, many economists spend their entire careers trying to predict the unemployment rate in advance of its official publication, with only limited success.

For the important but intractable problems faced by governments, predicting the future would not only be socially beneficial but also highly remunerative. Many smart and well-trained people are attacking these problems even as the state of the art remains somewhat disappointing. It is great to work on difficult problems, but it is not a good idea to learn on such problems.

Found Time Series in Government Data

It's not difficult to find a variety of timely and promising data sets from government websites, such as those just mentioned. However, there is also a lot of potential for found time series in government data. For example, imagine laying out parallel time series for economics and climate, or correlating different crimes with government spending patterns. That data would come from a variety of sources, and you would need to integrate them.

You should be wary of working with found time series in government data sets for a number of reasons. Recording conventions, column names, or even column definitions shift over time without accompanying documentation. Projects start and stop depending on politics, budgets, and other exogenous considerations. Also data formats from government websites can be very messy, and are usually messier and less continuous than comparable data sets in the private sector. This makes it particularly challenging to construct found time series.

Additional helpful sources

While we cannot extensively cover all good sources of time series data, there are several other repositories that you should explore:

CompEngine (https://comp-engine.org)
> This "self organizing database of time-series data" has more than 25,000 time series databases that total almost 140 million individual data points. The emphasis of this repository, and the associated software it offers on its web interface, is to facilitate and promote *highly comparative time-series analysis* (hctsa). The goal of such analysis is to generate high-level insights and understanding of how many kinds of temporal behavior can be understood without discipline-specific data.

Mcomp (https://cran.r-project.org/package=Mcomp) and M4comp2018 (https://github.com/carlanetto/M4comp2018) R packages
> These R packages provide the competition data from the 1982 M-competition (1,001 time series), the M3 competition in 2000 (3,003 time series), and the M4 competition in 2018 (100,000 time series). These time series forecasting competitions were previously discussed in the Chapter 1 mention of Professor Rob Hyndman's history of time series forecasting. Additional time series forecasting competition data is included in R's *tscompdata (https://github.com/robjhyndman/tscompdata)* package. Finally, more specialized time series data sets can also be found in a variety of packages described in the CRAN repository listing of time series packages (*https://perma.cc/2694-D79K*) under the "Time Series Data" header.

Found Time Series

Earlier in the chapter, we discussed the concept of a found time series, which is time series data we put together ourselves from data sources in the wild. More specifically, such time series would be put together from individual data points recorded without any special allowances for time series analysis but with enough information to construct a time series. Putting together a time series of transactions for a particular customer from a SQL database that stores a company's transactions is a clean example. In such a case, the time series could be constructed so long as a timestamp, or some proxy for a timestamp, was saved in the database.[2] We can also imagine other time series constructed from the same data, such as a time series of total transaction volume per day for the company or total dollar volume for female customers per week. We can even imagine generating multivariate time series data, such as a time series that would separately indicate total volume of all customers under 18 per week, total dollars spent by women over 65 per week, and total ad spend by the company per week. This would give us three indicators at each time step, a multivariate time series.

Finding time series data in structured data not explicitly stored as a time series can be easy in the sense that timestamping is ubiquitous. Here are a few examples of where you'll see timestamps in your database:

Timestamped recordings of events
> If there is a timestamp on your data, you have the potential to construct a time series. Even if all you do is record the time a file was accessed with no other information, you have a time series. For example, in that case, you could model the delta time between timestamps with each marked according to its later timestamp, so that your time series would consist of time on your temporal axis and delta time on your value axis. You could go further, aggregating these delta times as means or totals over larger periods, or you could keep them individually recorded.

"Timeless" measurements where another measurement substitutes for time
> In some cases, time is not explicit in data but is accounted for in the underlying logic of the data set. For example, you may think of your data as "distance versus value" when the distance is being caused by a known experimental parameter, such as retracting a sensor from a position at a known rate. If you can map one of your variables to time, you have a time series. Alternately, if one of your axes has a known distance and ordering relationship (such as wavelength), you are also looking at time series data such as the case of the wine spectra mentioned earlier.

---

2 SQL databases are traditional databases using a table-based mechanism of storing data. For example, if you wanted to store customer transactions in a SQL database, you might have a table with customer information, including a unique identifier, and another table with transactions, each one including one of those unique customer identifiers.

Physical traces

Many scientific disciplines record physical traces, be it for medicine, audiology, or weather. These used to be physically generated traces collected via analog processes, but nowadays they are stored in a digital format. These are also time series, although they may stored in a format that doesn't make this obvious, such as in an image file or in a single vector within a single field of a database.

Retrofitting a Time Series Data Collection from a Collection of Tables

The quintessential example of a found time series is one extracted from state-type and event-type data stored in a SQL database. This is also the most relevant example because so much data continues to be stored in traditional structured SQL databases.

Imagine working for a large nonprofit organization. You have been tracking a variety of factors that could lend themselves to time series analysis:

- An email recipient's reaction to emails over time: Did they open the emails or not?

- A membership history: Were there periods when a member let their membership lapse?

- Transaction history: When does an individual buy and can we predict this?

You could look at the data with several time series techniques:

- You can generate a 2D histogram of member responses to emails over time with a member-specific time line to get an idea of whether members develop fatigue from emails. (We'll illustrate the use of 2D histograms for understanding time series in Chapter 3.)

- You can turn donation predictions into a time series forecasting problem. (We'll discuss classic statistical forecasting in Chapter 4.)

- You could examine whether there are typical patterns of trajectories for member behavior in important situations. For example, is there a typical pattern of events that indicates when a member is about to leave your organization (perhaps three email deletes in a row)? In time series analysis, you could frame this as detecting a member's underlying state based on external actions. (We'll cover this when we discuss state space methods of time series analysis in Chapter 7.)

As we can see, there are many time series questions and answers in a simple SQL database. In many cases, organizations don't plan for time series analysis when they are designing their database schema. In such examples, we need to collect and assemble time series from disparate tables and sources.

A Worked Example: Assembling a Time Series Data Collection

If you are lucky enough to have several related data sources available, you will need to line them up together, possibly dealing with disparate timestamping conventions or different levels of granularity in the data. Let's create some numbers for the nonprofit example we were using. Suppose you have data shown in Table 2-1 through Table 2-3:

Table 2-1. The year each member joined and current status of membership

MemberId	YearJoined	MemberStatus
1	2017	gold
2	2018	silver
3	2016	inactive

Table 2-2. Number of emails you sent out in a given week that were opened by the member

MemberId	Week	EmailsOpened
2	2017-01-08	3
2	2017-01-15	2
1	2017-01-15	1

Table 2-3. Time a member donated to your organization

MemberId	Timestamp	DonationAmount
2	2017-05-22 11:27:49	1,000
2	2017-04-13 09:19:02	350
1	2018-01-01 00:15:45	25

You have likely worked with data in this tabular form. With such data, you can answer many questions, such as how the overall number of emails opened by a member correlates with the overall donations.

You can also answer time-oriented questions, such as whether a member donates soon after joining or long after. Without putting this data into a more time-series-friendly format, however, you cannot get at more granular behaviors that might help you predict when someone is likely to make a donation (say, based on whether they have recently been opening emails).

You need to put this data into a sensible format for time series analysis. There are a number of challenges you will need to address.

You should start by considering the temporal axes of the data we have. In the preceding tables we have three levels of temporal resolution:

- A yearly member status

- A weekly tally of emails opened
- Instantaneous timestamps of donations

You will also need to examine whether the data means what you think it means. For example, you would want to determine whether the member status is a yearly status or just the most recent status. One way to answer this is to check whether any member has more than one entry:

```python
## python
>>> YearJoined.groupby('memberId').count().
                  groupby('memberStats').count()

1000
```

Here we can see that all 1,000 members have only one status, so that the year they joined is indeed likely to be the YearJoined, accompanied by a status that may be the member's current status or status when they joined. This distinction affects how you would use the status variable, so if you were going to analyze this data further you'd want to clarify with someone who knows the data pipeline. If you were applying a member's current status to an analysis of past data, that would be a *lookahead* because you would be inputting something into a time series model that could not be known at the time. This is why you would not want to use a status variable, such as YearJoined, without knowing when it was assigned.

What Is a Lookahead?

The term *lookahead* is used in time series analysis to denote any knowledge of the future. You shouldn't have such knowledge when designing, training, or evaluating a model. A lookahead is a way, through data, to find out something about the future earlier than you ought to know it.

A lookahead is any way that information about what will happen in the future might propagate back in time in your modeling and affect how your model behaves earlier in time. For example, when choosing hyperparameters for a model, you might test the model at various times in your data set, then choose the best model and start at the beginning of your data to test this model. This is problematic because you chose the model for one time knowing things that would happen at a subsequent time—a lookahead.

Unfortunately, there is no automated code or statistical test for a lookahead, so it is something you must be vigilant and thoughtful about.

Looking at the emails table, both the column name week and its contents suggest that the data is a weekly timestamp or time period. This must be an aggregate over the week, so we should think of these timestamps as weekly periods rather than timestamps occurring one week apart.

You should assess some important characteristics. For example, you could start by asking how the weeks are reported in time. While we may not have information to restructure the table, if the week is divided in a somewhat strange way relative to our industry, we might want to know about this too. For analyzing human activities, it generally makes sense to look at the calendar week of Sunday through Saturday or Monday through Sunday rather than weeks that are less in line with the cycle of human activity. So, for example, don't arbitrarily start your week with January 1st.

You could also ask whether null weeks are reported? That is, do the weeks in which the member opened 0 emails have a place in the table? This matters when we want to do time-oriented modeling. In such cases we need to always have the null weeks present in the data because a 0 week is still a data point.

```python
## python
>>> emails[emails.EmailsOpened < 1]

Empty DataFrame
Columns: [EmailsOpened, member, week]
Index: []
```

There are two possibilities: either nulls are not reported or members always have at least one email event. Anyone who has worked with email data knows that it's difficult to get people to open emails, so the hypothesis that members always open at least one email per week is quite unlikely. In this case, we can resolve this by looking at the history of just one user:

```python
## python
>>> emails[emails.member == 998]
      EmailsOpened member   week
25464 1           998    2017-12-04
25465 3           998    2017-12-11
25466 3           998    2017-12-18
25467 3           998    2018-01-01
25468 3           998    2018-01-08
25469 2           998    2018-01-15
25470 3           998    2018-01-22
25471 2           998    2018-01-29
25472 3           998    2018-02-05
25473 3           998    2018-02-12
25474 3           998    2018-02-19
25475 2           998    2018-02-26
25476 2           998    2018-03-05
```

We can see that some weeks are missing. There aren't any December 2017 email events after December 18, 2017.

We can check this more mathematically by calculating how many weekly observations we should have between the first and last event for that member. First we calculate the length of the member's tenure, in weeks:

```python
## python
>>> (max(emails[emails.member == 998].week) -
                min(emails[emails.member == 998].week)).days/7
25.0
```

Then we see how many weeks of data we have for that member:

```python
## python
>>> emails[emails.member == 998].shape
(24, 3)
```

We have 24 rows here, but we should have 26. This shows some weeks of data are missing for this member. Incidentally, we could also run this calculation on all members simultaneously with group-by operations, but it's more approachable to think about just one member for example purposes.

Why 26 Rows?

You may be surprised that we need 26 instead of 25 given the subtraction we just performed, but that was an incomplete calculation. When you work with time series data, one thing you should always ask yourself after doing this kind of subtraction is whether you should add 1 to account for the offset at the end. In other words, did you subtract the positions you wanted to count?

Consider this example. Let's say I have information for April 7th, 14th, 21st, and 28th. I want to know how many data points I should have in total. Subtracting 7 from 28 and dividing by 7 yields 21/7 or 3. However, I should obviously have four data points. I subtracted out April 7th and need to put it back in, so the proper calculation is the difference between the first and last days divided by 7, *plus 1* to account for the subtracted start date.

We'll move on to filling in the blanks so that we have a complete data set now that we have confirmed we do indeed have missing weeks. We can't be sure of identifying all missing weeks, since some may have occurred before our earliest recorded date or after our last recorded date. What we can do, however, is fill in the missing values between the first and last time a member had a non-null event.

It's a lot easier to fill in all missing weeks for all members by exploiting Pandas' indexing functionality, rather than writing our own solution. We can generate a MultiIndex for a Pandas data frame, which will create all combinations of weeks and members—that is, a Cartesian product:

```python
## python
>>> complete_idx = pd.MultiIndex.from_product((set(emails.week),
                                               set(emails.member)))
```

We use this index to reindex the original table and fill in the missing values—in this case with 0 on the assumption that nothing recorded means there was nothing to record. We also reset the index to make the member and week information available as columns, and then name those columns:

```python
## python
>>> all_email = emails.set_index(['week', 'member']).
                       reindex(complete_idx, fill_value = 0).
                       reset_index()
>>> all_email.columns = ['week', 'member', 'EmailsOpened']
```

Let's take a look at member 998 again:

```python
## python
>>> all_email[all_email.member == 998].sort_values('week')
   week      member EmailsOpened
2015-02-09 998     0
2015-02-16 998     0
2015-02-23 998     0
2015-03-02 998     0
2015-03-09 998     0
```

Python's Pandas

Pandas is a data frame analysis package in Python that is used widely in the data science community. Its very name indicates its suitability for time series analysis: "Pandas" refers to "panel data," which is what social scientists call time series data.

Pandas is based on tables of data with row and column indices. It has SQL-like operations built in, such as group by, row selection, and key indexing. It also has time series–specific functionality, such as indexing by time period, downsampling, and time-based grouping operations.

If you are unfamiliar with Pandas, I strongly recommend looking at a brief overview, such as that provided in the official documentation (*https://perma.cc/7R9B-2YPS*).

Notice that we have a large number of zeros at the start. These are likely before the member joined the organization, so they would not have been on an email list. There are not too many kinds of analyses where we'd want to keep the member's truly null weeks around—specifically those weeks before the member ever indicated opening an email. If we had the precise date a member started receiving emails, we would have an objective cutoff. As it is, we will let the data guide us. For each member we determine the start_date and end_date cutoffs by grouping the email DataFrame per member and selecting the maximum and minimum week values:

```
## python
>>> cutoff_dates = emails.groupby('member').week.
                        agg(['min', 'max']).reset_index)
>>> cutoff_dates = cutoff_dates.reset_index()
```

We drop rows from the DataFrame that don't contribute sensibly to the chronology, specifically 0 rows before each member's first nonzero count:

```
## python
>>> for _, row in cutoff_dates.iterrows():
>>>     member     = row['member']
>>>     start_date = row['min']
>>>     end_date   = row['max']
>>>     all_email.drop(
                all_email[all_email.member == member]
                [all_email.week < start_date].index, inplace=True)
>>>     all_email.drop(all_email[all_email.member == member]
                [all_email.week > end_date].index, inplace=True)
```

< or <= ?

We use the < and > operations, without equality, because the start_date and end_date are inclusive of the meaningful data points and because we are dropping data, not retaining data, as our code is written. In this case we want to include those weeks in our analysis because they were the first and last meaningful data points.

You will do well to work with your data engineers and database administrators to convince them to store the data in a time aware manner, especially with respect to how timestamps are created and what they mean. The more you can solve problems upstream, the less work you have to do downstream in the data pipeline.

Now that we have cleaned up our email data, we can consider new questions. For example, if we want to think about the relationship of member email behavior to donations, we can do a few things:

- Aggregate the DonationAmount to weekly so that the time periods are comparable. Then it becomes reasonable to ask whether donations correlate in some way to member responses to emails.

- Treat the prior week's EmailsOpened as a predictor for a given week's DonationAmount. Note that we have to use the prior week because EmailsOpened is a summary statistic for the week. If we want to predict a Wednesday donation, and our EmailsOpened summarizes email opening behavior from Monday to Sunday, then using the same week's information will potentially give us information about what the member did subsequent to when we could have known it (for example, whether they opened an email on the Friday after the donation).

Constructing a Found Time Series

Consider how to relate the email and donations data to one another. We can down-sample the donation data to turn it into a weekly time series, comparable to the email data. As an organization, we are interested in the total weekly amounts, so we aggregate the timestamps into weekly periods by summing. More than one donation in a week is unlikely, so the weekly donation amounts will reflect the individual donation amounts for most donors.

```python
## python
>>> donations.timestamp = pd.to_datetime(donations.timestamp)
>>> donations.set_index('timestamp', inplace = True)
>>> agg_don = donations.groupby('member').apply(
            lambda df: df.amount.resample("W-MON").sum().dropna())
```

In this code we first convert a string character to a proper timestamped data class so as to benefit from Pandas' built-in date-related indexing. We set the timestamp as an index, as is necessary for resampling a data frame. Finally for the data frame that we obtain from subsetting down to each member, we group and sum donations by week, drop weeks that do not have donations, and then collect these together.

Note that we resampled with an anchored week so that we will match the same weekly dates we already have in our email table. Note also that a week anchored to "Monday" makes sense from a human perspective.

We now have donation information and email information sampled at the same frequency, and we can join them. Pandas makes this simple so long as we anchor the weeks to the same day of the week, as we've done already. We can iterate through each member and merge the data frames per member:

```python
## python
>>> for member, member_email in all_email.groupby('member'):
>>>     member_donations = agg_donations[agg_donations.member
                                        == member]

>>>     member_donations.set_index('timestamp', inplace = True)
>>>     member_email.set_index    ('week', inplace = True)

>>>     member_email = all_email[all_email.member == member]
>>>     member_email.sort_values('week').set_index('week')

>>>     df = pd.merge(member_email, member_donations, how = 'left',
                    left_index = True,
                    right_index = True)
>>>     df.fillna(0)

>>>     df['member'] = df.member_x
>>>     merged_df = merged_df.append(df.reset_index()
                        [['member', 'week', 'emailsOpened', 'amount']])
```

We now have our email and donations data lined up per member. For each member we include only meaningful weeks, and not weeks that appear to be before or after their membership period.

We might treat the email behavior as a "state" variable relative to the donation behavior, but we probably want to carry forward the state from the previous week so as to avoid a lookahead. Suppose, for example, we were building a model that uses email behavior to predict a member's next donation. In this case we might consider looking at the pattern of email opening over time as a possible indicator. We would need to line up a given week's donation with the previous week's email behavior. We can easily take our data, processed to align week-to-week, and then shift by the appropriate number of weeks. If, say, we want to shift the donation a week forward, we can easily do so with the shift operator, although we would need to be sure to do this on a per-member basis:

```python
## python
>>> df = merged_df[merged_df.member == 998]
>>> df['target'] = df.amount.shift(1)
>>> df = df.fillna(0)
>>> df
```

It's good practice to store this target in a new column rather than overwrite the old, particularly if you are not also bringing forward the timestamp for the donation amount separately. We have shifted the donation amount one week into the future using Pandas' built-in shift functionality. You can also shift back in time with negative numbers. Generally you will have more predictors than targets, so it makes sense to shift your targets. We can see the outcome of the code here:

```
amount    emailsOpened    member    week          target
0         1               998       2017-12-04    0
0         3               998       2017-12-11    0
0         3               998       2017-12-18    0
0         0               998       2017-12-25    0
0         3               998       2018-01-01    0
50        3               998       2018-01-08    0
0         2               998       2018-01-15    50
```

Now that we have filled in the missing rows, we have the desired 26 rows for member 998. Our data is now cleaner and more complete.

To recap, these are the time-series-specific techniques we used to restructure the data:

1. *Recalibrate the resolution of our data* to suit our question. Often data comes with more specific time information than we need.

2. Understand how we can *avoid lookahead* by not using data for timestamps that produce the data's availability.

---

3. Record *all relevant time periods* even if "nothing happened." A zero count is just as informative as any other count.

4. *Avoid lookahead* by not using data for timestamps that produce information we shouldn't yet know about.

So far, we have created raw found time series by aligning our donation and email time series to be sampled for the same points in time and at the same frequency. However, we haven't done a thorough job of cleaning up this data or fully exploring before analysis. That's something we'll do in Chapter 3.

Timestamping Troubles

Timestamps are quite helpful for time series analysis. From timestamps, we can extrapolate a number of interesting features, such as time of day or day of the week. Such features can be important to understanding your data, especially for data concerning human behavior. However, timestamps are tricky. Here we discuss some of the difficulties of timestamped data.

Whose Timestamp?

The first question you should ask when seeing a timestamp is what process generated the timestamp, how, and when. Often an event happening is not coincident with an event being recorded. For example, a researcher might write something down in their notebook and later transcribe it to a CSV file used as a log. Does the timestamp indicate when they wrote it down or when they transcribed it to the CSV file? Or a mobile app user may take actions that trigger logging when their phone is offline, so that the data is only later uploaded to your server with some combination of timestamps. Such timestamps could reflect when the behavior took place, when the action was recorded by the app, when the metadata was uploaded to the server, or when the data was last accessed for download from the server back to the app (or any number of other events along the data pipeline).

Timestamps may appear at first to offer clarity but this quickly falls away if proper documentation is lacking. When you're looking at novel timestamps, your first step should always be to find out as best you can what established the time of an event.

A concrete example will illustrate these difficulties. Suppose you are looking at data taken from a mobile app for weight loss and see a meal diary with entries such as those in Table 2-4.

Table 2-4. Sample meal diary from a weight loss app

Time	Intake
Mon, April 7, 11:14:32	pancakes
Mon, April 7, 11:14:32	sandwich
Mon, April 7, 11:14:32	pizza

It's possible this user had pancakes, a sandwich, and a pizza all at once, but there are more likely scenarios. Did the user specify this time or was it automatically created? Does the interface perhaps offer an automatic time that the user can adjust or choose to ignore? Some answers to these questions would explain the identical timestamps better than the possibility that a user trying to lose weight would have had pancakes with a side of pizza and a sandwich appetizer.

Even if the user did have all these foods at 11:14, where in the world was it 11:14? Was this the user's local time or a global clock? Even in the unlikely case that the user had all this food at one meal, we still don't know very much about the temporal aspect of the meal from these rows alone. We don't know if this was breakfast, lunch, dinner, or a snack. To say something interesting to the user, we'd want to be able to talk in concrete terms about local hour of the day, which we can't do without time zone information.

The best way to address these questions is to see all the code that collects and stores the data or talk to the people who wrote that code. After going through all available human and technical data specifications on the system, you should also try the full system out yourself to ensure that the data behaves as you have been told it does. The better you understand your data pipeline, the less likely you are to ask the wrong questions because your timestamps don't really mean what you think they do.

You bear the ultimate responsibility for understanding the data. People who work upstream in the pipeline don't know what you have in mind for analysis. Try to be as hands-on as possible in assessing how timestamps are generated. So, if you are analyzing data from a mobile app pipeline, download the app, trigger an event in a variety of scenarios, and see what your own data looks like. You're likely to be surprised about how your actions were recorded after speaking to those who manage the data pipeline. It's hard to track multiple clocks and contingencies, so most data sets will flatten the temporal realities. You need to know exactly how they do so.

Guesstimating Timestamps to Make Sense of Data

If you are dealing with legacy data pipelines or undocumented data, you may not have the option of exploring a working pipeline and talking to those who maintain it. You will need to do some empirical investigation to understand whether you can make inferences about what the timestamps mean:

- Reading the data as we did in the previous example, you can generate initial hypotheses about what the timestamps mean. In the preceding case, look at data for multiple users to see whether the same pattern (multiple rows with identical timestamps and improbable single meal contents) held or whether this was an anomaly.
- Using aggregate-level analyses, you can test hypotheses about what timestamps mean or probably mean. For the preceding data, there are a couple of open questions:
  — Is the timestamp local or universal time?
  — Does the time reflect a user action or some external constraint, such as connectivity?

Local or Universal Time?

Most timestamps are stored in universal (UTC) time or in a single time zone, depending on the server's location but independent of the user's location. It is quite unusual to store data according to local time. However, we should consider both possibilities, because both are found in "the wild."

We form the hypothesis that if the time is a local timestamp (local to each user), we should see daily trends in the data reflecting daytime and nighttime behavior. More specifically, we should expect not to see much activity during the night when our users are sleeping. For our mobile app example, if we create a histogram of meal time counts per hour, there should be hours with significantly fewer meals logged, because in most cultures people don't eat in the middle of the night.

If we did not see a day-oriented pattern in the hours as displayed, we could conclude that the data was most likely timestamped by some universal clock and that the user base must be spread out across many time zones. In such a case, it would be quite challenging to extrapolate individual users' local hours (assuming time zone information wasn't available). We might consider individual explorations per user to see if we could code heuristics to label users approximately by time zone, but such work is both computationally taxing and also not always accurate.

Even if you cannot pinpoint an exact user time zone, it's still useful to have a global timestamp available. For starters, you can determine likely usage patterns for your app's servers, knowing when meals are most frequently logged at a given time of day and day of the week. You can also calculate the time differential between each meal the user recorded, knowing that since the timestamps are in absolute time, you don't need to worry about whether the user changed time zones. In addition to sleuthing, this would be interesting as a form of feature generation:

```python
## python
>>> df['dt'] = df.time - df.time.shift(-1)
```

The dt column would be a feature you could pass forward in your analysis. Using this time differential could also give you an opportunity to estimate each user's time zone. You could look at the time of day when the user generally had a long dt, which could point to when nighttime occurs for that user. From there you could begin to figure out each individual's "nighttime" without having to do peak-to-peak-style analysis.

User behavior or network behavior?

To return to another question posed by our short data sample, we asked whether our user had a strange meal or whether our timestamps relate to upload activity.

The same analyses used to pinpoint a user's time zone are applicable to determining whether the timestamps are a function of user or network behavior. Once you had a dt column (as previously calculated), you could look for clusters of 0s and qualitatively determine whether they were likely a single behavioral event or a single network event. You could also look at whether the dts appeared to be periodic across different days. If they were a function of user behavior, they would be more likely to be periodic than if they were a function of network connectivity or other software-related behaviors.

To summarize, here are some questions you could likely address with the available data set, even with little or no information about how timestamps are generated:

- Use differences in timestamps per user to get a sense of spacing between meals or spacing between data entries (depending on your working hypothesis of what the times indicate, a user behavior or a network behavior).

- Describe aggregate user behavior to determine when your servers are most likely to be active in the 24-hour cycle.

Time Zones

In the best-case scenario, your data is timestamped in UTC time. Most databases and other storage systems will default to this. However, there are some situations in which you are likely to run into data that is not UTC timestamped:

- Timestamps created in ways that did not use date-specific data objects, such as API calls. Such API calls use strings rather than time-specific objects.

- Data created "by hand" in small, local organizations where time zones are not relevant, such as spreadsheets generated by business analysts or field biologists.

What's a Meaningful Time Scale?

You should take the temporal resolution of the timestamping you receive with a grain of salt, based on domain knowledge about the behavior you are studying and also based on the details you can determine relating to how the data was collected.

For example, imagine you are looking at daily sales data, but you know that in many cases managers wait until the end of the week to report figures, estimating rough daily numbers rather than recording them each day. The measurement error is likely to be substantial due to recall problems and innate cognitive biases. You might consider changing the resolution of your sales data from daily to weekly to reduce or average out this systematic error. Otherwise, you should build a model that factors in the possibility of biased errors across different days of the week. For example, it may be that managers systematically overestimate their Monday performance when they report the numbers on a Friday.

Psychological Time Discounting

Time discounting is a manifestation of a phenomenon known as *psychological distance*, which names our tendency to be more optimistic (and less realistic) when making estimates or assessments that are more "distant" from us. Time discounting predicts that data reported from further in the past will be biased systematically compared to data reported from more recent memory. This is distinct from the more general problem of forgetting and implies a nonrandom error. You should keep this in mind whenever you are looking at human-generated data that was entered manually but not contemporaneously with the event recorded.

Another situation involves physical knowledge of the system. For example, there is a limit to how quickly a person's blood glucose can change, so if you are looking at a series of blood glucose measurements within seconds of one another, you should probably average them rather than treating them as distinct data points. Any physician will tell you that you are studying the error of the device rather than the rate of change of blood glucose if you are looking at many measurements within a few seconds of one another.

Humans Know Time Is Passing

Whenever you are measuring humans, keep in mind that people respond in more than one way to the passage of time. For example, recent research shows how manipulating the speed of a clock that is in a person's view influences how quickly that person's blood glucose level changes.

Cleaning Your Data

In this section we will tackle the following common problems in time series data sets:

- Missing data
- Changing the frequency of a time series (that is, upsampling and downsampling)
- Smoothing data
- Addressing seasonality in data
- Preventing unintentional lookaheads

Handling Missing Data

Missing data is surprisingly common. For example, in healthcare, missing data in medical time series can have a number of causes:

- The patient didn't comply with a desired measurement.
- The patient's health stats were in good shape, so there was no need to take a particular measurement.
- The patient was forgotten or undertreated.
- A medical device had a random technical malfunction.
- There was a data entry error.

To generalize at great peril, missing data is even more common in time series analysis than it is in cross-sectional data analysis because the burden of longitudinal sampling is particularly heavy: incomplete time series are quite common and so methods have been developed to deal with holes in what is recorded.

The most common methods to address missing data in time series are:

Imputation
When we fill in missing data based on observations about the entire data set.

Interpolation
When we use neighboring data points to estimate the missing value. Interpolation can also be a form of imputation.

Deletion of affected time periods
When we choose not to use time periods that have missing data at all.

We will discuss imputations and interpolations as well as illustrate the mechanics of these methods soon. We focus on preserving data, whereas a method such as deleting time periods with missing data will result in less data for your model. Whether to preserve the data or throw out problematic time periods will depend on your use case and whether you can afford to sacrifice the time periods in question given the data needs of your model.

Preparing a data set to test missing data imputation methodologies

We will work with monthly unemployment data (*https://data.bls.gov/timeseries/ LNS14000000*) that has been released by the US government since 1948 (this is freely available for download). We will then generate two data sets from this baseline data: one in which data is truly missing at random, and one in which are the highest unemployment months in the history of the time series. This will give us two test cases to see how imputation behaves in the presence of both random and systematic missing data.

 We move to R for the next example. We will freely switch between R and Python throughout the book. I assume you have some background in working with data frames, as well as matrices, in both R and Python.

```
## R
> require(zoo)       ## zoo provides time series functionality
> require(data.table) ## data.table is a high performance data frame

> unemp <- fread("UNRATE.csv")
> unemp[, DATE := as.Date(DATE)]
> setkey(unemp, DATE)

> ## generate a data set where data is randomly missing
> rand.unemp.idx <- sample(1:nrow(unemp), .1*nrow(unemp))
> rand.unemp     <- unemp[-rand.unemp.idx]

> ## generate a data set where data is more likely
> ## to be missing when unemployment is high
> high.unemp.idx <- which(unemp$UNRATE > 8)
> num.to.select  <- .2 * length(high.unemp.idx)
> high.unemp.idx <- sample(high.unemp.idx,)
> bias.unemp     <- unemp[-high.unemp.idx]
```

Because we have deleted rows from our data tables to create a data set with missing data, we will need to read the missing dates and NA values, and for this the `data.table` package's *rolling join* functionality is quite useful.

R's data.table Package

R's `data.table` package is a highly performant but woefully underutilized alternative to R's mainstream `data.frame` functionality. While many users are familiar with the `tidyverse` selection of packages (*https://perma.cc/E4S8-RUHN*), `data.table` is a standalone package built only on top of `data.frame`. It has many helpful functions for time series analysis. This is likely due to its origins as a package developed by software engineers who worked in finance and who needed highly performant R operations to work with financial data, most often time series data.

While there is a steeper learning curve for `data.table` than for `data.frame`, I have found this package invaluable in working with time series data in R, and so I use it heavily throughout the book. I've provided a few quick examples of how `data.table` works and may differ from `data.frame` here:

```
dt[, new.col := old.col + 7]
```

In this example, we assume that the original `data.table` has an old column called `old.col`. We make a new column, in place, which means that we add a column to the `data.table` without copying the entire `data.table` over to a new object, thereby reducing memory use. You will see this operator `:=` used frequently throughout the book.

To access a subset of columns in a `data.table`, we will usually simply put them in a list in the column indexing argument, which is the second argument to a data.table:

```
dt[, .(col1, col2, col3)]
```

This will return a subsetted `data.table` including only the col1, col2, and col3 values. Notice the `.()` operator surrounding the column names. This is shorthand for writing out `list()`.

If we wish to do row selection rather than column selection, we use the first argument to the `data.table` to subset on rows like so:

```
dt[col1 < 3 & col2 > 5]
```

This will return all columns of the `data.table` for which the two logic conditions were met. And of course we can combine row and column selection, as follows:

```
dt[col1 < 3 & col2 > 5, .(col1, col2, col3)]
```

Finally, group-by operations are quite performant and fast. For example, to count the number of rows in each group among the groups designated in col1, we could use the following code.

```
dt[, .N, col1]
```

And if we wanted to do multiple operations on our grouped objects, we would again make a list and could even name the columns:

```
dt[, .(total = .N, mean = mean(col2), col1]
```

The above would, by groups as designated by col1, return the total number of rows in the group and the mean of col2 for each group.

The data.table package is under active development, and it's an invaluable tool for time series analysis, particularly for large data sets. I strongly recommend reading the official introduction (*https://perma.cc/3HEB-NE6A*) to data.table.

```
## R
> all.dates <- seq(from = unemp$DATE[1], to = tail(unemp$DATE, 1),
                                            by = "months")
> rand.unemp = rand.unemp[J(all.dates), roll=0]
> bias.unemp = bias.unemp[J(all.dates), roll=0]
> rand.unemp[, rpt := is.na(UNRATE)]
## here we label the missing data for easy plotting
```

With a rolling join, we generate the sequence of all dates that should be available between the start and end date of the data set. This gives us rows in the data set to fill in as NA.

Now that we have data sets with missing values, we will take a look at a few specific ways to fill in numbers for those missing values:

- Forward fill
- Moving average
- Interpolation

We will compare the performance of these methods in both the randomly missing and systematically missing data sets. Since we have generated these data sets from a complete data set, we can in fact determine how they did instead of speculating. In the real world, of course, we will never have the missing data to check our data imputation.

Rolling Joins

In data.table's rolling join, we enjoy the benefit of a special spatially aware SQL-like join designed for timestamps. Although tables will not always match exactly on their timestamps, rolling joins can deal with timestamps intelligently.

```
## R
> ## we have a small data.table of donation dates
> donations <- data.table(
>    amt = c(99, 100, 5, 15, 11, 1200),
>    dt = as.Date(c("2019-2-27", "2019-3-2", "2019-6-13",
>                   "2019-8-1", "2019-8-31", "2019-9-15"))
> )
```

```
> ## we also have information on the dates of each
> ## publicity campaign
> publicity <- data.table(
>                 identifier = c("q4q42", "4299hj", "bbg2"),
>                 dt         = as.Date(c("2019-1-1",
>                                        "2019-4-1",
>                                        "2019-7-1")))


> ## we set the primary key on each data.table
> setkey(publicity, "dt")
> setkey(donations, "dt")

> ## we wish to label each donation according to
> ## what publicity campaign most recently preceded it
> ## we can easily see this with roll = TRUE
> publicity[donations, roll = TRUE]
```

This yields the following delightful output:

```
## R
> publicity[donations, roll = TRUE]
   identifier         dt  amt
1:      q4q42 2019-02-27   99
2:      q4q42 2019-03-02  100
3:     4299hj 2019-06-13    5
4:       bbg2 2019-08-01   15
5:       bbg2 2019-08-31   11
6:       bbg2 2019-09-15 1200
```

Each donation now is paired with the publicity identifier that indicates the campaign that most immediately preceded the donation.

Forward fill

One of the simplest ways to fill in missing values is to carry forward the last known value prior to the missing one, an approach known as *forward fill*. No mathematics or complicated logic is required. Simply consider the experience of moving forward in time with the data that was available, and you can see that at a missing point in time, all you can be confident of is what data has already been recorded. In such a case, it makes sense to use the most recent known measurement.

Forward fill can be accomplished easily using na.locf from the zoo package:

```
## R
> rand.unemp[, impute.ff := na.locf(UNRATE, na.rm = FALSE)]
> bias.unemp[, impute.ff := na.locf(UNRATE, na.rm = FALSE)]
>
> ## to plot a sample graph showing the flat portions
> unemp[350:400, plot (DATE, UNRATE,
```

```
                          col = 1, lwd = 2, type = 'b')]
> rand.unemp[350:400, lines(DATE, impute.ff,
                          col = 2, lwd = 2, lty = 2)]
> rand.unemp[350:400][rpt == TRUE, points(DATE, impute.ff,
                          col = 2, pch = 6, cex = 2)]
```

This will result in a plot that looks natural except where you see repeated values to account for missing data, as in Figure 2-6. As you will notice in the plot, the forward-filled values usually do not deviate far from the true values.

Figure 2-6. The original time series plotted with a solid line and the time series with forward filled values for randomly missing points plotted with a dashed line. The forward filled values are marked with downward pointing triangles.

We can also compare the values in the series by plotting the values of the series against one another. That is, for each time step, we plot the true known value against the value at the same time from the series with imputed values. Most values should match exactly since most data is present. We see that manifested in the 1:1 line in Figure 2-7. We also see scattered points off this line, but they do not appear to be systematically off.

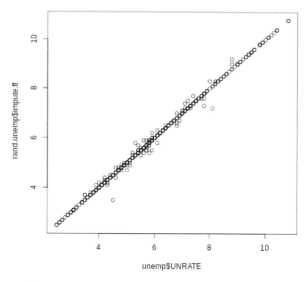

Figure 2-7. Plotting the true unemployment rate versus the forward-filled series. This plot shows that forward fill did not systematically distort the data.

Backward Fill

Just as you can bring values from the past forward to fill in missing data, you can also choose to propagate values backward. However, this is a case of a lookahead, so you should only do this when you are not looking to predict the future from the data and when, from domain knowledge, it makes more sense to fill in data backward rather than forward in time.

In some settings, forward fill makes sense as the best way to complete missing data, even if "fancier methods" are possible. For example, in medical settings, a missing value often indicates that a medical worker did not think it necessary to remeasure a value, likely because the patient's measurement was expected to be normal. In many medical cases, this means we could apply forward fill to missing values with the last known value since this was the presumption motivating the medical worker not to retake a measurement.

There are many advantages to forward fill: it is not computationally demanding, it can be easily applied to live-streamed data, and it does a respectable job with imputation. We will see an example soon.

Moving average

We can also impute data with either a rolling mean or median. Known as a *moving average*, this is similar to a forward fill in that you are using past values to "predict" missing future values (imputation can be a form of prediction). With a moving average, however, you are using input from *multiple* recent times in the past.

There are many situations where a moving average data imputation is a better fit for the task than a forward fill. For example, if the data is noisy, and you have reason to doubt the value of any individual data point relative to an overall mean, you should use a moving average rather than a forward fill. Forward filling could include random noise more than the "true" metric that interests you, whereas averaging can remove some of this noise.

To prevent a lookahead, use only the data that occurred before the missing data point. So, your implementation would like something like this:

```
## R
> ## rolling mean without a lookahead
> rand.unemp[, impute.rm.nolookahead := rollapply(c(NA, NA, UNRATE), 3,
>             function(x) {
>                         if (!is.na(x[3])) x[3] else mean(x, na.rm = TRUE)
>                         })]
> bias.unemp[, impute.rm.nolookahead := rollapply(c(NA, NA, UNRATE), 3,
>             function(x) {
>                         if (!is.na(x[3])) x[3] else mean(x, na.rm = TRUE)
>                         })]
```

We set the value of the missing data to the mean of the values that come before it (because we index off the final value and use this value to determine whether it is missing and how to replace it).

 A moving average doesn't have to be an arithmetic average. For example, exponentially weighted moving averages would give more weight to recent data than to past data. Alternately, a geometric mean can be helpful for time series that exhibit strong serial correlation and in cases where values compound over time.

When imputing missing data with a moving average, consider whether you need to know the value of the moving average with only forward-looking data, or whether you are comfortable building in a lookahead. If you are unconcerned about a lookahead, your best estimate will include points both before and after the missing data because this will maximize the information that goes into your estimates. In that case, you can implement a rolling window, as illustrated here with the zoo package's roll apply() functionality:

```
## R
> ## rolling mean with a lookahead
> rand.unemp[, impute.rm.lookahead := rollapply(c(NA, UNRATE, NA), 3,
>             function(x) {
>                         if (!is.na(x[2]))
>                             x[2]
```

```
>                        else
>                            mean(x, na.rm = TRUE)
>                        })]
```

Using both past and future information is useful for visualizations and recordkeeping applications, but, as mentioned before, it is not appropriate if you are preparing your data to be fed into a predictive model.

The results for both a forward-looking moving average and a moving average computed with future and past data are shown in Figure 2-8.

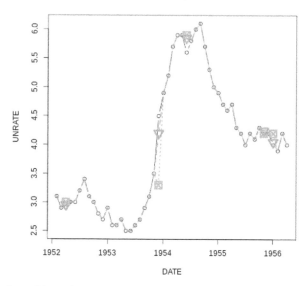

Figure 2-8. The dotted line shows the moving average imputation without a lookahead, while the dashed line shows the moving average imputation with a lookahead. Likewise, the squares show the nonlookahead imputed points while the upside down triangles show the moving average with a lookahead.

A rolling mean data imputation reduces variance in the data set. This is something you need to keep in mind when calculating accuracy, R^2 statistics, or other error metrics. Your calculation may overestimate your model's performance, a frequent problem when building time series models.

Using a Data Set's Mean to Impute Missing Data

In a cross-sectional context, it is common to impute missing data by filling in the mean or median for that variable where it is missing. While this can be done with time series data, it is not appropriate for most cases. Knowing the mean for the data set involves looking into the future…and that's a lookahead!

Interpolation

Interpolation is a method of determining the values of missing data points based on geometric constraints regarding how we want the overall data to behave. For example, a linear interpolation constrains the missing data to a linear fit consistent with known neighboring points.

Linear interpolation is particularly useful and interesting because it allows you to use your knowledge of how your system behaves over time. For example, if you know a system behaves in a linear fashion, you can build that knowledge in so that only linear trends will be used to impute missing data. In Bayesian speak, it allows you to inject a *prior* into your imputation.

As with a moving average, interpolation can be done such that it is looking at both past and future data or looking in only one direction. The usual caveats apply: allow your interpolation to have access to future data only if you accept that this creates a lookahead and you are sure this is not a problem for your task.

Here we apply interpolation using both past and future data points (see Figure 2-9):

```R
## R
> ## linear interpolation
> rand.unemp[, impute.li := na.approx(UNRATE)]
> bias.unemp[, impute.li := na.approx(UNRATE)]
>
> ## polynomial interpolation
> rand.unemp[, impute.sp := na.spline(UNRATE)]
> bias.unemp[, impute.sp := na.spline(UNRATE)]
>
> use.idx = 90:120
> unemp[use.idx, plot(DATE, UNRATE, col = 1, type = 'b')]
> rand.unemp[use.idx, lines(DATE, impute.li, col = 2, lwd = 2, lty = 2)]
> rand.unemp[use.idx, lines(DATE, impute.sp, col = 3, lwd = 2, lty = 3)]
```

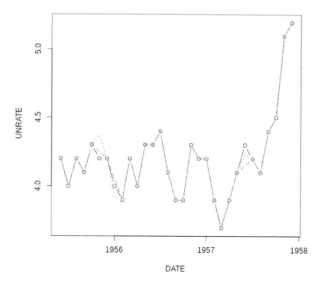

Figure 2-9. The dashed line shows the linear interpolation while the dotted line shows the spline interpolation.

There are many situations where a linear (or spline) interpolation is appropriate. Consider mean average weekly temperature where there is a known trend of rising or falling temperatures depending on the time of year. Or consider yearly sales data for a growing business. If the trend has been for business volume to increase linearly from year to year, it is a reasonable data imputation to fill in missing data based on that trend. In other words, we could use a linear interpolation, which would account for the trend, rather than a moving average, which would not. If there were a trend of increasing values, the moving average would systematically underestimate the missing values.

There are also plenty of situations for which linear (or spline) interpolation is not appropriate. For example, if you are missing precipitation data in a weather data set, you should not extrapolate linearly between the known days; as we all know, that's not how precipitation works. Similarly, if we are looking at someone's hours of sleep per day but are missing a few days of data, we should not linearly extrapolate the hours of sleep between the last known days. As an example, one of the known endpoints might include an all-nighter of studying followed by a 30-minute catnap. This is unlikely to estimate the missing data.

Overall comparison

Now that we have performed a few different kinds of imputations, we can take a look at the results to compare how the different data imputations behaved on this data set.

We generated two data sets with missing data, one in which data was missing at random and another in which unfavorable data points (high unemployment) were missing. When we compare the methods we employed to see which yields the best results, we can see that the mean square error can differ by a high percentage:

```R
## R
> sort(rand.unemp[ , lapply(.SD, function(x) mean((x - unemp$UNRATE)^2,
>                na.rm = TRUE)),
>           .SDcols = c("impute.ff", "impute.rm.lookahead",
>                      "impute.rm.nolookahead", "impute.li",
>                      "impute.sp")])
impute.li   impute.rm.lookahead   impute.sp   impute.ff   impute.rm.nolookahead
0.0017      0.0019                0.0021      0.0056      0.0080

> sort(bias.unemp[ , lapply(.SD, function(x) mean((x - unemp$UNRATE)^2,
>                na.rm = TRUE)),
>           .SDcols = c("impute.ff", "impute.rm.lookahead",
>                      "impute.rm.nolookahead", "impute.li",
>                      "impute.sp")])
impute.sp   impute.li   impute.rm.lookahead   impute.rm.nolookahead   impute.ff
0.0012      0.0013      0.0017                0.0030                  0.0052
```

Remember that many of the preceding methods include a lookahead. The only methods that do not are the forward fill and the moving average without a lookahead (there is also a moving average with a lookahead). For this reason, it's not surprising that there is a range of difference in errors and that the methods without a lookahead do not perform as well as the others.

Final notes

Here we covered the simplest and most often-used methods of missing data imputation for time series applications. Data imputation remains an important area of data science research. The more significant the decisions you are making, the more important it is to carefully consider to the potential reasons for data to be missing and the potential ramifications of your corrections. Here are a few cautionary tips you should keep in mind:

- It is impossible to prove that data is truly missing at random, and it is unlikely that missingness is truly random in most real-world occurrences.
- Sometimes the probability that a measurement is missing is explainable by the variables you have measured, but sometimes not. Wide data sets with many features are the best way to investigate possible explanations for patterns of missing data, but these are not the norm for time series analysis.
- When you need to understand the uncertainty introduced by imputing values to missing data, you should run through a variety of scenarios and also speak to as many people involved in the data collection process as possible.

- How you handle missing data should account for your downstream use of that data. You must guard carefully against lookaheads or decide how seriously a lookahead will affect the validity of your subsequent work.

Upsampling and Downsampling

Often, related time series data from different sources will not have the same sampling frequency. This is one reason, among many, that you might wish to change the sampling frequency of your data. Of course you cannot change the actual rate at which information was measured, but you can change the frequency of the timestamps in your data collection. This is called *upsampling* and *downsampling*, for increasing or decreasing the timestamp frequency, respectively.

We downsampled temporal data in "Retrofitting a Time Series Data Collection from a Collection of Tables" on page 26. Here, we address the topic more generally, learning the how's and why's of both downsampling and upsampling.

 Downsampling is subsetting data such that the timestamps occur at a lower frequency than in the original time series. Upsampling is representing data as if it were collected more frequently than was actually the case.

Downsampling

Anytime you reduce frequency of your data, you are downsampling. This is most often done in the following cases.

The original resolution of the data isn't sensible. There can be many reasons that the original granularity of the data isn't sensible. For example, you may be measuring something too often. Suppose you have a data set where someone had measured the outside air temperature every second. Common experience dictates that this measurement is unduly frequent and likely offers very little new information relative to the additional data storage and processing burden. In fact, it's likely the measurement error could be as large as the second-to-second air temperature variation. So, you likely don't want to store such excessive and uninformative data. In this case—that is, for regularly sampled data—downsampling is as simple as selecting out every *n*th element.

Focus on a particular portion of a seasonal cycle. Instead of worrying about seasonal data in a time series, you might choose to create a subseries focusing on only one season. For example, we can apply downsampling to create a subseries, as in this case, where we generate a time series of January measurements out of what was originally a monthly time series. In the process, we have downsampled the data to a yearly frequency.

```
## R
> unemp[seq.int(from = 1, to = nrow(unemp), by = 12)]
     DATE       UNRATE
1948-01-01    3.4
1949-01-01    4.3
1950-01-01    6.5
1951-01-01    3.7
1952-01-01    3.2
1953-01-01    2.9
1954-01-01    4.9
1955-01-01    4.9
1956-01-01    4.0
1957-01-01    4.2
```

Match against data at a lower frequency. You may want to downsample data so that you can match it with other low-frequency data. In such cases you likely want to aggregate the data or downsample rather than simply dropping points. This can be something simple like a mean or a sum, or something more complicated like a weighted mean, with later values given more weight. We saw earlier in the donation data the idea of summing all donations over a single week, since it was the total amount donated that was likely to be most interesting.

In contrast, for our economic data, what is most likely to be interesting is a yearly average. We use a mean instead of a rolling mean because we want to summarize the year rather than get the latest value of that year to emphasize recency. (Note the difference from data imputation.) We group by formatting the date into a string, representing its year as an example of how you can creatively exploit SQL-like operations for time series functionality:

```
## R
> unemp[, mean(UNRATE), by = format(DATE, "%Y")]
    format    V1
    1948    3.75
    1949    6.05
    1950    5.21
    1951    3.28
    1952    3.03
    1953    2.93
    1954    5.59
    1955    4.37
    1956    4.13
    1957    4.30
```

Upsampling

Upsampling is not simply the inverse of downsampling. Downsampling makes inherent sense as something that can be done in the real world; it's simple to decide to measure less often. In contrast, upsampling can be like trying to get something for free—that is, not taking a measurement but still somehow thinking you can get high-resolution data from infrequent measurements. To quote the author of the popular R time series package XTS (*https://perma.cc/83E9-4N79*):

> It is not possible to convert a series from a lower periodicity to a higher periodicity - e.g., weekly to daily or daily to 5 minute bars, as that would require magic.

However, there are legitimate reasons to want to label data at a higher frequency than its default frequency. You simply need to keep in mind the data's limitations as you do so. Remember that you are adding more time labels but not more information.

Let's discuss a few situations where upsampling can make sense.

Irregular time series. A very common reason to upsample is that you have an irregularly sampled time series and you want to convert it to a regularly timed one. This is a form of upsampling because you are converting all the data to a frequency that is likely higher than indicated by the lags between your data. If you are upsampling for this reason, you already know how to do it with a rolling join, as we did this in the case of filling in missing economic data, via R:

```R
## R
> all.dates <- seq(from = unemp$DATE[1], to = tail(unemp$DATE, 1),
>                   by = "months")
> rand.unemp = rand.unemp[J(all.dates), roll=0]
```

Inputs sampled at different frequencies. Sometimes you need to upsample low-frequency information simply to carry it forward with your higher-frequency information in a model that requires your inputs to be aligned and sampled contemporaneously. You must be vigilant with respect to lookahead, but if we assume that known states are true until a new known state comes into the picture, we can safely upsample and carry our data forward. For example, suppose we know it's (relatively) true that most new jobs start on the first of the month. We might decide we feel comfortable using the unemployment rate for a given month indicated by the jobs report for the *entire* month (not considering it a lookahead because we make the assumption that the unemployment rate stays steady for the month).

```R
## R
> daily.unemployment = unemp[J(all.dates), roll = 31]
> daily.unemployment
   DATE      UNRATE
1948-01-01   3.4
1948-01-02   3.4
1948-01-03   3.4
1948-01-04   3.4
1948-01-05   3.4
```

Knowledge of time series dynamics. If you have underlying knowledge of the usual temporal behavior of a variable, you may also be able to treat an upsampling problem as a missing data problem. In that case, all the techniques we've discussed already still apply. An interpolation is the most likely way to produce new data points, but you would need to be sure the dynamics of your system could justify your interpolation decision.

As discussed earlier, upsampling and downsampling will routinely happen even in the cleanest data set because you will almost always want to compare variables of different timescales. It should also be noted that Pandas has particularly handy upsampling and downsampling functionality with the `resample` method.

Smoothing Data

Smoothing data can be done for a variety of reasons, and often real-world time series data is smoothed before analysis, especially for visualizations that aim to tell an understandable story about the data. In this section we discuss further why smoothing is done, as well as the most common time series smoothing technique: exponential smoothing.

Purposes of smoothing

While outlier detection is a topic in and of itself, if you have reason to believe your data should be smoothed, you can do so with a moving average to eliminate measurement spikes, errors of measurement, or both. Even if the spikes are accurate, they may not reflect the underlying process and may be more a matter of instrumentation problems; this is why it's quite common to smooth data.

Smoothing data is strongly related to imputing missing data, and so some of those techniques are relevant here as well. For example, you can smooth data by applying a rolling mean, with or without a lookahead, as that is simply a matter of the point's position relative to the window used to calculate its smoothed value.

When you are smoothing, you want to think about a number of questions:

- Why are you smoothing? Smoothing can serve a number of purposes:

    Data preparation
      Is your raw data unsuitable? For example, you may know very high values are unlikely or unphysical, but you need a principled way to deal with them. Smoothing is the most straightforward solution.

    Feature generation
      The practice of taking a sample of data, be it many characteristics about a person, image, or anything else, and summarizing it with a few metrics. In this way a fuller sample is collapsed along a few dimensions or down to a few traits. Feature generation is especially important for machine learning.

Prediction

The simplest form of prediction for some kinds of processes is mean reversion, which you get by making predictions from a smoothed feature.

Visualization

Do you want to add some signal to what seems like a noisy scatter plot? If so, what is your intention in doing so?

- How will your outcomes be affected by smoothing or not smoothing?

  — Does your model assume noisy and uncorrelated data, whereby your smoothing could compromise this assumption?

  — Will you need to smooth in a live production model? If so, you need to choose a smoothing method that does not employ a lookahead.

  — Do you have a principled way to smooth, or will you simply do a hyperparameter grid search? If the latter, how will you make sure that you use a time-aware form of cross-validation such that future data does not leak backward in time?

What Are Hyperparameters?

The term *hyperparameters* can sound awfully scary, but they are simply the numbers you use to tune, or adjust, the performance of a statistical or machine learning model. To optimize a model, you try several variations of the hyperparameters of that model, often performing a *grid search* to identify these parameters. A grid search is exactly that: you construct a "grid" of all possible combinations of hyperparameters in a search space and try them all.

For example, imagine an algorithm takes hyperparameters A and B, where we think reasonable values for A are 0, 1, or 2, and we think reasonable values for B are 0.7, 0.75, or 0.8. Then we would have nine possible hyperparameter options for our grid: every possible combination of potential A values (of which there are 3) with every possible B value (of which there are 3) for a total of $3 \times 3 = 9$ possibilities. As you can see, hyperparameter tuning can quickly get quite labor-intensive.

Exponential smoothing

You often won't want to treat all time points equally when smoothing. In particular, you may want to treat more recent data as more informative data, in which case exponential smoothing is a good option. In contrast to the moving average we looked at before—where each point where data was missing could be imputed to the mean of its surrounding points—exponential smoothing is geared to be more temporally aware, weighting more recent points higher than less recent points. So, for a given window, the nearest point in time is weighted most heavily and each point earlier in time is weighted exponentially less (hence the name).

The mechanics of exponential smoothing work as follows. For a given time period t, you find the smoothed value of a series by computing:

Smoothed value at time $t = S_t = d \times S_{t-1} + (1 - d) \times x_t$

Think about how this propagates over time. The smoothed value at time $(t - 1)$ is itself a product of the same thing:

$$S_{t-1} = d \times S_{t-2} + (1 - d) \times x_{t-1}$$

So we can see a more complex expression for the smoothed value at time t:

$$d \times (d \times S_{t-2} + (1 - d) \times x_{t-1}) + (1 - d) \times x_t$$

Mathematically inclined readers will notice that we have a series of the form:

$$d^3 \times x_{t-3} + d^2 \times x_{t-2} + d \times x_{t-1}$$

In fact, it is thanks to this form that exponential moving averages are quite tractable. More details are widely available online and in textbooks; see my favorite summaries in "More Resources" on page 69.

I will illustrate smoothing in Python, as Pandas includes a variety of smoothing options. Smoothing options are also widely available in R, including base R, as well as many time series packages.

While we have been looking at US unemployment rate data, we will switch to another commonly used data set: the airline passenger data set (which dates back to Box and Jenkins's famous time series book and is widely available). The original data set is an account of the thousands of monthly airline passengers broken down by month:

```python
## python
>>> air
        Date  Passengers
0    1949-01         112
1    1949-02         118
2    1949-03         132
3    1949-04         129
4    1949-05         121
5    1949-06         135
6    1949-07         148
7    1949-08         148
```

```
8      1949-09      136
9      1949-10      119
10     1949-11      104
```

We can easily smooth the values of passengers using a variety of decay factors and applying the Pandas `ewma()` function, like so:

```
air['Smooth.5'] = air.ewm(alpha=0.5).mean().Passengers
air['Smooth.1'] = air.ewm(alpha=0.1,).mean().Passengers
air['Smooth.9'] = air.ewm(alpha=0.9,).mean().Passengers
```

As we can see, the level of the `alpha` parameter, also called the *smoothing factor*, affects how much the value is updated to its current value versus retaining information from the existing average. The higher the `alpha` value, the more quickly the value is updated closer to its current price. Pandas accepts a number of parameters, all of which plug into the same equation, but they offer multiple ways to think about specifying an exponential moving average.[3]

```
## python
>>> air
        Date  Passengers      Smooth.5   Smooth.9
0    1949-01         112    112.000000  112.000000
1    1949-02         118    116.000000  117.454545
2    1949-03         132    125.142857  130.558559
3    1949-04         129    127.200000  129.155716
4    1949-05         121    124.000000  121.815498
5    1949-06         135    129.587302  133.681562
6    1949-07         148    138.866142  146.568157
7    1949-08         148    143.450980  147.856816
8    1949-09         136    139.718200  137.185682
9    1949-10         119    129.348974  120.818568
10   1949-11         104    116.668295  105.681857
```

However, simple exponential smoothing does not perform well (for prediction) in the case of data with a long-term trend. Holt's Method and Holt–Winters smoothing , are two exponential smoothing methods applied to data with a trend, or with a trend and seasonality.

There are many other widely used smoothing techniques. For example, Kalman filters smooth the data by modeling a time series process as a combination of known dynamics and measurement error. LOESS (short for "locally estimated scatter plot smoothing") is a nonparametric method of locally smoothing data. These methods, and others, gradually offer more complex ways of understanding smoothing but at increased computational cost. Of note, Kalman and LOESS incorporate data both earlier and later in time, so if you use these methods keep in mind the leak of

---

3 You can specify the alpha smoothing factor, the halflife, the span, or the center of mass according to which you feel comfortable with. Details are available in the documentation (*https://perma.cc/4265-4U8L*).

information backward in time, as well as the fact that they are usually not appropriate for preparing data to be used in forecasting applications.

Smoothing is a commonly used form of forecasting, and you can use a smoothed time series (without lookahead) as one easy null model when you are testing whether a fancier method is actually producing a successful forecast.

How to Start a Smoothing Calculation

One important reason to use a packaged exponential smoothing function is that it is difficult to get the start of your smoothed series correct. Imagine that the first point in your series is equal to 3 and the second point is equal to 6. Also imagine that your discounting factor (which you set according to how quickly you want the average to adapt to the most recent information) is equal to 0.7, so that you would compute the second point in your series as:

$$3 \times 0.7 + 6 \times (1 - 0.7) = 3.9$$

In fact, you would be wrong. The reason is that multiplying your discounting factor of 0.7 by 3 implicitly assumes that 3 is the sum of your knowledge going back in time to infinity, rather than a mere single data point you just measured. Therefore, it unduly weighs 3 relative to 6 as though 3 has a significantly more longstanding set of knowledge than it actually does.

The algebra to derive the correct expression is too complicated for a short sidebar, but check the additional resources at the end of this book for more details. In the meantime, I will leave you with the general guidance that 6 should be weighed a little more heavily than it is here, and 3 a little less. As you get further along in a time series, your older values should be weighted closer to your true discounting factor of 0.7. How soon this happens can be seen in the algebra, so take a look at the additional resources if you are interested.

Seasonal Data

Seasonality in data is any kind of recurring behavior in which the frequency of the behavior is stable. It can occur at many different frequencies at the same time. For example, human behavior tends to have a daily seasonality (lunch at the same time every day), a weekly seasonality (Mondays are similar to other Mondays), and a yearly seasonality (New Year's Day has low traffic). Physical systems also demonstrate seasonality, such as the period the Earth takes to revolve around the sun.

Identifying and dealing with seasonality is part of the modeling process. On the other hand, it's also a form of data cleaning, such as the economically important US Jobs Report (*https://perma.cc/GX6J-QJG9*). Indeed, many government statistics, particularly economic ones, are deseasonalized in their released form.

To see what seasonal data smoothing can do, we return to the canonical data set on airline passenger counts. A plot quickly reveals that this is highly seasonal data, but only if you make the right plot.

Notice the difference between using R's default plot (which uses points; see Figure 2-10) versus adding the argument to indicate that you want a line (Figure 2-11).

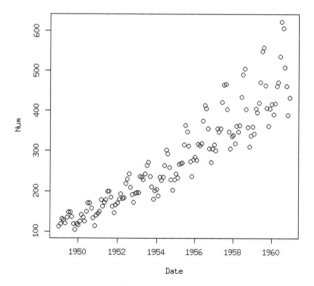

Figure 2-10. The increasing mean and variance of the data are apparent from a scatter plot, but we do not see an obvious seasonal trend.

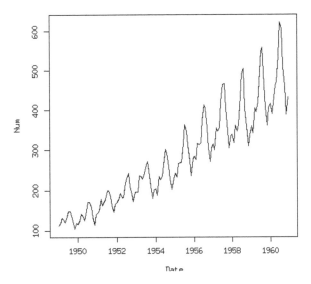

Figure 2-11. A line plot makes the seasonality abundantly clear.

If you were looking only at the default R plot, the seasonal nature of the data might elude you. Hopefully that wouldn't remain the case for long, as no doubt you would also do things to explore your data more fully, perhaps with an autocorrelation plot (discussed in Chapter 3) or other diagnostics.

Humans Are Creatures of Habit

With human behavior data there is almost always some form of seasonality, even with several cycles (an hourly pattern, a weekly pattern, a summer-winter pattern, etc.).

The scatter plot does show some information more clearly than the line plot. The variance of our data is increasing, as is the mean, which is most obvious when we see a cloud of data points beaming outward in a conical shape, tilted upward. This data clearly has a trend, and so we would likely transform it with a log transform or differencing, depending on the demands of our model. This data also clearly has a trend of increasing variance. We'll talk more about model-specific data transformations on seasonal data in the modeling chapters, so we won't elaborate here.

We also get useful information apart from the evidence for seasonality from the line plot. We get information about what *kind* of seasonality. That is, we see that the data is not only seasonal, but seasonal in a multiplicative way. As the overall values get larger, so do the season swings (think of this as sizing from peak to trough).

We can decompose the data into its seasonal, trend, and remainder components very easily as shown here with just one line of R:

```
## R
> plot(stl(AirPassengers, "periodic"))
```

The resulting plot seems very sensible based on the original data (see Figure 2-12). We can imagine adding the seasonal, trend, and remainder data back together to get the original series. We can also see that this particular decomposition did not take into account the fact that this series shows a multiplicative seasonality rather than an additive one because the residuals are greatest at the beginning and end of the time series. It seems like this decomposition settled on the average seasonal variance as the variance for the seasonal component.

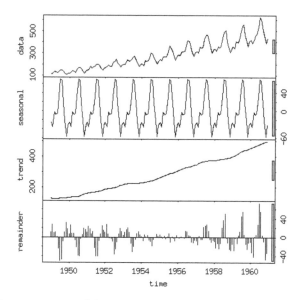

Figure 2-12. A decomposition of the original time series into a seasonal component, a trend, and the residuals. Pay attention to each plot's y-axis, as they are quite different. Note that this is the reason for the gray bars on the right side of each plot. These gray bars are all the same absolute size (in units of the y-axis), so that their relatively different display is a visual reminder of the different y-axis scales across different components.

To get an initial understanding of how this works, we can look at the official R documentation:

> The seasonal component is found by LOESS smoothing the seasonal subseries (the series of all January values, ...). If s.window = "periodic" smoothing is effectively replaced by taking the mean. The seasonal values are removed, the remainder smoothed to find the trend. The overall level is removed from the seasonal component and added to the

trend component. This process is iterated a few times. The remainder component is the residuals from the seasonal plus trend fit.

LOESS, introduced earlier, is a computationally taxing method for smoothing data points that involves a moving window to estimate the smoothed value of each point based on its neighbors (I hope your lookahead alarm is sounding!).

Seasonal Data and Cyclical Data

Seasonal time series are time series in which behaviors recur over a fixed period. There can be multiple periodicities reflecting different tempos of seasonality, such as the seasonality of the 24-hour day versus the 12-month calendar season, both of which exhibit strong features in most time series relating to human behavior.

Cyclical time series also exhibit recurring behaviors, but they have a variable period. A common example is a business cycle, such as the stock market's boom and bust cycles, which have an uncertain duration. Likewise, volcanoes show cyclic but not seasonal behaviors. We know the approximate periods of eruption, but these are not precise and vary over time.

Time Zones

Time zones are intrinsically tedious, painful, and difficult to get right even with a lot of effort. That is why you should never use your own solution. From their very invention, time zones have been complicated, and they have only gotten more so with the advent of personal computers. There are many reasons for this:

- Time zones are shaped by political and social decisions.
- There is no standard way to transport time zone information between languages or via an HTTP protocol.
- There is no single protocol for naming time zones or for determining start and end dates of daylight savings offsets.
- Because of daylight savings, some times occur twice a year in their time zones!

Most languages rely on the underlying operating system to get time zone information. Unfortunately, the built-in automatic time retrieval function in Python, `date time.datetime.now()`, does not return a time zone–aware timestamp. Part of this is by design. Some decisions made in the standard library include forbidding time zone information in the `datetime` module (because this information changes so often) and allowing `datetime` objects both with and without time zone information. However, comparing an object with a time zone to one without a time zone will cause a `TypeError`.

Some bloggers claim that a majority of libraries are written with the assumption that tzinfo==None. Although this is a difficult claim to substantiate, it is consistent with much experience. People also report difficulty pickling time zone–stamped objects, so this is also something to check on early if you plan to use pickling.[4]

So let's have a look at working with time zones in Python. The main libraries you are likely to use are datetime, pytz, and dateutil. Additionally Pandas offers convenient time zone–related functionality built on top of the latter two libraries.

We'll cover the most important time zone functionality next.

First, notice that when you retrieve a "now" from the datetime module, it does not come with time zone information, even though it will give you the appropriate time for your time zone. Note, for example, the difference in the responses for now() and utcnow():

```python
## python
>>> datetime.datetime.utcnow()
datetime.datetime(2018, 5, 31, 14, 49, 43, 187680)

>>> datetime.datetime.now()
datetime.datetime(2018, 5, 31, 10, 49, 59, 984947)
>>> # as we can see, my computer does not return UTC
>>> # even though there is no time zone attached

>>> datetime.datetime.now(datetime.timezone.utc)
datetime.datetime(2018, 5, 31, 14, 51, 35, 601355,
                  tzinfo=datetime.timezone.utc)
```

Notice that if we do pass in a time zone, we will get the correct information, but this is not the default behavior. To work with time zones in Python, we create a time zone object, such western as for the US Pacific time zone:

```python
## python
>>> western = pytz.timezone('US/Pacific')
>>> western.zone
'US/Pacific'
```

We can then use these objects to localize a time zone as follows:

```python
## python
>>> ## the API supports two ways of building a time zone-aware time,
>>> ## either via 'localize' or to convert a time zone from one locale
>>> ## to another
>>> # here we localize
>>> loc_dt = western.localize(datetime.datetime(2018, 5, 15, 12, 34, 0))
```

---

4 Pickling is the process of storing Python objects in byte format. This is done via the popular built-in pickle module. Most of the time pickling will work well, but sometimes time related objects can be difficult to pickle.

```
datetime.datetime(2018, 5, 15, 12, 34,
            tzinfo=<DstTzInfo 'US/Pacific' PDT-1 day, 17:00:00 DST>)
```

Note, however, that passing the time zone directly into the datetime constructor will often not produce the result we were expecting:

```python
## python
>>> london_tz = pytz.timezone('Europe/London')
>>> london_dt = loc_dt.astimezone(london_tz)

>>> london_dt
datetime.datetime(2018, 5, 15, 20, 34,
            tzinfo=<DstTzInfo 'Europe/London' BST+1:00:00 DST>)

>>> f = '%Y-%m-%d %H:%M:%S %Z%z'
>>> datetime.datetime(2018, 5, 12, 12, 15, 0,
                    tzinfo = london_tz).strftime(f)
'2018-05-12 12:15:00 LMT-0001'

>>> ## as highlighted in the pytz documentation using the tzinfo of
>>> ## the datetime.datetime initializer does not always lead to the
>>> ## desired outcome, such as with the London example

>>> ## according to the pytz documentation, this method does lead to
>>> ## the desired results in time zones without daylight savings
```

This is important, such as when you are calculating time deltas. The first example of the following three is the gotcha:

```python
## python
>>> # generally you want to store data in UTC and convert only when
>>> # generating human-readable output
>>> # you can also do date arithmetic with time zones
>>> event1 = datetime.datetime(2018, 5, 12, 12, 15, 0,
                            tzinfo = london_tz)
>>> event2 = datetime.datetime(2018, 5, 13, 9, 15, 0,
                            tzinfo = western)
>>> event2 - event1
>>> ## this will yield the wrong time delta because the time zones
>>> ## haven't been labelled properly

>>> event1 = london_tz.localize(
                datetime.datetime(2018, 5, 12, 12, 15, 0))
>>> event2 = western.localize(
                datetime.datetime(2018, 5, 13, 9, 15, 0))
>>> event2 - event1

>>> event1 = london_tz.localize(
                (datetime.datetime(2018, 5, 12, 12, 15, 0))).
                    astimezone(datetime.timezone.utc)
>>> event2 = western.localize(
                datetime.datetime(2018, 5, 13, 9, 15, 0)).
```

```
                     astimezone(datetime.timezone.utc)
>>> event2 - event1
```

pytz provides a list of common time zones and time zones by country, both of which can be handy references:

```
## python
## have a look at pytz.common_timezones
>>> pytz.common_timezones
(long output...)

## or country specific
>>> pytz.country_timezones('RU')
['Europe/Kaliningrad', 'Europe/Moscow', 'Europe/Simferopol',
 'Europe/Volgograd', 'Europe/Kirov', 'Europe/Astrakhan',
 'Europe/Saratov', 'Europe/Ulyanovsk', 'Europe/Samara',
 'Asia/Yekaterinburg', 'Asia/Omsk', 'Asia/Novosibirsk',
 'Asia/Barnaul', 'Asia/Tomsk', 'Asia/Novokuznetsk',
 'Asia/Krasnoyarsk', 'Asia/Irkutsk', 'Asia/Chita',
 'Asia/Yakutsk', 'Asia/Khandyga', 'Asia/Vladivostok',
 'Asia/Ust-Nera', 'Asia/Magadan', 'Asia/Sakhalin',
 'Asia/Srednekolymsk', 'Asia/Kamchatka', 'Asia/Anadyr']
>>>
>>> pytz.country_timezones('fr')
>>> ['Europe/Paris']
```

A particularly hairy issue is the matter of daylight savings. Certain human-readable times exist twice (falling behind in the autumn), while others do not exist at all (skip ahead in the spring):

```
## python
>>> ## time zones
>>> ambig_time = western.localize(
                    datetime.datetime(2002, 10, 27, 1, 30, 00)).
                        astimezone(datetime.timezone.utc)
>>> ambig_time_earlier = ambig_time - datetime.timedelta(hours=1)
>>> ambig_time_later = ambig_time + datetime.timedelta(hours=1)
>>> ambig_time_earlier.astimezone(western)
>>> ambig_time.astimezone(western)
>>> ambig_time_later.astimezone(western)

>>> #results in this output
datetime.datetime(2002, 10, 27, 1, 30,
          tzinfo=<DstTzInfo 'US/Pacific' PDT-1 day, 17:00:00 DST>)
datetime.datetime(2002, 10, 27, 1, 30,
          tzinfo=<DstTzInfo 'US/Pacific' PST-1 day, 16:00:00 STD>)
datetime.datetime(2002, 10, 27, 2, 30,
          tzinfo=<DstTzInfo 'US/Pacific' PST-1 day, 16:00:00 STD>)
>>> # notice that the last two timestamps are identical, no good!


>>> ## in this case you need to use is_dst to indicate whether daylight
>>> ## savings is in effect
```

```
>>> ambig_time = western.localize(
        datetime.datetime(2002, 10, 27, 1, 30, 00), is_dst = True).
          astimezone(datetime.timezone.utc)
>>> ambig_time_earlier = ambig_time - datetime.timedelta(hours=1)
>>> ambig_time_later = ambig_time + datetime.timedelta(hours=1)
>>> ambig_time_earlier.astimezone(western)
>>> ambig_time.astimezone(western)
>>> ambig_time_later.astimezone(western)

datetime.datetime(2002, 10, 27, 0, 30,
        tzinfo=<DstTzInfo 'US/Pacific' PDT-1 day, 17:00:00 DST>)
datetime.datetime(2002, 10, 27, 1, 30,
        tzinfo=<DstTzInfo 'US/Pacific' PDT-1 day, 17:00:00 DST>)
datetime.datetime(2002, 10, 27, 1, 30,
        tzinfo=<DstTzInfo 'US/Pacific' PST-1 day, 16:00:00 STD>)
## notice that now we don't have the same time happening twice.
## it may appear that way until you check the offset from UTC
```

Time zone concerns may not be important to your work, so the utility of this knowledge will depend on the nature of your data. There are certainly situations where getting this wrong could be catastrophic (say, generating weather forecasts for commercial airliners flying during the time change and suddenly finding their positions drastically altered).

Preventing Lookahead

Lookahead is dangerously easy to introduce into a modeling pipeline, especially with vectorized functional data manipulation interfaces, such as those offered by R and Python. It is easy to shift a variable in the wrong direction, shift it more or less than you intended, or otherwise find yourself with data that is not entirely "honest" in that you have data available before it would have been known in your system.

Unfortunately, there isn't a definitive statistical diagnosis for lookahead—after all, the whole endeavor of time series analysis is modeling the unknown. Unless a system is somewhat deterministic with known dynamical laws, it can be difficult to distinguish a very good model from a model with lookahead—that is, until you put a model into production and realize either that you are missing data when you planned to have it, or simply that your results in production do not reflect what you see during training.

The best way to prevent this embarrassment is constant vigilance. Whenever you are time-shifting data, smoothing data, imputing data, or upsampling data, ask yourself whether you could know something at a given time. Remember that doesn't just include calendar time. It also includes realistic time lags to reflect how long a delay there is between something happening and your organization having that data available. For example, if your organization scrapes Twitter only weekly to gather its sentiment analysis data, you need to include this weekly periodicity in your training and validation data segmentation. Similarly, if you can retrain your model only once a month, you need to figure out what model would apply to what data over time. You can't, for example, train a model for July and then apply it to July for testing, because in a real situation, you wouldn't have that model trained up in time if training takes you a long time.

Here are some other ideas to use as a general checklist. Keep them in mind both when planning to build a model and when auditing your process after the fact:

- If you are smoothing data or imputing missing data, think carefully about whether it might impact your results by introducing a lookahead. And don't just think about it—experiment as we did earlier and see how the imputations and smoothing work. Do they seem to be forward looking? If so, can you justify using them? (Probably not.)

- Build your entire process with a very small data set (only a few rows in a `data.table` or a few row time steps in whatever data format). Then, do random spot checks at each step in the process and see whether you accidentally shift any information temporally to an inappropriate place.

- For each kind of data, find out what the lag is for it relative to its own timestamp. For example, if the timestamp is when the data "happened" but not when it was uploaded to your servers, you need to know that. Different columns of a data frame may have different lags. To address this, you can either customize your lag per data frame or (better and more realistic) pick the biggest lag and apply that to everything. While you won't want to unduly pessimize your model, it's a good starting point after which you can relax these overly constrained rules one at a time, carefully!

- Use time-aware error (rolling) testing or cross-validation. This will be discussed in Chapter 11, but remember that randomizing your training versus testing data sets does not work with time series data. You do not want information from the future to leak into models for the past.

- Intentionally introduce a lookahead and see how your model behaves. Try various degrees of lookahead, so you have an idea how it shifts accuracy. If you have some idea of the accuracy with lookahead, you have an idea of what the ceiling on a real model without unfair knowledge of the future will do. Remember that many time series problems are extremely difficult, so a model with a lookahead may seem great until you realize you are dealing with a high-noise/low-signal data set.

- Add features slowly, particularly features you might be processing, so that you can look for jumps. One sign of a lookahead is when a particular feature is unexpectedly good, and there isn't a very good explanation. At the top of your explanation list should always be "lookahead."

Processing and cleaning time-related data can be a tedious and detail-oriented process.

There is tremendous danger in data cleaning and processing of introducing a *lookahead*! You should have lookaheads only if they are intentional, and this is rarely appropriate.

More Resources

- On missing data:

Steffen Moritz et al., "Comparison of Different Methods for Univariate Time Series Imputation in R," unpublished research paper, October 13, 2015, https://perma.cc/ M4LJ-2DFB.
> This thorough summary from 2015 outlines available methods for imputing time series data in the case of univariate time series data. Univariate time series data is a particular challenge because many advanced missing data imputation methods rely on looking at distributions among covariates, an option that is not available in the case of a univariate time series. This paper summarizes both the usability and performance of various R packages as well as empirical results of the methods available on a variety of data sets.

James Honaker and Gary King, "What to Do About Missing Values in Time-Series Cross-Section Data," American Journal of Political Science 54, no. 2 (2010): 561– 81, https://perma.cc/8ZLG-SMSX.
> This article explores best practices for missing data in time series with wide panels of covariates.

Léo Belzile, "Notes on Irregular Time Series and Missing Values," n.d. https:// perma.cc/8LHP-92FP.
> The author provides examples of working with irregular data as a missing data problem and an overview of some commonly used R packages.

- On time zones:

Tom Scott, The Problem with Time & Timezones, Computerphile video, December 30, 2013, https://oreil.ly/iKHkp.
> This 10-minute YouTube video with over 1.5 million views outlies the perils and challenges of dealing with time zones, particularly in the context of a web application.

Wikipedia, "Time Zone," https://perma.cc/J6PB-232C.
> This is a fascinating history in short form that gives some insight into how timekeeping worked before the last century and how technology advancements (beginning with the railway) led to the increasing need for people in

different locations to coordinate their time. There are also some fun maps of time zones.

Declan Butler, "GPS Glitch Threatens Thousands of Scientific Instruments,"
Nature, April 3, 2019, https://perma.cc/RPT6-AQBC.
This article is not directly related to time zones but to the problem of time-stamping more generally. It describes a recent problem in which a bug in the US Global Positioning System could cause a problem with timestamped data because it transmits a binary 10-digit "week number" that began counting from January 6, 1980. This system can cover only 1,024 weeks total (2 raised to the power of 10). This count was reached for the second time in April 2019. Devices that were not designed to accommodate this limitation reset to zero and incorrectly timestamped scientific and industrial data. This article describes some of the difficulties with a fairly limited representation of time combined with scientific devices that were not future-proofed to account for this problem.

- On smoothing and seasonality:

Rob J. Hyndman and George Athanasopoulos, "Exponential Smoothing," in Fore-
casting: Principles and Practices, 2nd ed. (Melbourne: OTexts, 2018), https://
perma.cc/UX4K-2V5N.
This chapter of Hyndman and Athanasopoulos's introductory academic textbook covers exponential smoothing methods for time series data, including a helpful taxonomy of exponential smoothing and extensions of exponential smoothing to forecasting applications.

David Owen, "The Correct Way to Start an Exponential Moving Average," For-
ward Motion blog, January 31, 2017, https://perma.cc/ZPJ4-DJJK.
As highlighted in a side note earlier, exponential moving averages are conceptually simple and easy to calculate, with the complication that how to "start" off the calculation is a little complicated. We want to make sure our moving average adapts to new information in a way that recognizes how long it has been recording information. If we start a moving average without giving consideration to this, even a new moving average will behave as though it has infinite lookback and unduly discount new information relative to what it should do. More details and a computational solution are presented in this blog post.

Avner Abrami, Aleksandr Arovkin, and Younghun Kim, "Time Series Using Exponential Smoothing Cells," unpublished research paper, last revised September 29, 2017, https://perma.cc/2JRX-K2JZ.

This is an extremely accessible time series research paper elaborating on the idea of simple exponential smoothing to develop "exponential smoothing cells."

- On functional data analysis:

Jane-Ling Wang, Jeng-Min Chiou, and Hans-Georg Müller, "Review of Functional Data Analysis," Annual Reviews of Statistics and its Application, 2015, https://perma.cc/3DNT-J9EZ.

This statistics article offers an approachable mathematical overview of important functional data analysis techniques. There are also helpful visualization techniques illustrated throughout the article.

Shahid Ullah and Caroline F. Finch, "Applications of Functional Data Analysis: A Systematic Review," BMC Medical Research Methodologyv, 13, no. 43 (2013), https://perma.cc/VGK5-ZEUX.

This review takes a multidisciplinary approach to surveying recently published analyses using functional data analysis. The authors argue that the techniques of functional data analysis have much wider applicability than is currently appreciated and make the case that the biological and health sciences could benefit from more use of these techniques to look at medical time series data.

## CHAPTER 3

Exploratory Data Analysis for Time Series

We will break down our discussion of exploratory data analysis for time series into two distinct sections. First, we discuss the application of commonly used data methods on time series. In particular, we discuss how histograms, plotting, and group-by operations can be applied to time series data.

Second, we highlight fundamentally temporal methods for time series analysis—that is, methods developed specifically for time series data and that make sense only in the context of data points that have a temporal relationship with one another (rather than a cross-sectional one).

Familiar Methods

We begin by thinking about how to apply commonly used data exploration techniques to time series data sets. The process is the same as what you have performed on non-time-series data to start with.

You will want to know the columns that are available, their value ranges, and what logical units of measurement work best. You will want to address the same exploratory questions you would ask about any new data set, such as:

- Are any of the columns strongly correlated with one another?
- What is the overall mean of an interesting variable? What is its variance?

To answer these, you can use familiar techniques such as plotting, taking summary statistics, applying histograms, and using targeted scatter plots. You'll also want to answer explicitly time-oriented questions such as:

- What is the range of values you see, and do they vary by time period or some other logical unit of analysis?

- Does the data look consistent and uniformly measured, or does it suggest changes in either measurement or behavior over time?

To address these, you will want the previously mentioned methods—histograms, scatter plots, and summary statistics—to take into account the temporal axis. To implement this behavior, we will incorporate time into our statistics, as an axis in our graphs or as a group in group-by operations, such as those applied to histograms or scatter plots.

In the rest of this section, we will look at examples of exploratory data analysis with a few different kinds of time series data to explore how we can use traditional nontemporal methods directly or with time-specific modifications. To demonstrate these exploratory methods, we will use European stock markets data, a time series data set available in base R.

Group-by Operations

With time series data, there are groups just as there are with nontemporal data, so group-by-style thinking is still applicable. For example, in cross-sectional data you might take group means based on age, gender, or neighborhood. In time series analysis, you might find exploratory group-by operations useful, such as computing a monthly average or weekly medians. You can also combine temporal and nontemporal groups, such as the monthly average calorie intake per gender in a population, or the weekly median amount of sleep per age group in a set of hospital patients. There are many ways that grouping data can allow the temporal axis to interact with other relationships within the data. For example:

- You may want to analyze different time periods separately, if you have exogenous reasons to expect a regime change. Use your exploratory analysis to find sensible demarcations for different meaningful time periods.

- You may discover that data presented as a single series (such as when donations come in) is clearer when broken down into many parallel processes (such as individual donation timelines).

Plotting

Let's explore a typical time series provided by R, the EuStockMarkets data set. To get a sense of it, we can look at the head of the data set:

```R
## R
> head(EuStockMarkets)
         DAX     SMI     CAC    FTSE
[1,] 1628.75  1678.1  1772.8  2443.6
[2,] 1613.63  1688.5  1750.5  2460.2
[3,] 1606.51  1678.6  1718.0  2448.2
```

```
[4,] 1621.04 1684.1 1708.1 2470.4
[5,] 1618.16 1686.6 1723.1 2484.7
[6,] 1610.61 1671.6 1714.3 2466.8
```

This is a built-in data set in base R, and it contains the daily closing prices of four major European stock indexes from 1991 to 1998. The data set only includes business days.

This data set is better prepared than the ones we have looked at so far because it's already nicely formatted and sampled for us. We don't need to worry about missing values, time zones, or incorrect measurements, so we can jump right into exploratory data analysis.

Our first steps for looking at this data will be very similar to a non-time-based time series, although we have a simpler option with temporal data than we would with cross-sectional data. We can plot each value individually, and such a plot makes sense. Contrast this with cross-sectional data, where a plot of any individual feature against its index in the data set will only reveal, if anything, the order in which data was entered in a data table or returned from a SQL server (in other words, an arbitrary order), but nothing about the underlying situation. However, in the case of time series, plotting is quite informative, as you can see in Figure 3-1.

```
## R
> plot(EuStockMarkets)
```

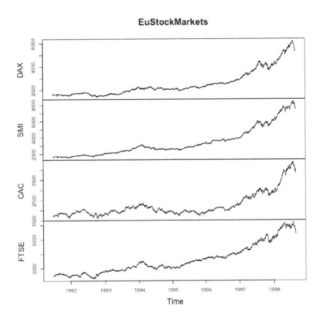

Figure 3-1. A simple plot of the time series data.

Notice that the image is automatically segmented into different time series with a simple plot() command. In fact, as shown here, we are using an R mts object (we would be using a ts object if there were only one time series, but we have several in parallel):

```
## R
> class(EuStockMarkets)
[1] "mts"     "ts"       "matrix"
```

A number of popular packages make heavy use of ts objects and derivative classes. These objects come with nice automatic calls to appropriate plotting functions, as we see in the previous example where a simple call to plot() created a nicely labeled, multipanel plot. ts objects also have a few convenience functions:

- frequency for finding out the yearly frequency of the data:

  ```
  ## R
  > frequency(EuStockMarkets)
  [1] 260
  ```

- start and end to find the first and last time represented in the series:

  ```
  ## R
  > start(EuStockMarkets)
  [1] 1991   130
  > end(EuStockMarkets)
  1] 1998   169
  ```

- window to take a temporal section of the data:

  ```
  ## R
  > window(EuStockMarkets, start = 1997, end = 1998)
  Time Series:
  Start = c(1997, 1)
  End = c(1998, 1)
  Frequency = 260          DAX SMI CAC    FTSE
  1997.000 2844.09 3869.8 2289.6 4092.5
  1997.004 2844.09 3869.8 2289.6 4092.5
  1997.008 2844.09 3869.8 2303.8 4092.5
  ...
  1997.988 4162.92 6115.1 2894.5 5168.3
  1997.992 4055.35 5989.9 2822.9 5020.2
  1997.996 4125.54 6049.3 2869.7 5018.2
  1998.000 4132.79 6044.7 2858.1 5049.8
  ```

There are pluses and minuses to the ts class. As mentioned earlier, ts and its derivative classes are used in many time series packages. Also, automatic setup of plotting parameters can be helpful. However, indexing can sometimes be tricky, and the process of accessing subsections of data with window can feel cumbersome over time. In this book, you will see several ways of containing and accessing time series data, and it will be up to you to choose the most convenient for your use cases.

Histograms

Let's continue our comparison of exploratory data analysis in time series and non-time-series data. We'd like to, for example, take a histogram of the data, just as with most exploratory data analysis. We add a wrinkle by also taking a histogram of the differenced data because we want to use our time axis (see Figure 3-2):

```
## R
> hist(    EuStockMarkets[, "SMI"], 30)
> hist(diff(EuStockMarkets[, "SMI"], 30))
```

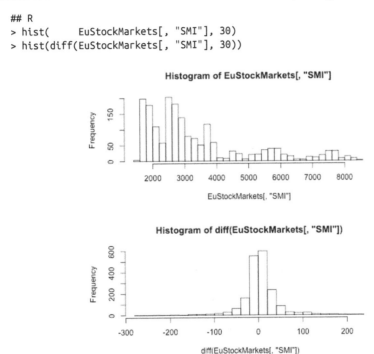

Figure 3-2. The histogram of the untransformed data (top) is extremely wide and does not show a normal distribution. This is to be expected given that the underlying data has a trend. We difference the data to remove the trend, and this transforms the data to a more normally shaped distribution (bottom).

In a time series context, a hist() of the difference of the data is often more interesting than a hist() of the untransformed data. After all, in a time series context, often (and particularly in finance) what is most interesting is how a value changes from one measurement to the next rather than the value's actual measurement. This is particularly true for plotting, because taking the histogram of data with a trend in it does not produce a very informative visualization.

Notice the new information we get from the histogram of the differenced series. While the original plots of the stocks in the previous section painted a very rosy economic picture with stocks inexorably rising, this histogram shows us the day-to-day experience of someone following the stocks. The histogram of the difference tells us that the value of the time series has gone both up (positive difference values) and down (negative difference values) about the same amount over time. The stock indexes did not go up and down exactly the same amount over time, because we know the stock indexes do have an increasing trend. However, we can see from this histogram that just a slight bias in favor of positive over negative differences is what accounts for that trend.

This is a good first example of why we need to pay attention to temporal scale when sampling, summarizing, and asking questions of data. Whether performance looks great (long-term plot) or just so-so (histogram of difference plots) will depend on what our temporal scale is: do we care about the day-to-day, or do we have a longer-term horizon? If we work in an organization that trades stock to make an annual profit, we need to think about the short-term experience we will have every time we report to our bosses a "diff" that leans toward negative values. However, if we are a large institutional investor—perhaps a university or a hospital—we may be able to afford to take the long view, which expects an upward climb. This latter scenario has its own challenges: how do we take long enough to maximize our profit but not so long that the party is over? These are the questions that keep the financial industry humming along with time series research and predictions.

Scatter Plots

The traditional method of using scatter plots is just as useful for time series data as it is for other kinds of data. We can use scatter plots to determine both how two stocks are linked at a specific time and how their price shifts are related over time.

In this example, we plot both cases (see Figure 3-3):

- The values of two different stocks over time
- The values of the daily changes in these two stocks over time (via differencing) with R's diff() function

```
## R
> plot(     EuStockMarkets[, "SMI"],        EuStockMarkets[, "DAX"])
> plot(diff(EuStockMarkets[, "SMI"]), diff(EuStockMarkets[, "DAX"]))
```

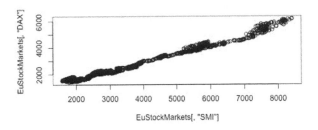

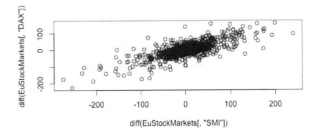

Figure 3-3. Simple scatter plots of two stock indexes suggest strong correlations. However, there are reasons to view these plots with suspicion.

As we've already seen, the actual values are less informative than the differences between adjacent time points, so we have plotted the diffs in a second scatter plot. These look like very strong correlations, but the relationships are not as strong as they appear. (To delve deeper, skip ahead to "Spurious Correlations" on page 102.)

The apparent correlations in Figure 3-3 are interesting, but even if they are true correlations (and there is reason to doubt they are), these are not correlations we can monetize as stock traders. By the time we know whether a stock is going up or down, the stocks it is correlated with will have also gone up or down, since we are taking correlations of values at identical time points. What we need to do is find out whether the change in one stock earlier in time can predict the change in another stock later in time. To do this, we shift one of the differences of the stocks back by 1 before looking at the scatter plot. Read the following code carefully; notice that we are still differencing, but now we are also applying a lag to one of the differenced time series (see Figure 3-4):

```
## R
> plot(lag(diff(EuStockMarkets[, "SMI"]), 1),
        diff(EuStockMarkets[, "DAX"]))
```

The lines in these code examples are easy to read because each section is aligned. It can be tempting to write long lines of unreadable code, particularly in a functional programming language such as R or Python, but you should avoid this whenever possible!

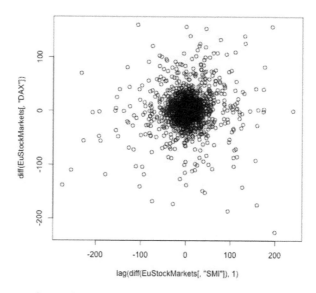

Figure 3-4. The correlation between the stocks disappears as soon as we put in a time lag, indicating that SMI does not appear to predict DAX.

This result tells us a number of important things:

- With time series data, while we may use the same exploratory techniques as with non-time-series data, mindless application won't work. We need to think about how to exploit the same techniques but with reshaped data.

- Oftentimes it's the relationship between data at different points or the change over time that is most informative about how your data behaves.

- The plot in Figure 3-4 shows why it can be difficult to be a stock trader. If you are a passive investor and wait it out, you might benefit from the long-term rising trend. However, if you are trying to make predictions about the future, you can see that it's not easy!

R's lag() function may not move data in the temporal direction you expect. The lag() function is a forward time shift. This is important to remember, as you wouldn't want to move data in the wrong temporal direction given a specific use case, and both forward and backward shifts in time are reasonable for different use cases.

Time Series–Specific Exploratory Methods

Several methods of analyzing time series data focus on relations of values at different times in the same series, and you likely won't have seen these before if you haven't worked with time series data. In the rest of this chapter, we walk through a few concepts and related techniques used to classify time series.

The concepts we will explore are:

Stationarity
> What it means for a time series to be stationary and a statistical test for stationarity

Self-correlation
> What it means to say that a time series correlates with itself and what such a correlation indicates about the underlying dynamics of the time series

Spurious correlations
> What it means for a correlation to be spurious and when you should expect to run into spurious correlations

The methods we will learn to apply are:

- Rolling and expanding window functions
- Self-correlation functions
  - The autocorrelation function
  - The partial autocorrelation function

We will cover the concepts and their resulting methods in order from stationarity to self-correlations to spurious correlations. Before we dive into the specifics, let's discuss the logic behind this particular ordering.

The first question you will likely ask about a time series is whether it appears to reflect a system that is "stable" or one that is constantly changing. The level of stability, or *stationarity*, is important to assess because we need to know how much we should expect the system's long-term past behavior to reflect its long-term future behavior. Once we have assessed the "stability" (this word is not used in a technical sense here) of a time series, we try to determine whether there are internal dynamics in that series (seasonal changes, for example). That is, we are looking for *self-correlations*, answering the fundamental question of how tightly past data, distant or recent, predicts future data. Finally, when we think we have found certain behavioral dynamics within the system, we need to make sure we are not identifying relationships based on dynamics that do not in any way imply the causal relationships we wish to discover; hence, we must look for *spurious correlations*.

Understanding Stationarity

Many traditional statistical time series models rely on a time series being stationary. Generally speaking, a stationary time series is one that has fairly stable statistical properties over time, particularly with respect to mean and variance. This seems relatively straightforward.

Nonetheless, stationarity can be a slippery concept, particularly when applied to real time series data. It is both too intuitive and too easy to fool yourself into relying on your natural intuition. We will walk through the concept both intuitively and with a somewhat formal definition before discussing a common test for stationarity and practical details for how to apply the concept.

Intuition

A stationary time series is one in which a time series measurement reflects a system in a steady state. Sometimes it is difficult to assert what exactly this means, and it can be easier to rule things out as *not* being stationary rather than saying something *is* stationary. An easy example of data that is not stationary is the airline passengers data set we examined in Chapter 2, which is plotted in Figure 3-5 (as a reminder, it is available in R as AirPassengers and also widely available for download on the internet).

There are several traits that show this process is not stationary. First, the mean value is increasing over time, rather than remaining steady. Second, the distance between peak and trough on a yearly basis is growing, so the variance of the process is increasing over time. Third, the process displays strong seasonal behavior, the antithesis of stationarity.

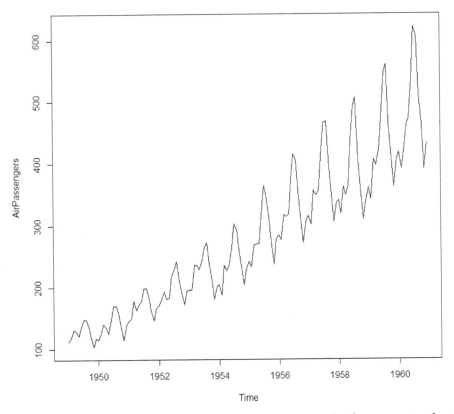

Figure 3-5. The airline passengers data set offers a clear example of a time series that is not stationary. As time passes, both the mean and the variance of the data are changing. We also see evidence of seasonality, which intrinsically reflects a nonstationary process.

Stationary definition and the Augmented Dickey–Fuller test

A simple definition of a stationary process is the following: a process is stationary if for all possible lags, k, the distribution of $y_t, y_{t+1},..., y_{t+k}$, does not depend on t.

Statistical tests for stationarity often come down to the question of whether there is a unit root—that is, whether 1 is a solution of the process's characteristic equation.[1] A linear time series is nonstationary if there is a unit root, although lack of a unit root does not prove stationarity. Addressing stationarity as a general question remains tricky, and determining whether a process has a unit root remains a current area of research.

---

1 If this doesn't ring a bell, don't worry about it. You can read more in "More Resources" on page 117, should you wish to pursue the topic.

Nonetheless, a simple intuition for what a unit root is can be gleaned from the example of a random walk:

$$y_t = \phi \times y_{t-1} + e_t$$

In this process, the value of a time series at a given time is a function of its value at the immediately preceding time and some random error. If ϕ is equal to 1, the series has a unit root, will "run away," and will not be stationary. Interesting to note is that the series not being stationary does not mean it has to have a trend. A random walk is a good example of a nonstationary time series that does not have an underlying trend.[2]

Tests for determining whether a process is stationary are called *hypothesis tests*. The Augmented Dickey–Fuller (ADF) test is the most commonly used metric to assess a time series for stationarity problems. This test posits a null hypothesis that a unit root is present in a time series. Depending on the results of the test, this null hypothesis can be rejected for a specified significance level, meaning the presence of a unit root test can be rejected at a given significance level.

Note that tests for stationarity focus on whether the mean of a series is changing. The variance is handled by transformations rather than formally tested. The test of whether a series is stationary is thus a test of whether a series is integrated. An integrated series of order d is a series that must be differenced d times to become stationary.

The framing of the Dickey–Fuller test is as follows:

$$\Delta y_t = y_t - y_{t-1} = (\phi - 1) \times y_{t-1} + e_t$$

Then the test of whether $\phi = 1$ is a simple t-test of whether the parameter on the lagged y_{t-1} is equal to 0. The difference that the ADF test makes is to account for more lags, so that the underlying model takes higher-order dynamics into account, which can be written as a series of differenced lags:

$$Y_t - \phi_1 \times y_{t-1} - \phi_2 \times y_{t-2} \ldots = e_t$$

This requires somewhat more algebra to write as a series of differenced lags, and the expected distribution against which to test the null hypothesis is somewhat different

---

[2] A given sample time series process produced by a random walk can appear to have a trend, and this inspires much debate, particularly in analyses of stock-price time series.

compared to the original Dickey–Fuller test. The ADF test is the most widely presented test for stationarity in the time series literature.

Unfortunately, these tests are far from a panacea to your stationarity problems for a number of reasons:

- These tests have low power in distinguishing *near* unit roots from unit roots.
- With low sample size, false positives for unit roots are fairly common.
- Most tests do not test for or against all kinds of problems that can lead to a nonstationary time series. For example, some times will look specifically to testing whether the mean or the variance (but not both) is stationary. Other tests will look more generally at the overall distribution. It is important to understand the limits of the test applied when using it and to ensure that the limits are consistent with your beliefs about your data.

Setting an Alternate Null Hypothesis: KPSS Test

While the ADF test posits a null hypothesis of a unit root, the Kwiatkowski-Phillips-Schmidt-Shin (KPSS) test posits a null hypothesis of a stationary process. Unlike the ADF, the KPSS is not available in base R, but is still widely implemented. There are some nuances as to what these tests are for and how to use them properly; these are beyond the scope of this text but widely discussed online (*https://perma.cc/D3F2-TATY*).

In practice

Stationarity matters in practice for a number of reasons. First, a large number of models assume a stationary process, such as traditional models with known strengths and statistical models. We will cover these classes of models in Chapter 6.

A broader point is that a model of a time series that is not stationary will vary in its accuracy as the metrics of the time series vary. That is, if you are seeking a model to help you estimate the mean of a time series with a nonstationary mean and variance, the bias and error in your model will vary over time, at which point the value of your model becomes questionable.

It is often the case that a time series can be made stationary enough with a few simple transformations. A log transformation and a square root transformation are two popular options, particularly in the case of changing variance over time. Likewise, removing a trend is most commonly done by differencing. Sometimes a series must be differenced more than once. However, if you find yourself differencing too much (more than two or three times) it is unlikely that you can fix your stationarity problem with differencing.

Log or sqrt?

While a square root tends to be less computationally complex than a logarithm, you should explore both options. Think about the range of your data and how you want to compress large values rather than prematurely optimizing (that is, unduly pessimizing) your code and analysis.

Stationarity is not the only assumption forecasting models make. Another common but distinct assumption is the normality of the distribution of the input variables or predicted variable. In such cases, other transformations may be necessary. A common one is the Box Cox transformation, which is implemented in the R forecast package and in `scipy.stats` in Python. The transformation makes non-normally distributed data (skewed data) more normal. However, just because you can transform your data doesn't mean you should. Think carefully about the meaning of the distances between data points in your original data set before transforming them, and make sure, regardless of your task, that transformations preserve the most important information.

Transformations Have Assumptions

You might think you're choosing only transformations that don't make underlying assumptions (`log` and `sqrt` seem simple, right?), but think carefully about whether that's true.

As noted, log and square root transformations are commonly used to reduce the variance *of variance* over time. These make a number of assumptions. One is that your data will always be positive. Alternately, if you choose to shift your data before taking the square root or log, you are again either adding a bias or assuming it doesn't matter. Finally, when you take the square root or log, you make larger values less different from one another, effectively compressing the space between larger values but not between smaller values, de-emphasizing the differences between outliers. This may or may not be appropriate.

Applying Window Functions

Let's review the most important time series exploratory graphs, which you are likely to use for most initial time series analysis.

Rolling windows

A common function distinct to time series is a window function, which is any sort of function where you aggregate data either to compress it (as we saw in the case of downsampling in the previous chapter) or to smooth it (also discussed in Chapter 2).

In addition to the applications already discussed, smoothed data and window-aggregated data make for informative exploratory visualizations.

We can calculate a moving average and other calculations that involve some linear function of a series of points with base R's `filter()` function, as follows:

```
## R
> ## calculate a rolling average using the base R
> ## filter function
> x  <- rnorm(n = 100, mean = 0, sd = 10) + 1:100
> mn <- function(n) rep(1/n, n)

> plot(x, type = 'l',               lwd = 1)
> lines(filter(x, mn( 5)), col = 2, lwd = 3, lty = 2)
> lines(filter(x, mn(50)), col = 3, lwd = 3, lty = 3)
```

This code produces the graph in Figure 3-6.

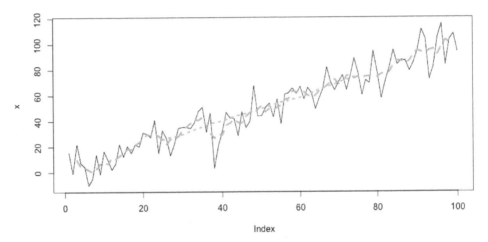

Figure 3-6. Two exploratory curves produced via a rolling mean smoothing. We might use these to look for a trend in particularly noisy data or to decide what sorts of deviations from linear behavior are interesting to investigate versus which are likely just noise.

If we are looking for functions that are not linear combinations of all points in the window, we can't use `filter()` because it relies on a linear transformation of the data. However, we can use `zoo`. The `rollapply()` function from the `zoo` package is quite handy (see Figure 3-7):

```
## R
> ## you can also do more 'custom' functionality
> require(zoo)

> f1 <- rollapply(zoo(x), 20, function(w) min(w),
```

```
>                    align = "left",  partial = TRUE)
> f2 <- rollapply(zoo(x), 20, function(w) min(w),
>                    align = "right", partial = TRUE)

> plot (x,            lwd = 1,           type = 'l')
> lines(f1, col = 2, lwd = 3, lty = 2)
> lines(f2, col = 3, lwd = 3, lty = 3)
```

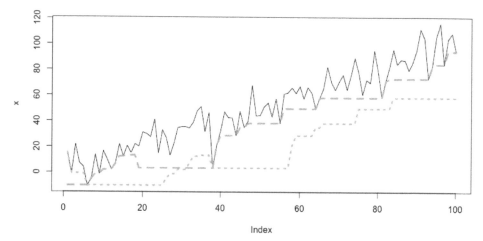

Figure 3-7. A rolling window minimum aligned either to the left (long dashes) or to the right (short dashes). The left alignment sees events in the future, whereas the right sees only events in the past. This is important to note for avoiding lookahead. However, sometimes a left alignment can be useful to ask exploratory questions such as "if I knew this in advance, would it even be useful?" Sometimes even a lookahead is not informative, which means that a particular variable is not informative. When knowing the future of a measure is not helpful, it's likely not a useful measure at any time in a time series context.

Use `zoo` objects in `zoo` functions. If you pass numeric vectors directly to `rollapply()`, the `align` argument will not take effect. You can confirm this by deleting the `zoo()` wrapper around x in the preceding code. You will see that the two curves are identical and, in fact, this is a silent failure, which is particularly dangerous in time series analysis because it can introduce an unintended lookahead. This is an example of why you need to sanity-check often, even with your exploratory data analysis. Unfortunately, silent failure is not unusual in many popular R packages and in other scripting languages, so stay alert!

It is also possible to "roll your own" for this kind of function, which can be a good idea to limit dependencies. In such cases, I'd strongly recommend starting with the source code (*https://perma.cc/5LTP-Q45T*) from an existing widely used package, such as zoo. This is because, even for a univariate time series, there are a surprising number of corner cases to think about, such as how to treat NAs, and how to treat the beginning and ending of a series where you will have fewer points than the specified size of the window.

A Number of R Options for Time Series

Earlier in this chapter, we considered the ts class, which is available in base R to handle time series data. In this section we use zoo objects. You should also be aware of xts objects. Here is a brief summary of how zoo and xts objects improve on ts objects:

- A ts object assumes a regularly spaced complete time series and for this reason does not store individual time indices but instead only the start, end, and frequency of the time series.
- ts objects support a recurring cycle, such as years or months.
- zoo objects store timestamps as an index attribute, so that they do not require regularly spaced, periodic time series.
- zoo objects can be printed horizontally or vertically.
- The data part of a zoo object can be a vector or a matrix.
- xts objects are an extension of zoo objects with even more options.

Expanding windows

Expanding windows are less commonly used in time series analysis than rolling windows because their proper application is more limited. Expanding windows make sense only in cases where you are estimating a summary statistic that you believe is a stable process rather than evolving over time or oscillating significantly. An expanding window starts with a given minimum size, but as you progress into the time series, it grows to include every point up to a given time rather than only a finite and constant size.

An expanding window offers greater certainty in your estimate of test statistics over time, allowing you to benefit from being "deeper" into a particular time series. However, it works only if you hypothesize that your underlying system is stationary. It can help you maintain "online" summary statistics as you would have estimated them in real time as you gathered more information.

If you look into base R, you will realize that many functions already exist as implementations of an expanding window, such as cummax and cummin. You can also easily repurpose cumsum to create a cumulative mean. In the following plot, we show both an expanding window max and an expanding window mean (see Figure 3-8):

```
## R
> # expanding windows
> plot(x, type = 'l', lwd = 1)
> lines(cummax(x),             col = 2, lwd = 3, lty = 2) # max
> lines(cumsum(x)/1:length(x), col = 3, lwd = 3, lty = 3) # mean
```

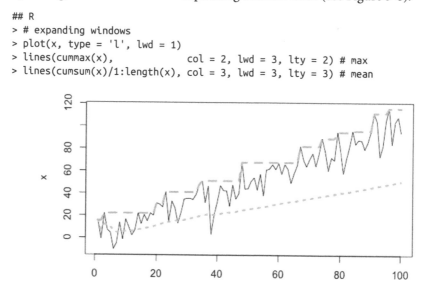

Figure 3-8. An expanding window "max" (long dashes) and an expanding window "mean" (short dashes). Use of an expanding window means that the max always reflects the global max up until that time, making it a monotonic function. Thanks to the long "memory" of the expanding window, the expanding window mean is lower than the rolling mean (Figure 3-6) because the underlying trend is less prominent in the expanding mean. Whether this is good, bad, or neutral depends on our assumptions and knowledge of the underlying system.

If you need a custom function with a rolling window, you can use rollapply() as we did for a rolling window. In this case, you need to specify a sequence of window sizes rather than a single scalar. Running the following code will produce an identical plot to the one in Figure 3-8, but this time it was created with rollapply() rather than built-in R functions:

```
## R
> plot(x, type = 'l', lwd = 1)
> lines(rollapply(zoo(x), seq_along(x), function(w) max(w),
>                       partial = TRUE, align = "right"),
>           col = 2, lwd = 3, lty = 2)
> lines(rollapply(zoo(x), seq_along(x), function(w) mean(w),
>                       partial = TRUE, align = "right"),
>           col = 3, lwd = 3, lty = 3)
```

Custom rolling functions

We only care about the option to apply a custom rolling function if we can actually imagine doing so. In practice, this is something you are likely to see when analyzing time series domains that have known underlying fundamental laws of behavior or useful heuristics that are necessary for proper analysis.

For example, we may be looking at windows that include a particular feature that is informative given domain knowledge. We might want to know when we are in a *monotonic* regime (say, blood sugar is increasing) versus an up-and-down regime that suggests instrumental noise rather than a trend. We could write a custom function for this scenario and apply it with either a moving or expanding window.

Understanding and Identifying Self-Correlation

At its most fundamental, self-correlation of a time series is the idea that a value in a time series at one given point in time may have a correlation to the value at another point in time. Note that "self-correlation" is being used here informally to describe a general idea rather than a technical one.

As an example of self correlation, if you take a yearly time series of daily temperature data, you may find that comparing May 15th of every year to August 15th of every year will give you some correlation, such that hotter May 15ths tend to correlate with hotter August 15ths (or tend to correlate with cooler August 15ths). You may feel you have learned a potentially interesting fact about the temperature system, indicating that there is a certain amount of long-term predictability. On the other hand, you may find the correlation closer to zero, in which case you will also have found something interesting, namely that knowing the temperature on May 15th does not alone give you any information about the likely range of temperatures on August 15th. That is self-correlation in an anecdotal nutshell.

From this simple example, we are going to expand into autocorrelation, which generalizes self-correlation by not anchoring to a specific point in time. In particular, autocorrelation asks the more general question of whether there is a correlation between any two points in a specific time series with a specific fixed distance between them. We'll look at this in more detail next, as well as final elaboration of partial autocorrelation.

The autocorrelation function

We begin with Wikipedia's (*https://perma.cc/U8JY-QD7U*) excellent definition of autocorrelation:

> Autocorrelation, also known as serial correlation, is the correlation of a signal with a delayed copy of itself as a function of the delay. Informally, it is the similarity between observations as a function of the time lag between them.

Let's translate that into plainer English. Autocorrelation gives you an idea of how data points at different points in time are linearly related to one another as a function of their time difference.

The *autocorrelation function* (ACF) can be intuitively understood with plotting. We can plot it easily in R (see Figure 3-9):

```R
## R
> x <- 1:100
> y <- sin(x × pi /3)
> plot(y, type = "b")
> acf(y)
```

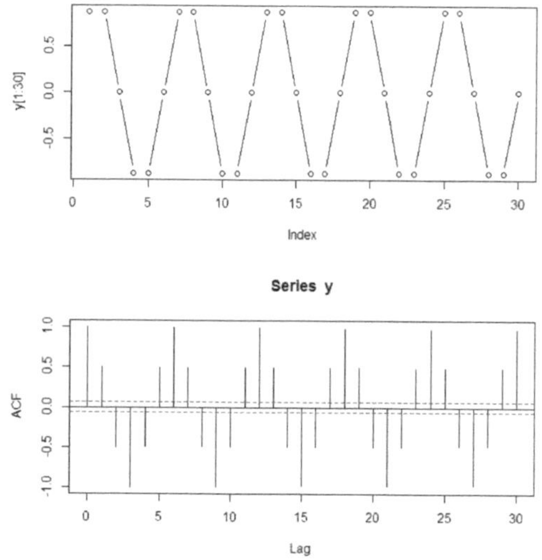

Figure 3-9. Plot of a sine function and its ACF.

Correlation in a Deterministic System

This sine series is a simple function and a fully determined system given a known input sequence. Nonetheless, we do not have a correlation of 1. Why is that? Inspect the series and think about what the ACF is calculating, and you will realize that for many values, the subsequent value can go either up or down depending on where you are in the cycle. If you know a few points in a row, you know which direction the process is going, but if you don't (as with the ACF, which is a 1:1 correlation measure), you will have a correlation of less than 1 because most values do not have a unique subsequent value but rather more than one. So, remember that a nonunitary correlation does not mean you necessarily have a probabilistic or noisy time series.

From the ACF, we see that points that have a lag between them of 0 have a correlation of 1 (this is true for every time series), whereas points separated by 1 lag have a correlation of 0.5. Points separated by 2 lags have a correlation of –0.5, and so on.

Calculating the ACF is straightforward. We can do it ourselves using data.table's shift() function:

```R
## R
> cor(y, shift(y, 1), use = "pairwise.complete.obs")
[1] 0.5000015
> cor(y, shift(y, 2), use = "pairwise.complete.obs")
[1] -0.5003747
```

Our calculations approximately match the graphical results in Figure 3-9.[3] While it is straightforward to calculate the ACF with custom code, it's usually better to use a prerolled version, such as R's acf function. There are several advantages to doing so:

- Automatic plotting with helpful labels
- A sensible (usually but not always) max number of lags for which to calculate ACF, as well as the option to override this max
- A graceful way to handle multivariate time series

There are a few important facts about the ACF, mathematically speaking:

- The ACF of a periodic function has the same periodicity as the original process. You can see this in the preceding sine example plots.
- The autocorrelation of the sum of periodic functions is the sum of the autocorrelations of each function separately. You can easily formulate an example of this with some simple code.
- All time series have an autocorrelation of 1 at lag 0.
- The autocorrelation of a sample of white noise will have a value of approximately 0 at all lags other than 0.
- The ACF is symmetric with respect to negative and positive lags, so only positive lags need to be considered explicitly. You can try plotting a manually calculated ACF to prove this.
- A statistical rule for determining a significant nonzero ACF estimate is given by a "critical region" with bounds at $+/-1.96 \times$ sqrt(n). This rule relies on a sufficiently large sample size and a finite variance for the process.

---

3 Sample correlations calculated with R's cor() function versus with R's acf() function will not match exactly because they use different divisors. For more information, see StackExchange (*https://perma.cc/M7V6-HN5Y*).

The partial autocorrelation function

The partial autocorrelation function (PACF) can be trickier to understand than the ACF. The partial autocorrelation of a time series for a given lag is the partial correlation of the time series with itself at that lag given all the information between the two points in time.

It's easy to nod your head here and think that sounds reasonable. But what exactly does it mean to account for the information between two points in time? It means you need to compute a number of conditional correlations and subtract these out of the total correlation. Computing the PACF is not straightforward, and there are a variety of methods for estimating it. We will not discuss these here, but you can find discussions in the related R and Python documentation.

The PACF is easier to understand graphically than conceptually. Its utility is also much more obvious in a graph than in a discussion (see Figure 3-10):

```
## R
> y <- sin(x × pi /3)
> plot(y[1:30], type = "b")
> pacf(y)
```

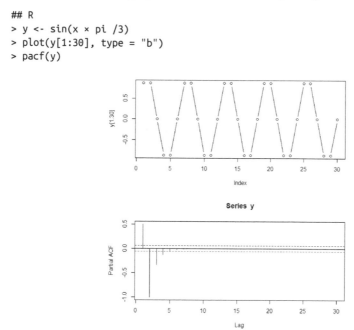

Figure 3-10. Plot and PACF of a seasonal, noiseless process.

For the case of a sine series, the PACF provides a striking contrast to the ACF. The PACF shows which data points are informative and which are harmonics of shorter time periods.

For a seasonal and noiseless process, such as the sine function, with period T, the same ACF value will be seen at T, 2T, 3T, and so on up to infinity. An ACF fails to weed out these redundant correlations. The PACF, on the other hand, reveals which correlations are "true" informative correlations for specific lags rather than redundancies. This is invaluable for knowing when we have collected enough information to get a sufficiently long window at a proper temporal scale for our data.

The critical region for the PACF is the same as for the ACF. The critical region has bounds at $\pm 1.96\sqrt{n}$. Any lags with calculated PACF values falling inside the critical region are effectively zero.

We have so far looked only at examples of perfectly noiseless single-frequency processes. Now we look at a slightly more complicated example. We'll consider the sum of two sine curves under no noise, low noise, and high noise conditions.

First, let's look at the plots with no noise, each individually (see Figure 3-11):

```R
## R
> y1 <- sin(x × pi /3)
> plot(y1, type = "b")
> acf (y1)
> pacf(y1)

> y2 <- sin(x × pi /10)
> plot(y2, type = "b")
> acf (y2)
> pacf(y2)
```

 The ACF of stationary data should drop to zero quickly. For non-stationary data the value at lag 1 is positive and large.

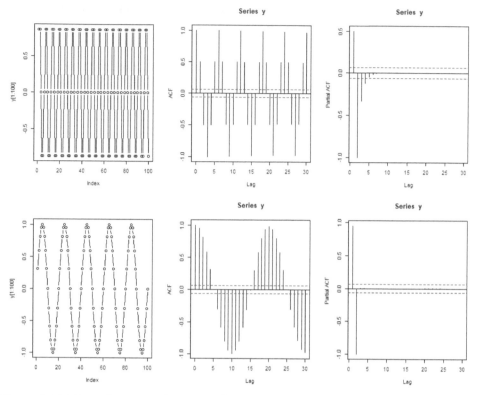

Figure 3-11. Plots of two sine functions, their ACF, and their PACF.

We combine the two series by summing and create the same plots for the summed series (see Figure 3-12):

```R
## R
> y <- y1 + y2
> plot(y, type = "b")
> acf (y)
> pacf(y)
```

As we can see, our ACF plot is consistent with the aforementioned property; the ACF of the sum of two periodic series is the sum of the individual ACFs. You can see this most clearly by noticing the positive → negative → positive → negative sections of the ACF correlating to the more slowly oscillating ACF. Within these waves, you can see the faster fluctuation of the higher-frequency ACF.

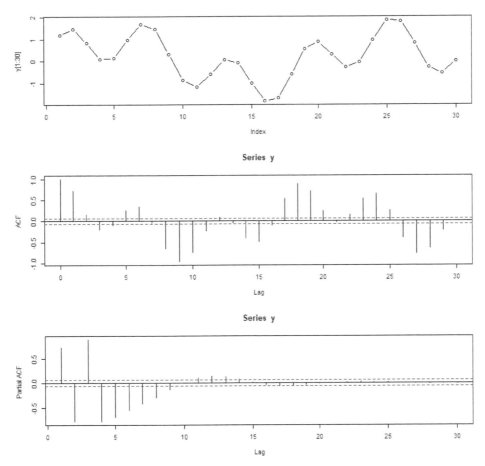

Figure 3-12. Plot, ACF, and PACF for the sum of the two sine series.

The PACF is not a straightforward sum of the PACF functions of the individual components. A PACF is simple enough to understand once it is calculated, but it's not so easy to generate or predict. This PACF indicates that partial autocorrelation is more substantial in the summed series than in either of the original series. That is, the correlation between points separated by a certain lag, when accounting for the values of the points between them, is more informative in the summed series than in the original series. This is related to the two different periods of the series, which result in any given point being less determined by the values of neighboring points since the location within the cycle of the two periods is less fixed now as the oscillations continue at different frequencies.

Let's look at the same situation, but with more noise (see Figure 3-13):

```
## R
> noise1 <- rnorm(100, sd = 0.05)
> noise2 <- rnorm(100, sd = 0.05)

> y1 <- y1 + noise1
> y2 <- y2 + noise2
> y  <- y1 + y2

> plot(y1, type = 'b')
> acf (y1)
> pacf(y1)

> plot(y2, type = 'b')
> acf (y2)
> pacf(y2)

> plot(y, type = 'b')
> acf (y)
> pacf(y)
```

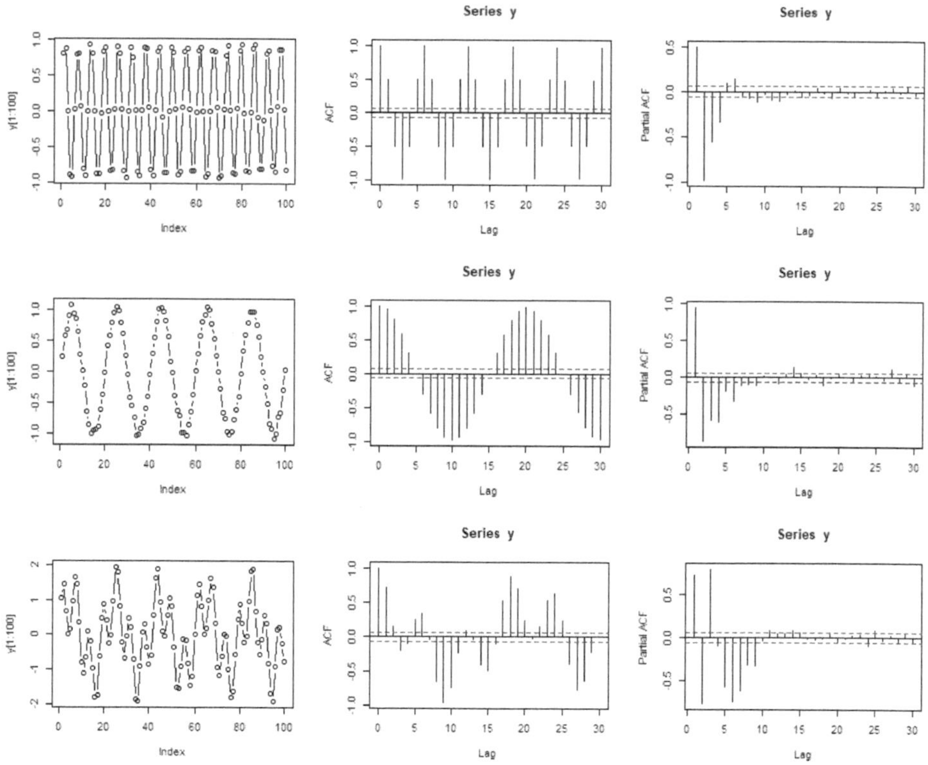

Figure 3-13. Plot, ACF, and PACF of two noisy sine processes and their sum.

Finally we add more noise to the time series so that the initial data itself does not even look particularly sine-like. (We omit the code example because it is the same as the previous example, simply with a larger sd parameter for rnorm.) We can see that this adds further interpretation difficulties, particularly to the PACF. The only difference between the plots in Figure 3-14 and the previous ones is a larger sd value for the noise variables.

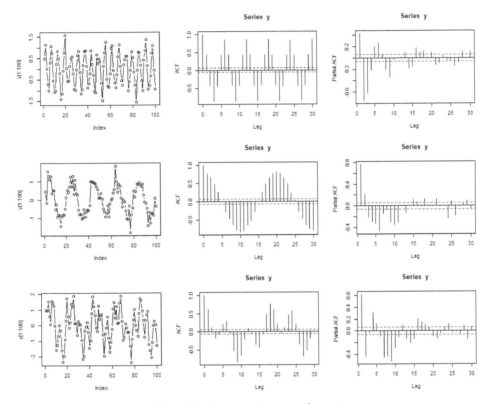

Figure 3-14. Plot, ACF, and PACF of very noisy sum of two sine processes.

Nonstationary Data

Let's consider how the ACF and PACF would look in the event of a series with a trend but no cycle (see Figure 3-15):

```R
## R
> x <- 1:100
> plot(x)
> acf (x)
> pacf(x)
```

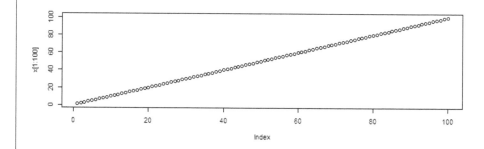

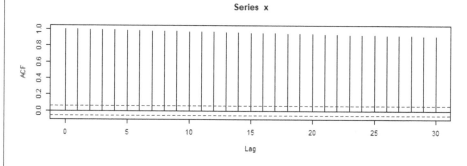

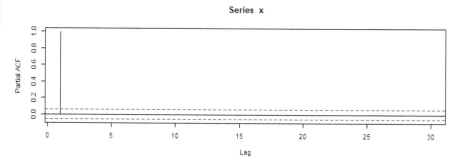

Figure 3-15. Plot, ACF, and PACF of a linearly trending process.

The ACF is not informative. It has a similar value for every lag, seemingly implying that all lags are equally correlated to the data. It's not clear what this means, and it's more a case of being able to compute *something*, not being able to make that computed quantity meaningful or sensible.

Luckily, the PACF is not as difficult to contemplate and gives us the information we need, which is that the only significant PACF correlation is at lag 1. Why is that? For a given point in time, once you know the point just before it, you know all the information the series can possibly give you about your point in time. This is due to the next point in time being 1 plus the old point.

Let's conclude with a look at the ACF and PACF of a real-world data set. Below, we examine the `AirPassengers` data. Given what we have seen so far, think about why the ACF has so many "critical" values (answer: it has a trend) and why the PACF has a critical value for a large lag (answer: the yearly seasonal cycle, which is identifiable even with a trend in the data).

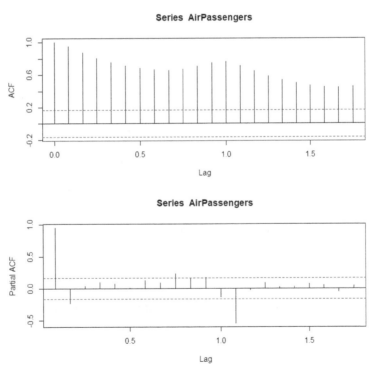

Figure 3-16. Plot of the ACF and PACF of the AirPassengers data. The lags here are not unit numbers because the lags are expressed as fractions of a year. This is because the AirPassengers data set comes in the form of a ts object, which has a built-in frequency that can be used for plotting (and other purposes).

Spurious Correlations

Those new to time series analysis will often begin with standard exploratory data practices, such as plotting two variables against each other and calculating their correlation. The new analyst will be very excited early on in the data exploration process when they notice what appears to be a very strong correlation and helpful relationship. They will continue looking and find other surprisingly high correlations; what an amazing system! They will wish they had started working with time series data sooner in their career. They will think to themselves, "All I do is win."

Then the analyst will do a little reading on time series analysis (best-case scenario), or present their findings to someone else and realize that it doesn't all make sense (not the best-case scenario). A skeptic will point out to the analyst that the correlations are too high. It can seem like any two values are related. More problems will emerge as the analyst reruns their analysis with other sets of variables and finds that they too have surprisingly high correlations. At some point it will become clear that there can't possibly be so many truly high correlations.

This trajectory is very much like the early history of econometrics. In the 19th century, when economists first began thinking of the idea of a business cycle, some of them went looking for external drivers of the cycle, such as sun spots (an 11-year cycle) or various meteorological cycles (such as a posited 4-year precipitation cycle). They invariably got very positive and strongly correlated results, even when they had no causal hypothesis to explain these results.

Many economists and statisticians remained skeptical, and rightly so. Udny Yule investigated the problem formally with a paper titled "Why Do We Sometimes Get Nonsense Correlations?" (*https://www.jstor.org/stable/2341482*), and an area of research was born and continues to give trouble and pleasure to academics. Spurious correlations remain an important problem to guard against, and they are hotly debated in litigation situations where one side asserts a relationship and the other side tries to discredit it. Similarly, one attempt to discredit climate change data relies on an argument that the correlation between increasing carbon emissions and warming global temperatures is a spurious correlation due to the trends in the two data sets (I do not find this argument convincing).

Economists have learned over time that data with an underlying trend is likely to produce spurious correlations. Here's a simple way to think about this: there is more information in a trending time series than in a stationary time series, so there are more opportunities for data points to move together.

In addition to trends, some other common features of time series can introduce spurious correlations:

- Seasonality—for example, think of a spurious correlation between hot dog consumption and death by drowning (summer).
- Level or slope shifts in data from regime changes over time (producing a dumbbell-like distribution with meaningless high correlation).
- Cumulatively summed quantities (this is a trick used in certain industries to make models or correlations look better than they are).

Cointegration

Cointegration refers to a real relationship between two time series. A commonly used example is a drunk pedestrian and their dog. Their individually measured walks might appear random taken alone, but they never stray too far from each other.

In the case of cointegration you will see high correlations. The difficulty will be in assessing whether two processes are cointegrated or whether you are seeing a spurious correlation because in both cases you will see surprisingly high correlations. The important difference is that there need not be any relationship in the case of a spurious correlation, whereas cointegrated time series are strongly related to one another.

There is a well-known blog (and now book) full of wonderful examples of spurious correlations, and I share one in Figure 3-17. Whenever you are tempted to think you have found a special and very strong relationship, make sure to check your data for obvious causes of trouble such as trends.

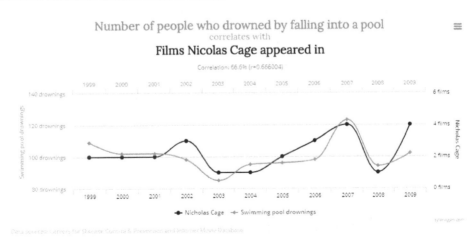

Figure 3-17. Some spurious correlations can look surprisingly convincing. This plot was taken from Tyler Vigen's website (https://perma.cc/6UYH-FPBX) of spurious correlations.

Some Useful Visualizations

Graphs are fundamental to a thorough exploratory analysis of time series. You will certainly want to visualize data with respect to the temporal axis—preferably in a way that answers the general questions you have about the data set, such as the behavior of a particular variable or the overall temporal distribution of the data points.

Earlier in this chapter, we looked at some plotting techniques familiar to any data analyst, such as a plot of values against time or a scatter plot of different columns' values over time. In this final section on exploratory data analysis, we discuss several visualizations that are particularly helpful for offering novel insights about time series behavior.

We will look at visualizations of varying degrees of complexity:

- A one-dimensional visualization to understand the overall temporal distribution of individuals with a found time series we assembled in Chapter 2

- A two-dimensional histogram to understand the typical trajectory of a value over time in the case of many parallel measurements (say many years measured or many time series of the same phenomenon measured)

- A three-dimensional visualization where time can take up as many as two of the dimensions or as few as none of the dimensions, but still be implicitly present

1D Visualizations

In the cases of many units of measurement (many users, members, etc.) we consider multiple time series in parallel. It can be interesting to stack these visually, emphasizing individual units of analysis and their respective time frames. We ignore the values measured and rather take the existence of data over a given range as the information of interest. The time span itself becomes the unit of analysis. Here we use R's timevis package, but there are many other options available. We look at a small subset of the donations data we prepared in Chapter 2 (see Figure 3-18):

```R
## R
> require(timevis)
> donations <- fread("donations.csv")
> d          <- donations[, .(min(timestamp), max(timestamp)), user]
> names(d)   <- c("content", "start", "end")
> d          <- d[start != end]
> timevis(d[sample(1:nrow(d), 20)])
```

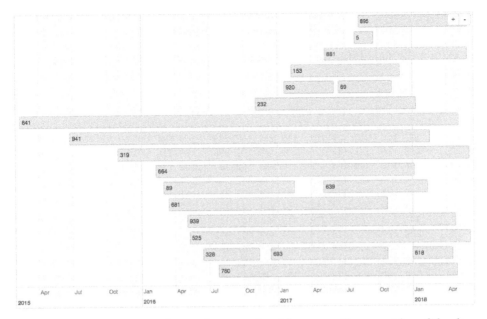

Figure 3-18. A Gantt chart of a random sample of data can offer some idea of the distribution of the range of "active" time periods for the users/donors.

The chart in Figure 3-18 helps us see that we probably have "busy" periods globally across the member population. We also glean some sense of the distribution of active donation spans in a member's "lifetime" in our organization.

Gantt charts have been used for over a century, most often for project management tasks. They came about independently in many different industries, and the idea is intuitive as soon as you see one. Despite the project management origins, Gantt charts can be useful in time series analysis where there are many independent actors, rather than a single process being measured. The plot in Figure 3-18 quickly answered questions I had about the relative overlap among the entire user base with respect to their donation history, a distribution I found difficult to understand when merely reading through the tabular data.

2D Visualizations

Now we'll use the `AirPassengers` data to see the seasonality and the trend, but we shouldn't think of time as linear. In particular, time happens on more than one axis. There is, of course, the axis of time going forward from day to day and year to year, but we can also consider laying time out along the axis of hour of the day or day of the week, and so on. In this way, we can more easily think about seasonality, such as certain behaviors happening at a certain time of the day or month of the year. We think in particular about how to understand our data in a seasonal fashion rather than just according to linear, chronological time visualizations.

We extract the data from the `AirPassengers` ts object and put it into appropriate matrix form:

```
## R
> t(matrix(AirPassengers, nrow = 12, ncol = 12))
      [,1] [,2] [,3] [,4] [,5] [,6] [,7] [,8] [,9] [,10] [,11] [,12]
 [1,]  112  118  132  129  121  135  148  148  136   119   104   118
 [2,]  115  126  141  135  125  149  170  170  158   133   114   140
 [3,]  145  150  178  163  172  178  199  199  184   162   146   166
 [4,]  171  180  193  181  183  218  230  242  209   191   172   194
 [5,]  196  196  236  235  229  243  264  272  237   211   180   201
 [6,]  204  188  235  227  234  264  302  293  259   229   203   229
 [7,]  242  233  267  269  270  315  364  347  312   274   237   278
 [8,]  284  277  317  313  318  374  413  405  355   306   271   306
 [9,]  315  301  356  348  355  422  465  467  404   347   305   336
[10,]  340  318  362  348  363  435  491  505  404   359   310   337
[11,]  360  342  406  396  420  472  548  559  463   407   362   405
[12,]  417  391  419  461  472  535  622  606  508   461   390   432
```

Notice we had to transpose the data so that it will line up as presented by the ts object.

Column Major Versus Row Major

R is column major (*https://perma.cc/L4BH-DKB8*) by default, which is unusual and different from Python's NumPy (row major) and also different from most SQL databases. It's good to be aware of what behaviors are default and available in a given language, not just for display purposes but to think about how to manage and access memory effectively later on.

We plot each year on a set of axes that reflect the progression of the months across the year (see Figure 3-19):

```
## R
> colors <- c("green",  "red",            "pink",   "blue",
>                "yellow","lightsalmon", "black",  "gray",
>                "cyan",  "lightblue",   "maroon", "purple")
> matplot(matrix(AirPassengers, nrow = 12, ncol = 12),
>           type = 'l', col = colors,  lty = 1, lwd = 2.5,
>           xaxt = "n", ylab = "Passenger Count")
> legend("topleft", legend = 1949:1960, lty = 1, lwd = 2.5,
>           col = colors)
> axis(1, at = 1:12, labels = c("Jan", "Feb", "Mar", "Apr",
>                                "May", "Jun", "Jul", "Aug",
>                                "Sep", "Oct", "Nov", "Dec"))
```

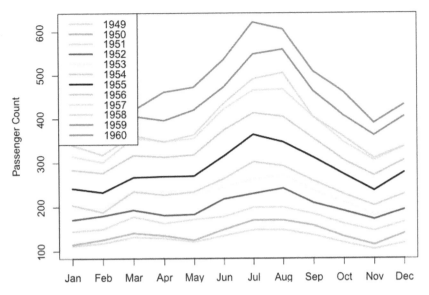

Figure 3-19. Per-year month-by-month counts.[4]

We can produce the same plot more easily with the forecast package (see Figure 3-20):

```R
## R
> require(forecast)
> seasonplot(AirPassengers)
```

The x-axis is month of the year for all years. Every year, the number of airline passengers peaks in July or August (months 7 and 8). We also see a local peak in March (month 3) most years. This graph can thus show us more details regarding the seasonality behavior.

---

4 Visit the GitHub repository (*https://github.com/PracticalTimeSeriesAnalysis/BookRepo*) to view the original figure, or plot it yourself for a more detailed look.

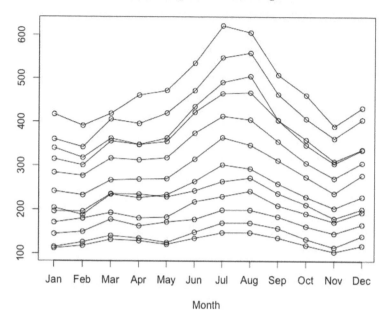

Seasonal plot: AirPassengers

Figure 3-20. We produce a similar seasonal plot more easily with the seasonplot() function.

The different years' curves rarely cross. There was such robust growth that it was rarely the case that different years had the same number of passengers in the same month. There are a few exceptions, but not during the peak months. From these observations alone, we can already produce advice for an air travel company looking to make decisions about how to plan for growth.

An alternate plot of per-month curves against years is less standard but also helpful (see Figure 3-21):

```R
## R
> months <- c("Jan", "Feb", "Mar", "Apr", "May", "Jun",
>             "Jul", "Aug", "Sep", "Oct", "Nov", "Dec")

> matplot(t(matrix(AirPassengers, nrow = 12, ncol = 12)),
>            type = 'l', col = colors, lty = 1, lwd = 2.5)
> legend("left", legend = months,
>                col = colors, lty = 1, lwd = 2.5)
```

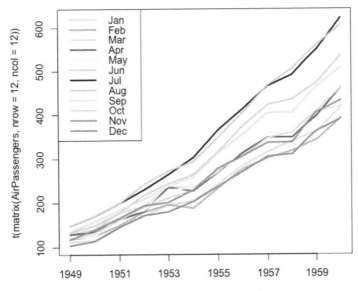

Figure 3-21. Per-month curves of year-to-year time series.[5]

Through the years, the growth trend is accelerating; that is, the rate of increase is itself increasing. Also, two months are growing faster than the others, namely July and August. We can get a similar visualization and similar insights with an easy visualization function supplied by the `forecast` package (see Figure 3-22):

```
## R
> monthplot(AirPassengers)
```

There are two general observations we can make from these plots:

- Time series have more than one useful set of temporal axes against which to plot. We used both an axis of month of the year (January through December) and an axis of years of the data set (the first year up through the last/twelfth year).

- We can glean a lot of useful information and predictive details from visualizations that stack time series data rather than linearly plotting.

---

5 Visit the GitHub repository (*https://github.com/PracticalTimeSeriesAnalysis/BookRepo*) to view the original figure, or plot it yourself for a more detailed look.

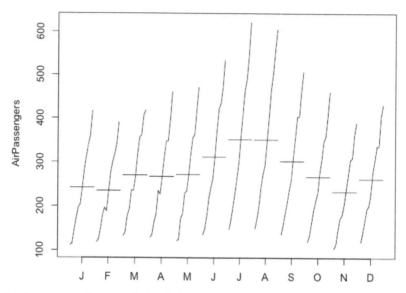

Figure 3-22. By using the monthplot() function, we can see how performance per month changes as the years pass.

We next consider a proper two-dimensional histogram. In a time series context, we can think of a two-dimensional histogram as having one axis for time (or a proxy for time) and another axis for a unit of interest. The "stacked" plots we just did are well on their way to being two-dimensional histograms, but some changes would be helpful:

- We need to bin data both on the time axis and on the number of passengers.
- We need more data. A 2D histogram doesn't make sense until the stacked curves run into one another; they can't properly be seen alone. Otherwise, the 2D histogram fails to convey any additional information.

We generate the 2D histogram on this small data set for illustration and then move on to a more meaningful example. We build our own 2D histogram function from scratch, like so:

```
## R
> hist2d <- function(data, nbins.y, xlabels) {
>    ## we make ybins evenly spaced to include
>    ## minimum and maximum points
>    ymin = min(data)
>    ymax = max(data) × 1.0001
>    ## the lazy way out to avoid worrying about inclusion/exclusion

>    ybins = seq(from = ymin, to = ymax, length.out = nbins.y + 1 )
```

```
>    ## make a zero matrix of the appropriate size
>    hist.matrix = matrix(0, nrow = nbins.y, ncol = ncol(data))

>    ## data comes in matrix form where each row
>    ## represents one data point
>    for (i in 1:nrow(data)) {
>        ts = findInterval(data[i, ], ybins)
>        for (j in 1:ncol(data)) {
>          hist.matrix[ts[j], j] = hist.matrix[ts[j], j] + 1
>        }
>    }
>    hist.matrix
> }
```

We make a histogram with heat map coloring, as follows:

```
## R
> h = hist2d(t(matrix(AirPassengers, nrow = 12, ncol = 12)), 5, months)
> image(1:ncol(h), 1:nrow(h), t(h), col = heat.colors(5),
>       axes = FALSE, xlab = "Time", ylab = "Passenger Count")
```

However, the resulting image (Figure 3-23) is not very satisfying.

Figure 3-23. *Heat map constructed from our homemade 2D histogram of the AirPas-sengers data.*

This plot is not useful because there is not enough data. We only have 12 curves, and we've divided them across 5 buckets. However, an even more important issue is that we don't have stationary data. The use of a histogram assumes a stationary data set. In this case there is a trend, so although we'd like to see seasonality, the trend will necessarily get in the way.

Now we look at a data set that features a larger number of samples and is not polluted by a trend. This is a subset of the FiftyWords data set taken from the UCR Time Series Classification Archive (*https://perma.cc/Y982-9FPS*). This data set includes a representation of 50 different words as recorded by a univariate time series, each time series being of equal length. The subset of the data used to plot Figure 3-24 is from the data set I used in a segment on time series classification in a general tutorial on the subject. You can download that subset (*https://oreil.ly/M6T-u*), although for the purposes of this exercise you can use either one:

```R
## R
> require(data.table)

> words <- fread(url.str)
> w1     <- words[V1 == 1]

> h = hist2d(w1, 25, 1:ncol(w1))

> colors <- gray.colors(20, start = 1, end = .5)
> par(mfrow = c(1, 2))
> image(1:ncol(h), 1:nrow(h), t(h),
>       col = colors, axes = FALSE, xlab = "Time",  ylab = "Projection Value")
> image(1:ncol(h), 1:nrow(h), t(log(h)),
>       col = colors, axes = FALSE, xlab = "Time", ylab = "Projection Value")
```

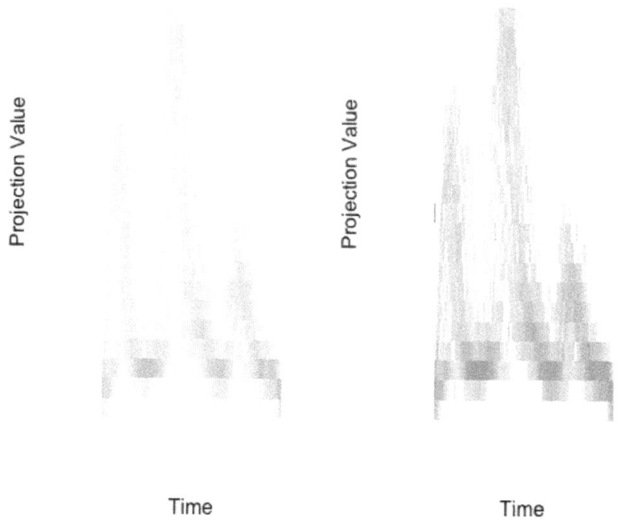

Figure 3-24. Two-dimensional histogram of an audio metric for a single word. The left plot has a linear count scale, while the right plot has a log count scale.

The plot on the right of Figure 3-24 looks better because the counts are colored according to a log transformation of the count rather than the count directly.

This is an application of the same idea as taking the log of a time series to decrease variance and to reduce the importance of, and distance between, outliers. Using a log transformation improves our visualization by not wasting a large portion of the range on the relatively sparse high-count values.

Our homemade option won't be as visually appealing as many precanned options, so we should take a look at these to see what is different. To take advantage of the pre-canned solutions, we need to reshape our data because these options expect pairs of x-y values to be turned into a 2D histogram. Unlike our homegrown solution, the precanned options for 2D histograms are not designed specifically for time series data. They nonetheless offer excellent visualization solutions (see Figure 3-25):

```R
## R
> w1 <- words[V1 == 1]

> ## melt the data to the pairs of paired-coordinates
> ## expected by most 2d histogram implementations
> names(w1) <- c("type", 1:270)
> w1          <- melt(w1, id.vars = "type")

> w1          <- w1[, -1]
> names(w1) <- c("Time point", "Value")

> plot(hexbin(w1))
```

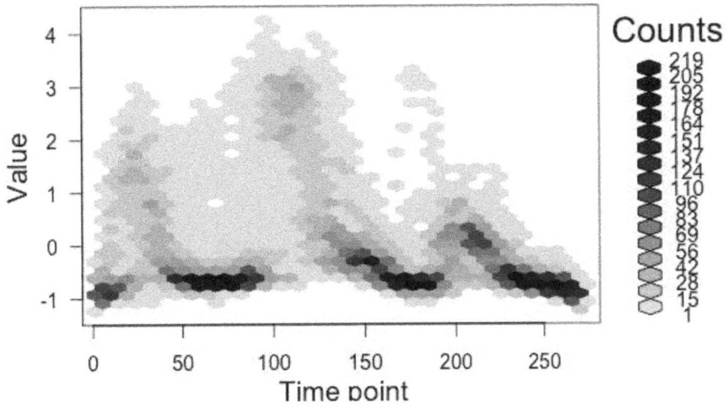

Figure 3-25. An alternative 2D histogram visualization of the same data.

3D Visualizations

3D visualizations do not come with base R, but there are many packages available for them. Here I present some fast plots made with plotly, which I chose because it produces plots that can be easily rotated in RStudio and exported to web interfaces. Also, downloading and installing plotly tends to be straightforward, which is not true of all visualization packages.

Let's consider the AirPassengers data. We plot it in three dimensions, using two dimensions for time (month and year) and one dimension for data values:

```
## R
> require(plotly)
> require(data.table)

> months = 1:12
> ap = data.table(matrix(AirPassengers, nrow = 12, ncol = 12))
> names(ap) = as.character(1949:1960)
> ap[, month := months]
> ap = melt(ap, id.vars = 'month')
> names(ap) = c("month", "year", "count")

> p <- plot_ly(ap, x = ~month, y = ~year, z = ~count,
>              color = ~as.factor(month)) %>%
>    add_markers() %>%
>    layout(scene = list(xaxis = list(title = 'Month'),
>                        yaxis = list(title = 'Year'),
>                        zaxis = list(title = 'PassengerCount')))
```

This 3D visualization helps us get a sense of the overall shape of the data. We have
seen much of this before, but expanding to a three-dimensional scatter plot proves to
be notably better than a two-dimensional histogram, perhaps because of the paucity
of the data (see Figures 3-26 and 3-27).

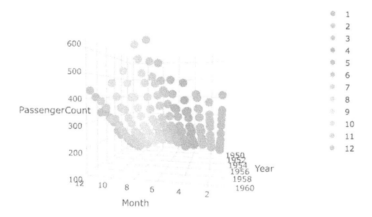

*Figure 3-26. A 3D scatterplot of the AirPassenger data. This perspective highlights the
seasonality.*[6]

---

6 Visit the GitHub repository (*https://github.com/PracticalTimeSeriesAnalysis/BookRepo*) to view the original
figure, or plot it yourself for a more detailed look.

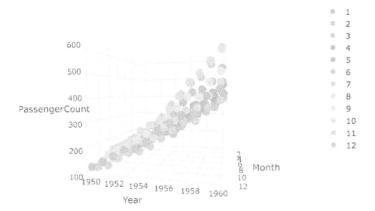

Figure 3-27. Another perspective on the same data more clearly illustrates the increasing trend from one year to the next. I highly recommend running the code on your own computer, where you can rotate it yourself.[7]

We do not necessarily need to expend two axes on time. Instead, we might use two axes for location and one for time. We can visualize a two-dimensional random walk as follows, with a lightly modified example of some `plotly` demonstration code (see Figures 3-28 and 3-29):

```
## R
> file.location <- 'https://raw.githubusercontent.com/plotly/datasets/master/\
                    _3d-line-plot.csv'
> data <- read.csv(file.location)
> p <- plot_ly(data, x = ~x1, y = ~y1, z = ~z1,
>               type = 'scatter3d', mode = 'lines',
>               line = list(color = '#1f77b4', width = 1))
```

The interactive nature of the plot is key. Different perspectives can be misleading or illuminating in ways we can't know until we are able to rotate the data.

---

7 Visit the GitHub repository (*https://github.com/PracticalTimeSeriesAnalysis/BookRepo*) to view the original figure, or plot it yourself for a more detailed look.

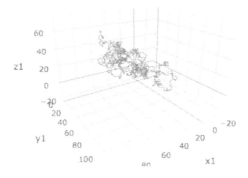

Figure 3-28. One perspective on a two-dimensional random walk over time.

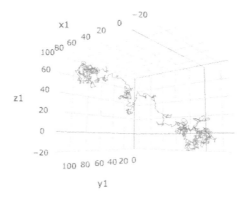

Figure 3-29. This perspective of the same random walk is much more revealing. Again, I encourage you to try this code out for yourself!

A good exercise would be to generate noisy seasonal motion data in two dimensions and visualize this in the same way we have visualized the random walk here. There should be a substantial difference in what you see in that plot compared to the random walk data. Packages like plotly can help you experiment quickly and with comprehensive visual feedback.

More Resources

- On spurious correlations:

Ai Deng, "A Primer on Spurious Statistical Significance in Time Series Regressions," Economics Committee Newsletter 14, no. 1 (2015), https://perma.cc/9CQR-RWHC.

> This industry write-up of what spurious correlations are and how they appear in data is useful for developing some practical insights into when and where to look for this problem in your own data sets. The material is written at a very approachable level.

Tyler Vigen, Spurious Correlations (New York: Hachette, 2015), https://perma.cc/YY6R-SKWA.

> This collection of ridiculous time series correlations is essential reading for any time series analyst or thinking person.

Antonio Noriega and Daniel Ventosa-Santaulària, "Spurious Regression Under Broken-Trend Stationarity," Journal of Time Series Analysis 27, no. 5 (2006): 671–84, https://perma.cc/V993-SF4F.

> The authors develop both theory and simulation data to show that shifts in the level or trend of independently and randomly generated data sets influence the presence of spurious correlations.

C.W.J. Granger and P. Newbold, "Spurious Regressions in Econometrics," Journal of Econometrics 2, no. 2 (1974): 111–20, https://perma.cc/M8TE-AL6U.

> This econometrics article led to a Nobel Prize for identifying the difficulties inherent in dealing with spurious correlations and arguing for a more robust approach to identifying related time series.

- On exploratory data analysis:

David R. Brillinger and Mark A. Finney, "An Exploratory Data Analysis of the Temperature Fluctuations in a Spreading Fire," Environmetrics 25, no. 6 (2014): 443–53, https://perma.cc/QB3D-APKM.

> This is a very thorough example of how some real-world laboratory data with a geotemporal grid was analyzed by academic and government researchers.

Robert H. Shumway and David S. Stoffer, "Time Series Regression and Exploratory Data Analysis," in Time Series Analysis and Its Applications with R Examples (New York: Springer, 2011), https://perma.cc/UC5B-TPVS.

> This is a chapter on exploratory data analysis from the authors' canonical treatise on time series analysis for graduate students.

- More visualizations:

  Christian Tominski and Wolfgang Aigner, "The TimeViz Browser," https:// perma.cc/94ND-6ZA5.
  > This stunning catalog shows examples of, and source code for, many interesting time series visualizations from both academic research papers and industry use cases.

  Oscar Perpiñán Lamigueiro, "GitHub Repository for Displaying Time Series, Spatial, and Space-time Data with R," https://perma.cc/R69Y-5JPL.
  > This includes source code for a variety of R-based time series visualizations, including geospatial time series data.

  Myles Harrison, "5 Ways to Do 2D Histograms in R," R-bloggers, September 1, 2014, https://perma.cc/ZCX9-FQQY.
  > This is a practical guide to a variety of options offered by R packages to construct 2D histograms and give them meaningful coloring and binning. In addition to a basic overview, a related segment (*https://edav.info/tidy quant.html*) also provides a walkthrough of the `tidyquant` package for visualizing stock market data, an important source of time series.

- On assorted trends:

  Halbert White and Clive W.J. Granger, "Considerations of Trends in Time Series," Journal of Time Series Econometrics 3, no. 1 (2011), https://perma.cc/WF2H-TVTL.
  > This recent academic article points out that while trends are ubiquitous in data, even traditional statistics does not have a definitive set of definitions for describing different kinds of trends in data. In an approachable style, the authors offer statistical insights into different ways that nonstationary data may have an underlying trend and guidance on how to improve statistical methods for data.

## CHAPTER 4

Simulating Time Series Data

Up to this point, we have discussed where to find time series data and how to process it. Now we will look at how to create times series data via simulation.

Our discussion proceeds in three parts. First, we compare simulations of time series data to other kinds of data simulations, noting what new areas of particular concern emerge when we have to account for time passing. Second, we look at a few code-based simulations. Third, we discuss some general trends in the simulation of time series.

The bulk of this chapter will focus on specific code examples for generating various kinds of time series data. We will run through the following examples:

- We simulate email opening and donation behavior of members of a nonprofit organization over the course of several years. This is related to the data we examined in "Retrofitting a Time Series Data Collection from a Collection of Tables" on page 26.

- We simulate events in a taxicab fleet of 1,000 vehicles with various shift start times and hour-of-the-day-dependent passenger pickup frequencies over the course of a single day.

- We simulate step-by-step state evolution of a magnetic solid for a given temperature and size using relevant laws of physics.

These three code examples correlate to three classes of time series simulations:

Heuristic simulations
> We decide how the world should work, ensure it makes sense, and code it up, one rule at a time.

Discrete event simulations
> We build individual actors with certain rules in our universe and then run those actors to see how the universe evolves over time.

Physics-based simulations
> We apply physical laws to see how a system evolves over time.

Simulating time series can be a valuable analytical exercise and one we will also demonstrate in later chapters as it relates to specific models.

What's Special About Simulating Time Series?

Simulating data is an area of data science that is rarely taught, but which is a particularly useful skill for time series data. This follows from one of the downsides of having temporal data: no two data points in the same time series are exactly comparable since they happen at different times. If we want to think about *what could have happened at a given time*, we move into the world of simulation.

Simulations can be simple or complex. On the simpler side, you will encounter synthetic data in any statistics textbook on time series, such as in the form of a random walk. These are usually generated as cumulative sums of a random process (such as R's rnorm) or by a periodic function (such as a sine curve). On the more complex side, many scientists and engineers make their careers out of simulating time series. Time series simulations remain an active area of research—and a computationally demanding one—in many fields, including:

- Meteorology
- Finance
- Epidemiology
- Quantum chemistry
- Plasma physics

In some of these cases, the fundamental rules of behavior are well understood, but it can still be difficult to account for everything that can happen due to the complexity of the equations (meteorology, quantum chemistry, plasma physics). In other cases, not all of the predictive variables can ever be known, and experts aren't even sure that perfect predictions can be made due to the stochastic nonlinear nature of the systems studied (finance, epidemiology).

Simulation Versus Forecasting

Simulation and forecasting are similar exercises. In both cases you must form hypotheses about underlying system dynamics and parameters, and then extrapolate from these hypotheses to generate data points.

Nonetheless, there are important differences to keep in mind when learning about and developing simulations rather than forecasts:

- It can be easier to integrate qualitative observations into a simulation than into a forecast.

- Simulations are run at scale so that you can see many alternative scenarios (thousands or more), whereas forecasts should be more carefully produced.

- Simulations have lower stakes than forecasts; there are no lives and no resources on the line, so you can be more creative and exploratory in your initial rounds of simulations. Of course, you eventually want to make sure you can justify how you build your simulations, just as you must justify your forecasts.

Simulations in Code

Next we look at three examples of coding up simulations of time series. As you read these examples, consider what a wide array of data can be simulated to produce a "time series," and how the temporal element can be very specific and human-driven, such as days of the week and times of day of donations, but can also be very nonspecific and essentially unlabeled, such as the "nth step" of a physics simulation.

The three examples of simulation we will discuss in this section are:

- Simulating a synthetic data set to test our hypotheses about how members of an organization may (or may not) have correlated behavior between receptiveness to organizational email and willingness to make donations. This is the most DIY example in that we hardcode relationships and generate tabular data with `for` loops and the like.

- Simulating the synthetic data set to explore aggregate behavior in a fleet of taxis, complete with shift times and time-of-day-dependent frequency of passengers. In this data set, we make use of Python's object-oriented attributes as well as generators, which are quite helpful when we want to set a system going and see what it does.

- Simulating the physical process of a magnetic material gradually orienting its individual magnetic elements, which begin in disarray but ultimately coalesce into a well-ordered system. In this example, we see how physical laws can drive a time series simulation and insert natural temporal scaling into a process.

Doing the Work Yourself

When you are programming simulations, you need to keep in mind the logical rules that apply to your system. Here we walk through an example where the programmer does most of the work of making sure the data makes sense (for example, by not specifying events that happen in an illogical order).

We start by defining the membership universe—that is, how many members we have and when each joined the organization. We also pair each member with a member status:

```python
## python
>>> ## membership status
>>> years       = ['2014', '2015', '2016', '2017', '2018']
>>> memberStatus = ['bronze', 'silver', 'gold', 'inactive']

>>> memberYears = np.random.choice(years, 1000,
>>>              p = [0.1, 0.1, 0.15, 0.30, 0.35])
>>> memberStats = np.random.choice(memberStatus, 1000,
>>>              p = [0.5, 0.3, 0.1, 0.1])

>>> yearJoined = pd.DataFrame({'yearJoined': memberYears,
>>>                            'memberStats': memberStats})
```

Notice that there are already many rules/assumptions built into the simulation just from these lines of code. We impose specific probabilities of the years the members joined. We also make the status of the member entirely independent on the year they joined. In the real world, it's likely we can already do better than this because these two variables should have some connection, particularly if we want to incentivize people to remain members.

We make a table indicating when members opened emails each week. In this case, we define our organization's behavior: we send three emails a week. We also define different patterns of members behavior with respect to email:

- Never opening email
- Constant level of engagement/email open rate
- Increasing or decreasing level of engagement

We can imagine ways to make this more complex and nuanced depending on anecdotal observations from veterans or novel hypotheses we have about unobservable processes affecting the data:

```python
## python
>>> NUM_EMAILS_SENT_WEEKLY = 3

>>> ## we define several functions for different patterns
>>> def never_opens(period_rng):
```

```
>>>     return []

>>> def constant_open_rate(period_rng):
>>>     n, p = NUM_EMAILS_SENT_WEEKLY, np.random.uniform(0, 1)
>>>     num_opened = np.random.binomial(n, p, len(period_rng))
>>>     return num_opened

>>> def increasing_open_rate(period_rng):
>>>     return open_rate_with_factor_change(period_rng,
>>>                                         np.random.uniform(1.01,
>>>                                                           1.30))

>>> def decreasing_open_rate(period_rng):
>>>     return open_rate_with_factor_change(period_rng,
>>>                                         np.random.uniform(0.5,
>>>                                                           0.99))

>>> def open_rate_with_factor_change(period_rng, fac):
>>>     if len(period_rng) < 1 :
>>>         return []
>>>     times = np.random.randint(0, len(period_rng),
>>>                               int(0.1 * len(period_rng)))
>>>     num_opened = np.zeros(len(period_rng))
>>>     for prd in range(0, len(period_rng), 2):
>>>         try:
>>>             n, p = NUM_EMAILS_SENT_WEEKLY, np.random.uniform(0,
>>>                                                              1)
>>>             num_opened[prd:(prd + 2)] = np.random.binomial(n, p,
>>>                                                            2)
>>>             p = max(min(1, p * fac), 0)
>>>         except:
>>>             num_opened[prd] = np.random.binomial(n, p, 1)
>>>     for t in range(len(times)):
>>>         num_opened[times[t]] = 0
>>>     return num_opened
```

We have defined functions to simulate four distinct kinds of behavior:

Members who never open the emails we send them
   (never_opens())

Members who open about the same number of emails each week
   (constant_open_rate())

Members who open a decreasing number of emails each week
   (decreasing_open_rate())

Members who open an increasing number of emails each week
   (increasing_open_rate())

We ensure that those who grow increasingly engaged or disengaged over time are simulated in the same way with the open_rate_with_factor_change() function via the functions increasing_open_rate() and decreasing_open_rate().

We also need to come up with a system to model donation behavior. We don't want to be totally naive, or our simulation will not give us insights into what we should expect. That is, we want to build into the model our current hypotheses about member behavior and then test whether the simulations based on those hypotheses match what we see in our real data. Here, we make donation behavior loosely but not deterministically related to the number of emails a member has opened:

```python
## python
>>> ## donation behavior
>>> def produce_donations(period_rng, member_behavior, num_emails,
>>>                        use_id, member_join_year):
>>>     donation_amounts = np.array([0, 25, 50, 75, 100, 250, 500,
>>>                                  1000, 1500, 2000])
>>>     member_has = np.random.choice(donation_amounts)
>>>     email_fraction = num_emails /
>>>                         (NUM_EMAILS_SENT_WEEKLY * len(period_rng))
>>>     member_gives = member_has * email_fraction
>>>     member_gives_idx = np.where(member_gives
>>>                                 >= donation_amounts)[0][-1]
>>>     member_gives_idx = max(min(member_gives_idx,
>>>                             len(donation_amounts) - 2),
>>>                         1)
>>>     num_times_gave = np.random.poisson(2) *
>>>                         (2018 - member_join_year)
>>>     times = np.random.randint(0, len(period_rng), num_times_gave)
>>>     dons = pd.DataFrame({'member'   : [],
>>>                          'amount'   : [],
>>>                          'timestamp': []})

>>>     for n in range(num_times_gave):
>>>         donation = donation_amounts[member_gives_idx
>>>                     + np.random.binomial(1, .3)]
>>>         ts = str(period_rng[times[n]].start_time
>>>                 + random_weekly_time_delta())
>>>         dons = dons.append(pd.DataFrame(
>>>                 {'member'   : [use_id],
>>>                  'amount'   : [donation],
>>>                  'timestamp': [ts]}))
>>>
>>>     if dons.shape[0] > 0:
>>>         dons = dons[dons.amount != 0]
>>>         ## we don't report zero donation events as this would not
>>>         ## be recorded in a real world database
>>>
>>>     return dons
```

There are a few steps we have taken here to make sure the code produces realistic behavior:

- We make the overall number of donations dependent on how long someone has been a member.
- We generate a wealth status per member, building in a hypothesis about behavior that donation amount is related to a stable amount a person would have earmarked for making donations.

Because our member behaviors are tied to a specific timestamp, we have to choose which weeks each member made donations and also when during that week they made the donation. We write a utility function to pick a random time during the week:

```python
## python
>>> def random_weekly_time_delta():
>>>     days_of_week = [d for d in range(7)]
>>>     hours_of_day = [h for h in range(11, 23)]
>>>     minute_of_hour = [m for m in range(60)]
>>>     second_of_minute = [s for s in range(60)]
>>>     return pd.Timedelta(str(np.random.choice(days_of_week))
>>>                         + " days" ) +
>>>         pd.Timedelta(str(np.random.choice(hours_of_day))
>>>                         + " hours" )   +
>>>         pd.Timedelta(str(np.random.choice(minute_of_hour))
>>>                         + " minutes") +
>>>         pd.Timedelta(str(np.random.choice(second_of_minute))
>>>                         + " seconds")
```

You may have noticed that we only draw the hour of the timestamp from the range of 11 to 23 (`hours_of_day = [h for h in range(11, 23)]`). We are postulating a universe with people in a very limited range of time zones or even in just a single time zone, as we do not allow hours outside the range given. Here we are building in more of our underlying model as to how users behave.

We thus expect to see unified behavior from our users as though they are all in one or a few adjoining time zones, and we are further postulating that reasonable donation behavior is for people to donate from late morning to late evening, but not overnight and not first thing when they wake up.

Finally, we put all the components just developed together to simulate a certain number of members and associated events in a way that ensures that events happen only once a member has joined and that a member's email events have some relation (but not an unrealistically small relation) to their donation events:

```python
## python
>>> behaviors      = [never_opens,
>>>                    constant_open_rate,
>>>                    increasing_open_rate,
>>>                    decreasing_open_rate]
>>> member_behaviors = np.random.choice(behaviors, 1000,
>>>                                      [0.2, 0.5, 0.1, 0.2])

>>> rng = pd.period_range('2015-02-14', '2018-06-01', freq = 'W')
>>> emails = pd.DataFrame({'member'      : [],
>>>                        'week'        : [],
>>>                        'emailsOpened': []})
>>> donations = pd.DataFrame({'member'   : [],
>>>                           'amount'   : [],
>>>                           'timestamp': []})

>>> for idx in range(yearJoined.shape[0]):
>>>     ## randomly generate the date when a member would have joined
>>>     join_date = pd.Timestamp(yearJoined.iloc[idx].yearJoined) +
>>>                 pd.Timedelta(str(np.random.randint(0, 365)) +
>>>                 ' days')
>>>     join_date = min(join_date, pd.Timestamp('2018-06-01'))
>>>
>>>     ## member should not have action timestamps before joining
>>>     member_rng = rng[rng.start_time > join_date]
>>>
>>>     if len(member_rng) < 1:
>>>         continue
>>>
>>>     info = member_behaviors[idx](member_rng)
>>>     if len(info) == len(member_rng):
>>>         emails = emails.append(pd.DataFrame(
>>>             {'member': [idx] * len(info),
>>>              'week': [str(r.start_time) for r in member_rng],
>>>              'emailsOpened': info}))
>>>         donations = donations.append(
>>>             produce_donations(member_rng, member_behaviors[idx],
>>>                               sum(info), idx, join_date.year))
```

We then look at the temporal behavior of the donations to get a sense of how we might try this for further analysis or forecasting. We plot the total sum of donations we received for each month of the data set (see Figure 4-1):

```python
## python
>>> df.set_index(pd.to_datetime(df.timestamp), inplace = True)
>>> df.sort_index(inplace = True)
>>> df.groupby(pd.Grouper(freq='M')).amount.sum().plot()
```

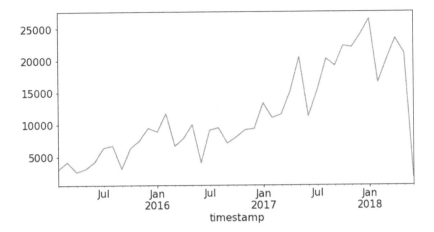

Figure 4-1. Total sum of donations received for each month of the data set.

It looks as though the number of donations and of emails opened rose over time from 2015 through 2018. This is not surprising, since the number of members also rose over time, as indicated in the cumulative sum of members and the year they joined. In fact, one built-in assumption of our model was that we got to keep a member indefinitely after they joined. We made no provision for termination other than allowing for members to open a decreasing number of emails. Even in that case, however, we left open the possibility of continued donations. We see this assumption of indefinitely continuing membership (and correlated donation behavior) in Figure 4-1. We should probably go back and refine our code, as indefinite membership and donation is not a realistic scenario.

This is not a classic time series simulation, so it may feel quite a bit more like an exercise in generating tabular data. It absolutely is that as well, but we did have to be time series–aware:

- We had to make decisions about how many time series our users were in.
- We had to make decisions about what kinds of trends we would model over time:
  — In the case of email, we decided to have three trends: stable, increasing, and decreasing email open rates.
  — In the case of donations, we made donations a stable behavioral pattern related to how many emails the member had ever opened in their lifetime. This included a lookahead, but since we were generating data, this was a way of deciding that a member's overall affinity in the organization, which would lead to more emails opened, would also increase the frequency of donations.

- We had to be careful to make sure we did not have emails opened or donations made before the member joined the organization.

- We had to make sure our data did not go into the future, to make it more realistic for consumers of the data. Note that for a simulation it is fine if our data goes into the future.

But it's not perfect. The code presented here is ungainly, and it doesn't produce a realistic universe. What's more, since only the programmer checked the logic, they could have missed edge cases such that events take place in an illogical order. It would be good to establish external metrics and standards of validity before running the simulation as one protection against such errors.

We need software that enforces a logical and consistent universe. We will look at Python generators as a better option in the next section.

Building a Simulation Universe That Runs Itself

Sometimes you have a specific system and you want to set up the rules for that system and see how it rolls along. Perhaps you want to envision what a universe of independent members accessing your application will use, or you want to attempt to validate an internal theory of decision making based on posited external behavior. In these cases, you are looking to see how individual agents contribute to your aggregate metrics over time. Python is an especially good fit for this job thanks to the availability of generators. When you start building software rather than staying purely in analysis, it makes sense to move to Python even if you are more comfortable in R.

Generators allow us to create a series of independent (or dependent!) actors and wind them up to watch what they do, without too much boilerplate code to keep track of everything.

In the next code example, we explore a taxicab simulation.[1] We want to imagine how a fleet of taxis, scheduled to begin their shifts at different times, might behave in aggregate. To do so, we want to create many individual taxis, set them loose in a cyber city, and have them report their activities back.

Such a simulation could be exceptionally complicated. For demonstration purposes, we accept that we will build a simpler world than what we imagine to truly be the case ("All models are wrong…"). We start by trying to understand what a Python generator is.

Let's first consider a method I wrote to retrieve a taxi identification number:

---

1 This example is heavily inspired by Luciano Ramalho's book, *Fluent Python* (*https://oreil.ly/fluent-python*) (O'Reilly 2015). I highly recommend reading the full simulation chapter in that book to improve your Python programming skills and see more elaborate opportunities for agent-based simulation.

```
## python
>>> import numpy as np

>>> def taxi_id_number(num_taxis):
>>>     arr = np.arange(num_taxis)
>>>     np.random.shuffle(arr)
>>>     for i in range(num_taxis):
>>>         yield arr[i]
```

For those who are not familiar with generators, here is the preceding code in action:

```
## python
>>> ids = taxi_id_number(10)
>>> print(next(ids))
>>> print(next(ids))
>>> print(next(ids))
```

which might print out:

```
7
2
5
```

This will iterate until it has emitted 10 numbers, at which point it will exit the `for` loop held within the generator and emit a `StopIteration` exception.

The `taxi_id_number()` produces single-use objects, all of which are independent of one another and keep their own state. This is a generator function. You can think of generators as tiny objects that maintain their own small bundle of state variables, which is useful when you want many objects parallel to one another, each one minding its own variables.

In the case of this simple taxi simulation, we compartmentalize our taxis into different shifts, and we also use a generator to indicate shifts. We schedule more taxis in the middle of the day than in the evening or overnight shifts by setting different probabilities for starting a shift at a given time:

```
## python
>>> def shift_info():
>>>     start_times_and_freqs = [(0, 8), (8, 30), (16, 15)]
>>>     indices               = np.arange(len(start_times_and_freqs))
>>>     while True:
>>>         idx   = np.random.choice(indices, p = [0.25, 0.5, 0.25])
>>>         start = start_times_and_freqs[idx]
>>>         yield (start[0], start[0] + 7.5, start[1])
```

Pay attention to `start_times_and_freqs`. This is our first bit of code that will contribute to making this a time series simulation. We are indicating that different parts of the day have different likelihoods of having a taxi assigned to the shift. Additionally, different times of the day have a different mean number of trips.

Now we create a more complex generator that will use the preceding generators to establish individual taxi parameters as well as create individual taxi timelines:

```python
## python
>>> def taxi_process(taxi_id_generator, shift_info_generator):
>>>     taxi_id = next(taxi_id_generator)
>>>     shift_start, shift_end, shift_mean_trips =
>>>                             next(shift_info_generator)
>>>     actual_trips = round(np.random.normal(loc   = shift_mean_trips,
>>>                                           scale = 2))
>>>     average_trip_time = 6.5 / shift_mean_trips * 60
>>>     # convert mean trip time to minutes
>>>     between_events_time = 1.0 / (shift_mean_trips - 1) * 60
>>>     # this is an efficient city where cabs are seldom unused
>>>     time = shift_start
>>>     yield TimePoint(taxi_id, 'start shift', time)
>>>     deltaT = np.random.poisson(between_events_time) / 60
>>>     time += deltaT
>>>     for i in range(actual_trips):
>>>         yield TimePoint(taxi_id, 'pick up    ', time)
>>>         deltaT = np.random.poisson(average_trip_time) / 60
>>>         time += deltaT
>>>         yield TimePoint(taxi_id, 'drop off   ', time)
>>>         deltaT = np.random.poisson(between_events_time) / 60
>>>         time += deltaT
>>>     deltaT = np.random.poisson(between_events_time) / 60
>>>     time += deltaT
>>>     yield TimePoint(taxi_id, 'end shift  ', time)
```

Here the taxi accesses generators to determine its ID number, shift start times, and mean number of trips for its start time. From there, it departs on its own individual journey as it runs through a certain number of trips on its own timeline and emits those to the client calling next() on this generator. In effect, this generator produces a time series of points for an individual taxi.

The taxi generator yields TimePoints, which are defined as follows:

```python
## python
>>> from dataclasses import dataclass

>>> @dataclass
>>> class TimePoint:
>>>     taxi_id:   int
>>>     name: str
>>>     time: float

>>>     def __lt__(self, other):
>>>         return self.time < other.time
```

We use the relatively new dataclass decorator to simplify the code (this requires Python 3.7). I recommend that all Python-using data scientists familiarize themselves with this new and data-friendly addition to Python.

Python's Dunder Methods

Python's *dunder* methods, whose names begin and end with two underscores, are a set of built-in methods for every class. Dunder methods are called automatically in the natural course using a given object. There are predefined implementations that can be overridden when you define them for your class yourself. There are many reasons you might want to do this, such as in the case of the preceding code, where we want `TimePoints` to be compared only based on their time and not based on their `taxi_id` or `name` attributes.

Dunder originated as an abbreviation of "double under."

In addition to the automatically generated initializer for `TimePoint`, we need only two other dunder methods, `__lt__` (to compare `TimePoints`) and `__str__` (to print out `TimePoints`, not shown here). We need comparison because we will take all `Time Points` produced into a data structure that will keep them in order: a priority queue. A *priority queue* is an abstract data type into which objects can be inserted in any order but which will emit objects in a specified order based on their priority.

Abstract Data Type

An *abstract data type* is a computational model defined by its behavior, which consists of an enumerated set of possible actions and input data and what the results of such actions should be for certain sets of data.

One commonly known abstract data type is a first-in-first-out (FIFO) data type. This requires that objects are emitted from the data structure in the same order in which they were fed into the data structure. How the programmer elects to accomplish this is a matter of implementation and not a definition.

We have a simulation class to run these taxi generators and keep them assembled. This is not merely a `dataclass` because it has quite a bit of functionality, even in the initializer, to arrange the inputs into a sensible array of information and processing. Note that the only public-facing functionality is the `run()` function:

```python
## python
>>> import queue

>>> class Simulator:
>>>     def __init__(self, num_taxis):
>>>         self._time_points = queue.PriorityQueue()
>>>         taxi_id_generator = taxi_id_number(num_taxis)
>>>         shift_info_generator = shift_info()
>>>         self._taxis = [taxi_process(taxi_id_generator,
```

```
>>>                                          shift_info_generator) for
>>>                                              i in range(num_taxis)]
>>>          self._prepare_run()

>>>     def _prepare_run(self):
>>>          for t in self._taxis:
>>>              while True:
>>>                  try:
>>>                      e = next(t)
>>>                      self._time_points.put(e)
>>>                  except:
>>>                      break

>>>     def run(self):
>>>          sim_time = 0
>>>          while sim_time < 24:
>>>              if self._time_points.empty():
>>>                  break
>>>              p = self._time_points.get()
>>>              sim_time = p.time
>>>              print(p)
```

First, we create the number of taxi generators that we need to represent the right number of taxis. Then we run through each of these taxis while it still has TimePoints and push all these TimePoints into a priority queue. The priority of the object is determined for a custom class such as TimePoint by our implementation of a Time Point's __lt__, where we compare start time. So, as the TimePoints are pushed into the priority queue, it will prepare them to be emitted in temporal order.

We run the simulation:

```
## python
>>> sim = Simulator(1000)
>>> sim.run()
```

Here's what the output looks like (your output will be different, as we haven't set a seed—and every time you run the code it will be different from the last iteration):

```
id: 0539 name: drop off    time: 23:58
id: 0318 name: pick up     time: 23:58
id: 0759 name: end shift   time: 23:58
id: 0977 name: pick up     time: 23:58
id: 0693 name: end shift   time: 23:59
id: 0085 name: end shift   time: 23:59
id: 0351 name: end shift   time: 23:59
id: 0036 name: end shift   time: 23:59
id: 0314 name: drop off    time: 23:59
```

Setting a Seed When Generating Random Numbers

When you write code that is generating random numbers, you might want to ensure that it's reproducible (e.g., if you wanted to set up unit tests for code that is normally random or if you were trying to debug and wanted to narrow down sources of variation to make debugging easier). To ensure that random numbers come out in the same nonrandom order, you set a seed. This is a common operation, so there are guides on how to set a seed in any computer language.

We have rounded to the nearest minute for display simplicity, although we do have more fine-grained data available. What temporal resolution we use will depend on our purposes:

- If we want to make an educational display for people in our city of how the taxi fleet affects traffic, we might display hourly aggregates.
- If we are a taxicab app and need to understand load on our server, we likely want to look at minute-by-minute data or even more highly resolved data to think about our infrastructure design and capacity.

We made the decision to report taxi TimePoints as they are "happening." That is, we report the start of a taxi ride ("pick up") without the time when the ride will end, even though we easily could have condensed this. This is one way of making the time series more realistic, in the sense that you likely would have recorded events in this way in a live stream.

Note that, as in the previous case, our time series simulation has not yet produced a time series. We have produced a log and can see our way through to making this a time series in a number of ways, however:

- Output to a CSV file or time series database as we run the simulation.
- Run some kind of online model hooked up to our simulation to learn how to develop a real-time streaming data processing pipeline.
- Save the output down to a file or database and then do more post-processing to package the data in a convenient (but possibly risky vis-à-vis lookahead) form, such as pairing together start and end times of a given ride to study how the length of a taxi ride behaves at different times of the day.

There are several advantages to simulating this data in addition to being able to test hypotheses about the dynamics of a taxi system. Here are a couple of situations where this synthetic time series data could be useful:.

- Testing the merits of various forecasting models relative to the known underlying dynamics of the simulation.

- Building a pipeline for data you eventually expect to have based on your synthetic data while you await the real data.

You will be well served as a time series analyst by your ability to make use of generators and object-oriented programming. This example offers just one example of how such knowledge can simplify your life and improve the quality of your code.

For Extensive Simulations, Consider Agent-Based Modeling

The solution we coded here was all right, but it was a fair amount of boilerplate to ensure that logical conditions would be respected. If a simulation of discrete events based on the actions of discrete actors would be a useful source of simulated time series data, you should consider a simulation-oriented module. The SimPy (*https://simpy.readthedocs.io/en/latest*) module is a helpful option, with an accessible API and quite a bit of flexibility to do the sorts of simulation tasks we handled in this section.

A Physics Simulation

In another kind of simulation scenario, you may be in full possession of the laws of physics that define a system. This doesn't have to be physics per se, however; it can also apply to a number of other areas:

- Quantitative researchers in finance often hypothesize the "physical" rules of the market. So do economists, albeit at different timescales.

- Psychologists posit the "psychophysical" rules of how humans make decisions. These can be used to generate "physical" rules about expected human responses to a variety of options over time.

- Biologists research rules about how a system behaves over time in response to various stimuli.

One case of knowing some rules for a simple physical system is that of modeling a magnet. This is the case we are going to work on, via an oft-taught statistical mechanics model called the Ising model.[2] We will look at a simplified version of how to simulate its behavior over time. We will initialize a magnetic material so that its

---

2 The Ising model is a well-known and commonly taught classical statistical mechanical model of magnets. You can find many code examples and further discussion of this model online in both programming and physics contexts if you are interested in learning more.

individual magnetic components are pointing in random directions. We will then watch how this system evolves into order where all the magnetic components point in the same direction, under the action of known physical laws and a few lines of code.

Next we discuss how such a simulation is accomplished via a Markov Chain Monte Carlo (MCMC) method, discussing both how that method works in general and as applied to this specific system.

Monte Carlo Simulations and Markov Chains

The idea behind a Monte Carlo simulation is to find clever ways to apply random numbers to situations that should in theory be solvable exactly, but in practice are much easier to solve probabilistically.

The Markov chain is a helpful addition to a general Monte Carlo simulation and is particularly applicable to time series simulation. A Monte Carlo simulation will help you figure out what a particular distribution or series of terms looks like, but not how those terms should evolve over time. This is where a Markov chain comes in. It calculates a probability of transitioning between states, and when we factor that in, we take "steps" rather than simply calculating a global integral. Now we can have a time series simulation rather than merely the calculation of an integral.

In physics, an MCMC simulation can be used, for example, to understand how quantum transitions in individual molecules can affect aggregate ensemble measurements of that system over time. In this case, we need to apply a few specific rules:

1. In a Markov process, the probability of a transition to a state in the future depends only on the present state (not on past information).

2. We will impose a physics-specific condition of requiring a Boltzmann distribution for energy; that is, $T_{ij} / T_{ji} = e^{-b(E_j - E_i)}$. For most of us, this is just an implementation detail and not something nonphysicists need to worry about.

We implement an MCMC simulation as follows:

1. Select the starting state of each individual lattice site randomly.

2. For each individual time step, choose an individual lattice site and flip its direction.

3. Calculate the change in energy that would result from this flip given the physical laws you are working with. In this case this means:

- If the change in energy is negative, you are transitioning to a lower energy state, which will always be favored, so you keep the switch and move on to the next time step.

- If the change in energy is not negative, you accept it with the acceptance probability of $e^{(-\text{energy change})}$. This is consistent with rule 2.

Continue steps 2 and 3 indefinitely until convergence to determine the most likely state for whatever aggregate measurement you are making.

Let's take a look at the specific details of the Ising model. Imagine we have a two-dimensional material composed of a grid of objects, each one having what boils down to a mini-magnet that can point up or down. We put those mini-magnets randomly in an up or down spin at time zero, and we then record the system as it evolves from a random state to an ordered state at low temperature.[3]

First we configure our system, as follows:

```python
## python
>>> ### CONFIGURATION
>>> ## physical layout
>>> N          = 5 # width of lattice
>>> M          = 5 # height of lattice
>>> ## temperature settings
>>> temperature = 0.5
>>> BETA       = 1 / temperature
```

Then we have some utility methods, such as random initialization of our starting block:

```python
>>> def initRandState(N, M):
>>>     block = np.random.choice([-1, 1], size = (N, M))
>>>     return block
```

We also calculate the energy for a given center state alignment relative to its neighbors:

```python
## python
>>> def costForCenterState(state, i, j, n, m):
>>>     centerS = state[i, j]
>>>     neighbors = [((i + 1) % n, j), ((i - 1) % n, j),
>>>                  (i, (j + 1) % m), (i, (j - 1) % m)]
>>>     ## notice the % n because we impose periodic boundary cond
>>>     ## ignore this if it doesn't make sense - it's merely a
>>>     ## physical constraint on the system saying 2D system is like
>>>     ## the surface of a donut
```

---

3 The Ising model is more often used to understand what a ferromagnet's equilibrium state is rather than to consider the temporal aspect of how a ferromagnet might make its way into an equilibrium state. However, we treat the evolution over time as a time series.

```python
>>>     interactionE = [state[x, y] * centerS for (x, y) in neighbors]
>>>     return np.sum(interactionE)
```

And we want to determine the magnetization of the entire block for a given state:

```python
## python
>>> def magnetizationForState(state):
>>>     return np.sum(state)
```

Here's where we introduce the MCMC steps discussed earlier:

```python
## python
>>> def mcmcAdjust(state):
>>>     n = state.shape[0]
>>>     m = state.shape[1]
>>>     x, y = np.random.randint(0, n), np.random.randint(0, m)
>>>     centerS = state[x, y]
>>>     cost = costForCenterState(state, x, y, n, m)
>>>     if cost < 0:
>>>         centerS *= -1
>>>     elif np.random.random() < np.exp(-cost * BETA):
>>>         centerS *= -1
>>>     state[x, y] = centerS
>>>     return state
```

Now to actually run a simulation, we need some recordkeeping as well as repeated calls to the MCMC adjustment:

```python
## python
>>> def runState(state, n_steps, snapsteps = None):
>>>     if snapsteps is None:
>>>         snapsteps = np.linspace(0, n_steps, num = round(n_steps / (M * N * 100)),
>>>             dtype = np.int32)
>>>     saved_states = []
>>>     sp = 0
>>>     magnet_hist = []
>>>     for i in range(n_steps):
>>>         state = mcmcAdjust(state)
>>>         magnet_hist.append(magnetizationForState(state))
>>>         if sp < len(snapsteps) and i == snapsteps[sp]:
>>>             saved_states.append(np.copy(state))
>>>             sp += 1
>>>     return state, saved_states, magnet_hist
```

And we run the simulation:

```python
## python
>>> ### RUN A SIMULATION
>>> init_state = initRandState(N, M)
>>> print(init_state)
>>> final_state = runState(np.copy(init_state), 1000)
```

We can get some insights from this simulation by looking at the beginning and ending states (see Figure 4-2).

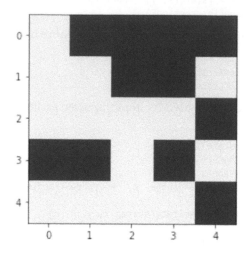

Figure 4-2. Initial state of a 5 × 5 simulated ferromagnetic material, initialized with each state randomly selected to be spin up or spin down with equal probability.

In Figure 4-2 we examine one randomly generated initial state. While you might expect to see the two states more mixed up, remember that probabilistically it's not that likely to get a perfect checkerboard effect. Try generating the initial state many times, and you will see that the seemingly "random" or "50/50" checkerboard state is not at all likely. Notice, however, that we start with approximately half our sites in each state. Also realize that any patterns you find in the initial states is likely your brain following the very human tendency to see patterns even where there aren't any.

We then pass the initial state into the `runState()` function, allow 1,000 time steps to pass, and then examine the outcome in Figure 4-3.

This is a snapshot of the state taken at step 1,000. There are at least two interesting observations at this point. First, the dominant state has reversed compared to step 1,000. Second, the dominant state is no more dominant numerically than was the other dominant state at step 1,000. This suggests that the temperature may continue to flip sites out of the dominant state even when it might otherwise be favored. To better understand these dynamics, we should consider plotting overall aggregate measurements, such as magnetization, or make movies where we can view our two-dimensional data in a time series format.

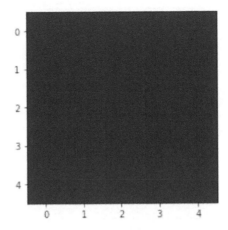

Figure 4-3. Final low temperature state in one run of our simulation, as seen at 1,000 time steps.

We do this with magnetization over time for many independent runs of the previous simulation, as pictured in Figure 4-4:

```python
## python
>>> we collect each time series as a separate element in results list
>>> results = []
>>> for i in range(100):
>>>     init_state = initRandState(N, M)
>>>     final_state, states, magnet_hist = runState(init_state, 1000)
>>>     results.append(magnet_hist)
>>>
>>> ## we plot each curve with some transparency so we can see
>>> ## curves that overlap one another
>>> for mh in results:
>>>     plt.plot(mh,'r', alpha=0.2)
```

The magnetization curves are just one example of how we could picture the system evolving over time. We might also consider recording 2D time series, as the snapshot of the overall state at each point in time. Or there might be other interesting aggregate variables to measure at each step, such as a measure of layout entropy or a measure of total energy. Quantities such as magnetization or entropy are related quantities, as they are a function of the geometric layout of the state at each lattice site, but each quantity is a slightly different measure.

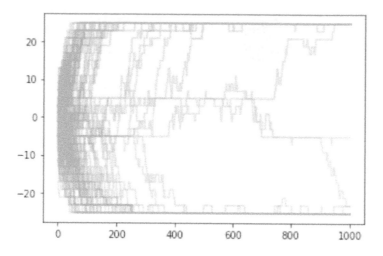

Figure 4-4. 100 independent simulations of potential ways the system could enter a magnetized state at a low temperature even when each original lattice site was initialized randomly.

We can use this data in similar ways to what we discussed with the taxicab data, even though the underlying system is quite different. For example, we could:

- Use the simulated data as the impetus to set up a pipeline.
- Test machine learning methods on this synthetic data to see if they can be helpful on physical data before we go to the trouble of cleaning up real-world data for such modeling.
- Watch the movie-like imagery of important metrics to develop better physical intuitions about the system.

Final Notes on Simulations

We have looked at a number of very different examples of simulating measurements that describe behavior over time. We have looked at simulating data related to consumer behavior (NGO membership and donation), city infrastructure (taxicab pickup patterns), and the laws of physics (the gradual ordering of a randomized magnetic material). These examples should leave you feeling comfortable enough to begin reading code examples of simulated data and also come up with ideas for how your own work could benefit from simulations.

Chances are that, in the past, you have made assumptions about your data without knowing how to test those or alternate possibilities. Simulations give you a route to do so, which means your conversations about data can expand to include hypothetical examples paired with quantitative metrics from simulations. This will ground your discussions while opening new possibilities, both in the time series domain and in other branches of data science.

Statistical Simulations

Statistical simulations are the most traditional route to simulated time series data. They are particularly useful when we know the underlying dynamics of a stochastic system and want to estimate a few unknown parameters or see how different assumptions would impact the parameter estimation process (we will see an example of this later in the book). Even for physical systems, sometimes the statistical simulation is better.

Statistical simulations of time series data are also quite valuable when we need to have a definitive quantitative metric to define our own uncertainty about the accuracy of our simulations. In traditional statistical simulations, such as an ARIMA model (to be discussed in Chapter 6), the formulas for the error are well established, meaning that to understand a system with a posited underlying statistical model, you do not need to run many simulations to make numerical assertions about error and variance.

Deep Learning Simulations

Deep learning simulations for time series are a nascent but promising field. The advantages of deep learning are that very complicated, nonlinear dynamics can be captured in time series data even without the practitioner fully understanding the dynamics. This is also a disadvantage, however, in that the practitioner has no principled basis for understanding the dynamics of the system.

Deep learning simulations also offer promise where privacy is a concern. For example, deep learning has been used to generate synthetic heterogeneous time series data for medical applications based on real time series data but without the potential to leak private information. Such a data set, if it can truly be produced without any privacy leaks, would be invaluable because researchers could have access to a large array of (otherwise expensive and privacy-violating) medical data.

More Resources

*Cristóbal Esteban, Stephanie L. Hyland, and Gunnar Rätsch, "Real-Valued (Medical)
Time Series Generation with Recurrent Conditional GANs," unpublished manuscript,
last revised December 4, 2017, https://perma.cc/Q69W-L44Z.*

> The authors demonstrate how generative adversarial networks can be used to
> produce realistic-looking heterogenous medical time series data. This is an
> example of how deep learning simulation can be used to create ethical, legal, and
> (hopefully) privacy-preserving medical data sets to enable wider access to useful
> data for machine learning and deep learning in the healthcare context.

*Gordon Reikard and W. Erick Rogers, "Forecasting Ocean Waves: Comparing a
Physics-based Model with Statistical Models," Coastal Engineering 58 (2011): 409–16,
https://perma.cc/89DJ-ZENZ.*

> This article offers an accessible and practical comparison of two drastically dif-
> ferent ways of modeling a system, with physics or with statistics. The researchers
> conclude that for the particular problem they address, the timescale of interest to
> the forecaster should drive decisions about which paradigm to apply. While this
> article is about forecasting, simulation is strongly related and the same insights
> apply.

*Wolfgang Härdle, Joel Horowitz, and Jens-Peter Kreiss, "Bootstrap Methods for Time
Series," International Statistical Review / Revue Internationale de Statistique 71, no. 2
(2003): 435–59, https://perma.cc/6CQA-EG2E.*

> A classic 2005 review of the difficulties of statistical simulation of time series data
> given temporal dependencies. The authors explain, in a highly technical statistics
> journal, why methods to bootstrap time series data lag behind methods for other
> kinds of data, as well as what promising methods were available at the time of
> writing. The state of the art has not changed too much, so this is a useful, if chal-
> lenging, read.

## CHAPTER 5

Storing Temporal Data

Often, the value of time series data comes in its retrospective, rather than live streaming, scenarios. For this reason, storage of time series data is necessary for most time series analysis.

A good storage solution is one that enables ease of access and reliability of data without requiring a major investment of computing resources. In this chapter, we will discuss what aspects of a data set you should consider when designing for time series data storage. We will also discuss the advantages of SQL databases, NoSQL databases, and a variety of flat file formats.

Designing a general time series storage solution is a challenge because there are so many different kinds of time series data, each with different storage, read/write, and analysis patterns. Some data will be stored and examined repeatedly, whereas other data is useful only for a short period of time, after which it can be deleted altogether.

Here are a few use cases for time series storage that have different read, write, and query patterns:

1. You are collecting performance metrics on a production system. You need to store these performance metrics for years at a time, but the older the data gets, the less detailed it needs to be. Hence you need a form of storage that will automatically downsample and cull data as information ages.

2. You have access to a remote open source time series data repository, but you need to keep a local copy on your computer to cut down on network traffic. The remote repository stores each time series in a folder of files available for download on a web server, but you'd like to collapse all these files into a single database for simplicity. The data should be immutable and able to be stored indefinitely, as it is intended to be a reliable copy of the remote repository.

3. You created your own time series data by integrating a variety of data sources at different timescales and with different preprocessing and formatting. The data collection and processing was laborious and time-consuming. You'd like to store the data in its final form rather than running a preprocessing step on it repeatedly, but you'd also like to keep the raw data in case you later investigate preprocessing alternatives. You expect to revisit the processed and raw data often as you develop new machine learning models, refitting new models on the same data and also adding to your data over time as new, more recent raw data becomes available. You will never downsample or cull data in storage even if you do so for your analyses.

These use cases are quite varied in what their major demands on a system will be:

Importance of how performance scales with size
In the first use case, we would look for a solution that could incorporate automated scripting to delete old data. We would not be concerned with how the system scaled for large data sets because we would plan to keep the data set small. By contrast, in the second and third use cases we would expect to have a stable, large collection of data (use case 2), or a large and growing collection of data (use case 3).

Importance of random versus sequential access of data points
In use case 2, we would expect all of the data to be accessed in equal amounts, as this time series data would all be the same "age" on insertion and would all reference interesting data sets. In contrast, in use cases 1 and 3 (and especially 1, given the preceding description) we would expect the most recent data to be accessed more often.

Importance of automation scripts
Use case 1 seems as though it might be automated, whereas use case 2 would not require automation (since that data would be immutable). Use case 3 suggests little automation but a great deal of data fetching and processing of all portions of the data, not just the most recent. In use case 1, we would want a storage solution that could be integrated with scripting or stored procedures, whereas in use case 3, we would want a solution that permitted a great deal of easy customization of data processing.

Just three examples already offer a good idea of the many use cases an all-purpose time series solution needs to meet.

In real-world use cases, you will be able to tailor your storage solution and not worry about finding a tool that fits all use cases. That said, you will always be choosing from a similar range of available technologies, which tend to boil down to:

- SQL databases

- NoSQL databases
- Flat file formats

In this chapter, we cover all three options and discuss the advantages and disadvantages of each. Of course, the specifics will depend on the use case at hand, but this chapter will equip you with the foundation you'll need when you begin your search for a time series storage option that works for your use case.

We first discuss what questions you should ask at the outset when picking a storage solution. Then we look at the great SQL versus NoSQL debate and examine some of the most popular time series storage solutions. Finally, we consider setting up policies to let old time series data expire and be deleted.

Defining Requirements

When you are thinking about storage for time series data, you need to ask yourself a number of questions:

- *How much time series data will you be storing? How quickly will that data grow?* You want to choose a storage solution appropriate to the growth rate of the data you expect. Database administrators moving into time series work from transaction-oriented data sets are often surprised at just how quickly time series data sets can grow.

- *Do your measurements tend toward endless channels of updates (e.g., a constant stream of web traffic updates) or distinct events (e.g., an hourly air traffic time series for every major US holiday in the last 10 years)?* If your data is like an endless channel, you will mostly look at recent data. On the other hand, if your data is a collection of distinct time series split into separate events, then events more distant in time may still be fairly interesting. In the latter case, random access is a more likely pattern.

- *Will your data be regularly or irregularly spaced?* If your data is regularly spaced, you can calculate more accurately in advance how much data you expect to collect and how often that data will be inputted. If your data is irregularly spaced, you must prepare for a less predictable style of data access that can efficiently allow for both dead periods and periods of write activity.

- *Will you continuously collect data or is there a well-defined end to your project?* If you have a well-defined end to the data collection, this makes it easier to know how large a data set you need to accommodate. However, many organizations find that once they start collecting a particular kind of time series, they don't want to stop!

- *What will you be doing with your time series? Do you need real-time visualizations? Preprocessed data for a neural network to iterate over thousands of times? Sharded data highly available to a large mobile user base?* Your primary use case will indicate whether you are more likely to need sequential or random access to your data and how important a factor latency should be to your storage format selection.

- *How will you cull or downsample data? How will you prevent infinite growth? What should be the lifecycle of an individual data point in a time series?* It is impossible to store all events for all time. It is better to make decisions about data deletion policies systematically and in advance rather than on an ad hoc basis. The more you can commit to up front, the better a choice you can make regarding storage formats. More on this in the next section.

The answers to these questions will indicate whether you should store raw or processed data, whether the data should be located in memory according to time or according to some other axis, and whether you need to store your data in a form that makes it easy to both read and write. Use cases will vary, so you should do a fresh inventory with every new data set.

 Sharded data is data that is part of a large data system but spread into smaller, more manageable chunks, often across several servers on a network.

Live Data Versus Stored Data

When thinking about what storage options are desirable for your data, it's key to understand your data's lifecycle. The more realistic you can be about your actual use cases for data, the less data you will need to save and the less time you will need to worry about finding the optimal storage system because you won't quickly size up to an intractable amount of data. Often organizations over-record events of interest and are afraid to lose their data stores, but having more data stored in an intractable form is far less useful than having aggregated data stored at meaningful timescales.

For short-lived data, such as performance data that will be looked at only to ensure nothing is going wrong, it's possible you may never need to store the data in the form in which it's collected, at least not for very long. This is most important for event-driven data, where no single event is important and instead aggregate statistics are the values of interest.

Suppose you are running a web server that records and reports to you the amount of time it took every single mobile device to fully load a given web page. The resulting irregularly spaced time series might look something like Table 5-1.

Table 5-1. Web server time series

Timestamp	Time to load page
April 5, 2018 10:22:24 pm	23s
April 5, 2018 10:22:28 pm	15s
April 5, 2018 10:22:41 pm	14s
April 5, 2018 10:23:02 pm	11s

For many reasons, you would not be interested in any individual measurement of the time to load a page. You would want to aggregate the data (say, time to load averaged per minute), and even the aggregate statistics would only be interesting for a short while. Suppose you were on call overnight for that server. You'd want to make sure you could show performance was good while you were in charge. You could simplify this to a data point for your 12 hours of being on call, and you would have most of the information you could ever want, as shown in Table 5-2.

Table 5-2. A simplified data point for on-call hours

Time period	Most popular hour of access	Num loads	Mean time to load	Max time to load
April 5, 2018 8pm–8 am	11 pm	3,470	21s	45s

In such a case, you should not plan to indefinitely store the individual events. Rather, you should build a storage solution that provides for staging of the individual events only as temporary storage until the data goes into its ultimate form. You will save yourself and your colleagues a lot of grief by preventing runaway data growth before it can even get started. Instead of having 3,470 individual events, none of which interests anyone, you will have readily accessible and compact figures of interest. You should simplify data storage via aggregation and deduplication whenever possible.

Next we consider a few opportunities for reducing data without losing information.

Slowly changing variables

If you are storing a state variable, consider recording only those data points where the value has changed. For example, if you are recording temperature at five-minute increments, your temperature curve might look like a step function, particularly if you only care about a value such as the nearest degree. In such a case, it's not necessary to store repetitive values, and you will save storage space by not doing so.

Noisy, high-frequency data

If your data is noisy, there are reasons for not caring very much about any particular data point. You might consider aggregating data points before recording them since the high level of noise makes any individual measurement less valuable. This will, of course, be quite domain specific and you will need to ensure that downstream users are still able to assess the noise in the measurements for their purposes.

Stale data

The older data is, the less likely your organization is going to use it, except in a very general way. Whenever you are beginning to record a new time series data set, you should think preemptively about when the time series data is likely to become irrelevant:

- Is there a natural expiration date?
- If not, can you look through past research from your analytics department and see how far back they realistically go? When did any script in a Git repository actually access the oldest data in your data set?

If you can automate data deletion in a way that will not impede data analysis efforts, you will improve your data storage options by reducing the importance of scalability or the drag of slow queries on bloated data sets.

Legal Considerations

If you work in an organization that is large or handles data required by outside parties (such as for regulatory audits), legal considerations should factor into your system requirements. You should assess how often you will likely get requests from outside parties for data, how much data they would require, and whether you would be able to choose the format or whether regulations impose a specific form.

It may seem unduly cautious to cite legal requirements when assessing data storage specifications, but recent developments in Europe and the United States suggest increased interest in the data sets used to power machine learning (e.g., the European Union's General Data Protection Regulation).

So far, we have discussed the general range of use cases for time series storage. We have also reviewed a general set of queries related to how a time series data set will be produced and analyzed so that these queries can inform our selection of a storage format. We now review two common options for storing time series: databases and files.

Database Solutions

For almost any data analyst or data engineer, a database is an intuitive and familiar solution to the question of how to store data. As with relational data, a database will often be a good storage choice for time series data. This is particularly true if you want an out-of-the-box solution with any of these classic database characteristics:

- A storage system that can scale up to multiple servers

- A low-latency read/write system

- Functions already in place for computing commonly used metrics (such as computing the mean in a group-by query, where the group-by can apply to time metrics)

- Troubleshooting and monitoring tools that can be used to tune system performance and analyze bottlenecks

These are good reasons among many (*https://perma.cc/K994-RXE9*) to opt for a database rather than a filesystem, and you should always consider a database solution for data storage, particularly when working with a new data set. A database, particularly a NoSQL database, can help you preserve flexibility. Also, a database will get your project up and running sooner than if you are going to work with individual files because much of the boilerplate you will need is already in place. Even if you ultimately decide on a file storage solution (and most people won't), working with a database first can help you determine how to structure your own files when your new data processes mature.

In the remainder of this section, we will cover the respective advantages of SQL and NoSQL databases for time series and then discuss currently popular database options for time series applications.

The good news is that time series graphs appear to be the fastest growth category of database (*https://perma.cc/RQ79-5AX7*) at this time, so you can expect to see more and even better options for time series database solutions in the future.

SQL Versus NoSQL

SQL versus NoSQL is just as lively a debate in the time series database community as it is more widely. Many expert database administrators insist that SQL is the only way to go and that there is no data in any shape that cannot be well described by a good set of relational tables. Nonetheless, in practice there is often a drop in performance when organizations try to scale SQL solutions to accommodate large amounts of time series data, and for this reason it's always worth considering a NoSQL solution as well, particularly if you are seeking an open-ended solution that can scale to accommodate cases where time series data collection begins with no finite time horizon in sight.

While both SQL and NoSQL solutions can be good for time series data, we first motivate our discussion of the difficulties of applying database logic to time series data by exploring how time series data differs from the kind of data for which SQL databases were developed.

Characteristics of the data that originally inspired SQL databases

We can best understand the mismatch between SQL-like thinking and time series data by reviewing the history of SQL solutions. SQL solutions were based on transactional data, which is whatever data is needed to completely describe a discrete event. A transaction is composed of attributes that reflect many primary keys, such as product, participants, time, and value of a transaction. Notice that time can be present as a primary key but only as one among many, not as a privileged axis of information.

There are two important features of transactional data that are quite different from time series needs:

- Existing data points are often updated.
- The data is accessed somewhat randomly as there is no underlying ordering required.

Characteristics of time series data

Time series data elaborates the entire history of something, whereas a transaction recording tells us only the final state. Hence, time series data does not usually require updating, meaning that random access for write operations is a low priority.

This means that the performance objectives that have been key to decades of SQL database design are not very important for time series databases. In fact, when we consider objectives in designing a database for time series, we have quite different priorities because of how we will use our time series data. The primary features of our time series data use case are:

- Write operations predominate over read operations.
- Data is neither written nor read nor updated in random order, but rather in the order related to temporal sequencing.
- Concurrent reads are far more likely than they are for transaction data.
- There are few, if any, primary keys other than time itself.
- Bulk deletes are far more common than individual data point deletion.

These features support the use of a NoSQL database, because many general-application NoSQL databases offer much of what we want for time series databases, particularly the emphasis on write operations over read operations. Conceptually, NoSQL databases map well to time series data in that they natively reflect aspects of time series data collection, such as not all fields being collected for all data points. The flexible schemas of NoSQL are natural with time series data. Much of the data that motivated the current popularity of NoSQL databases is time series data.

For this reason, out-of-the-box NoSQL databases tend to perform better than SQL databases for write operations. Figure 5-1 plots performance of a commonly used SQL database versus a commonly used NoSQL database for the case of time series data point insertions (write operations).

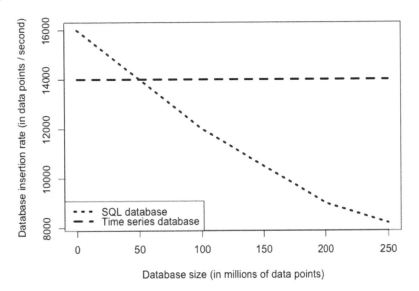

Figure 5-1. One common feature of a database designed for time series use is that the data insertion rate is constant with respect to the database size. This is a striking advantage in comparison to a traditional SQL storage solution.

How to choose between SQL and NoSQL

It may seem that I am attempting to push you toward the NoSQL use case, but there are also many good use cases for SQL databases. When thinking about your own data, keep in mind the principle that always applies to data storage, be it in a SQL database, a NoSQL database, or a plain old text file: *data that tends to be requested at the same time should be stored in the same location.* That's the most important factor regardless of your use case.

Many conference presentations and blog posts herald NoSQL solutions as precisely catering to the situation of a high-write low-update scenario where data is not ordered randomly. Nonetheless, SQL databases are a viable and regularly used time series storage option. In particular, with some architectural changes in emphasis to traditional SQL databases and their internal memory structure, these challenges can be met while keeping some of the advantages of SQL (*https://perma.cc/3ZUQ-B5WC*). For example, something as simple and seemingly obvious as structuring SQL tables' internal memory representation to account for time can make a substantial performance difference.

Ultimately, the distinctions between NoSQL and SQL are highly implementation-dependent and not as systematic or important as they are made out to be. Let your data drive your selection of a specific implementation of one of these technologies. As you consider the attributes of your time series data and access use patterns, you can keep some general limitations in mind:

Pluses of SQL for time series

- If your time series is stored in a SQL database, you can easily relate it to relevant cross-sectional data also stored in that database.

- Hierarchical time series data is a natural fit with relational tables. An appropriate set of SQL schema will help you group related time series and clearly delineate the hierarchy, whereas these could be easily scattered less systematically in a NoSQL solution.

- If you are crafting a time series based on transactional data, where the data is best stored in a SQL database, it will be advantageous to also store your time series in the same database for ease of verification, cross-referencing, and so on.

Pluses of NoSQL for time series

- Writes are fast.

- They're good if you don't know enough about future data to design and intelligent and robust schema.

- For the inexpert user, these databases are often the more performant out-of-the-box solution because you are less likely to design an awkward schema or to be locked into a mediocre schema design.

Popular Time Series Database and File Solutions

Now we'll discuss some popular database solutions for time series data. These give you a sense of what is available in addition to traditional SQL solutions. Note that the technologies discussed here occupy a crowded and fragmented technology landscape. What is commonly used this year may not be so popular next year. Thus, this discussion should not be viewed as a specific technology recommendation so much as a set of samples helping to illustrate the current state of the market.

Time series–specific databases and related monitoring tools

First we discuss tools that are built specifically for the purpose of storing and monitoring time series data. In particular, we take a look at a time series database (InfluxDB) and another product that is a performance monitoring tool, which can double as a time series storage solution (Prometheus). The advantages of each tool necessarily reflect its distinct emphasis and use patterns.

InfluxDB. InfluxDB is a time series–specific database, per its own description (*https:// oreil.ly/6qmVH*) on the GitHub project's web page:

> InfluxDB is an open source time series database...useful for recording metrics, events, and performing analytics.

In InfluxDB, data is organized by time series. A data point in InfluxDB consists of:

- A timestamp
- A label indicating what the measurement consists of
- One or more key/value fields (such as `temperature=25.3`)
- Key/value pairs containing metadata tags

InfluxDB, as a time-aware database, automatically timestamps any data point that comes in without a timestamp. Also, InfluxDB uses SQL-like querying, such as:

```
SELECT * FROM access_counts WHERE value > 10000
```

Other advantages of InfluxDB include:

- Data retention options that allow you to easily automate the designation and deletion of stale data
- High data ingestion speed and aggressive data compression
- Ability to tag individual time series to allow for fast indexing of time series that match a specific criterion
- Membership in the mature TICK stack (*https://oreil.ly/KLjGo*), which is a platform for capturing, storing, monitoring, and displaying time series data

There are many other time series–specific databases; right now InfluxDB happens to be the most popular, so it is the one you are most likely to encounter. The options it offers are those commonly offered by time series–specialized databases, as these options reflect the most commonly desired attributes for storage of time series data.

As a database, InfluxDB is a *push-based* system, meaning that when you use it you push data into the database for ingestion. This is different from the next option we'll discuss, Prometheus.

Given these specifications, a solution such as InfluxDB provides all general functionality right out of the box in a time-aware manner. This means you can use your existing SQL skills while also benefiting from advantages related to the need to capture but control the growth of time series data. Finally, you have the fast write so often necessary when capturing time series data as it is produced.

Prometheus. Prometheus describes itself (*https://github.com/prometheus/prometheus*) as a "monitoring system and time series database" that works via HTTP. This description indicates its general emphasis: monitoring first and storage second. The great advantage of Prometheus is that it is a *pull-based* system, which means that the logic for how data is pulled for creating a time series and how often it is centralized can be easily adjusted and inspected.

Prometheus is a great resource to have during emergencies because it is atomic and self-reliant. However, it is also not guaranteed to be entirely up-to-date or accurate, due to its pull-based architecture. While it should be the go-to technology for quick-and-dirty performance monitoring, it is not appropriate for applications where data has to be 100% accurate.

Prometheus uses a functional expression language called PromQL for its querying:

```
access_counts > 10000
```

Prometheus also offers, via PromQL, an API for many common time series tasks, even including sophisticated functions such as making a prediction (`predict_lin ear()`) and calculating the per-unit-of-time rate of increase for a time series (`rate()`). Aggregation over periods of time is also offered with a simple interface. Prometheus tends to emphasize monitoring and analysis over maintenance, so compared to InfluxDB there are fewer automated data-curating functionalities.

Prometheus is a helpful time series storage solution particularly for live streaming applications and when data availability is paramount. It has a steeper learning curve due to its custom scripting language and the less database-like architecture and API, but it is nonetheless widely used and enjoyed by many developers.

General NoSQL databases

While time series–specific databases offer many advantages, you can also consider the case of more general-use NoSQL databases. These sorts of databases are based on a *document* structure rather than a *table* structure, and they do not usually have many explicit functions specifically for time series.

Nonetheless, the flexible schemas of NoSQL databases remain useful for time series data, particularly for new projects where the rhythm of data collection and the number of input channels may change through the course of a data set's lifetime. For example, a time series may initially start out as only one channel of data but gradually grow to include more kinds of data all timestamped together. Later, it may be decided that some of the input channels are not particularly useful, and they may be discontinued.

In such a case, storing the data in a SQL table would be difficult for several reasons, and would result in a great many NaNs where data was not available. A NoSQL database would simply leave the channels absent when they were not available rather than marking up a rectangular data store with a bunch of NaNs.

One popular and performant NoSQL time series database is MongoDB. Mongo is particularly aware of its value as a time series database and has a strong push to develop IoT-friendly architecture and instructions. It offers high-level aggregation features that can be applied to aggregation operations by time and time-related groupings, and it also offers many automated processes to divide time into human-relevant markings, such as day of the week or day of the month:

- $dayOfWeek
- $dayOfMonth
- $hour

Also, Mongo has devoted extensive documentation efforts to demonstrating how time series can be handled.[1] A ready set of documentation and institutional focus on time series data explicitly mean that users can expect this database to continue developing even more time-friendly features.

More useful than all this functionality, however, is Mongo's flexibility in evolving schema over time. This schema flexibility will save you a lot of grief if you work on a rapidly evolving time series data set with ever-changing data collection practices.

For example, imagine a sequence of time series data collection practices such as what might occur in a healthcare startup:

1. Your startup launches as a blood pressure app and collects only two metrics, systolic and diastolic blood pressure, which you encourage users to measure several times daily…

2. But your users want enhanced lifestyle advice and they are happy to give you more data. Users provide everything from their birth dates to monthly weight recordings to hourly step counts and calorie counts. Obviously, these different kinds of data are collected at drastically different rhythms…

3. But later you realize some of this data is not very useful so you stop collecting it…

4. But your users miss your display of the data even if you don't use it, so you restart collection of the most popular data…

1 See, for example, "Schema Design for Time Series Data in MongoDB" (*https://perma.cc/E8XL-R3SY*), "Time Series Data and MongoDB: Part 1 – An Introduction" (*https://perma.cc/A2D5-HDFB*), and "Time Series Data and MongoDB: Part 2 – Schema Design Best Practices" (*https://perma.cc/2LU7-YDHX*). For a contrary view, see "How to store time-series data in MongoDB, and why that's a bad idea" (*https://perma.cc/T4KM-Z2B4*).

5. And then the government of one of your largest markets passes a law regarding the storage of health data and you need to purge that data or encrypt it, so you need a new encrypted field…

6. And the changes continue.

When you need the querying and schema flexibility described in an application like this, you would be well served by a general NoSQL database that offers a reasonable balance between time-specific and more general flexibility.

Another advantage of a general-purpose NoSQL database rather than a time series–specific database is you can more easily integrate non-time series data into the same database for ease of cross-referencing the related data sets. Sometimes a general-purpose NoSQL database is just the right mix of performance considerations and SQL-like functionality without the cleverness needed to optimize a SQL database schema for time series functionality.

In this section, we have examined NoSQL database solutions[2] and surveyed some options now in popular use. While we can expect the particular technologies that dominate the market to evolve over time, the general principles on which these technologies operate and the advantages they offer will remain the same. As a quick recap, some of the advantages we find in different varieties of databases are:

- High read or write capacities (or both)
- Flexible data structures
- Push or pull data ingestion
- Automated data culling processes

A time series–specific database will offer the most out-of-the-box automation for your time series–specific tasks, but it will offer less schema flexibility and fewer opportunities to integrate related cross-sectional data than a more general, all-purpose NoSQL database.

Generally speaking, databases will offer more flexibility than flat file storage formats, but this flexibility means that databases are less streamlined and less I/O performant than simple flat files, which we discuss next.

---

2 The optimization of SQL databases for time series data is beyond the scope of this book and tends to be very specific to use case. Some resources at the end of the chapter address this question.

File Solutions

At the end of the day, a database is a piece of software that integrates both scripting and data storage. It is essentially a flat file wrapped in a special piece of software responsible for making that file as safe and easy to use as possible.

Sometimes it makes sense to take away this outer layer and take on the full responsibility of the data storage ourselves. While this is not very common in business applications, it is frequently done in scientific research and in the rare industrial application (such as high-frequency trading) in which speed is paramount. In such cases, it is the analyst who will design a much more intricate data pipeline that involves allocating storage space, opening files, reading files, closing files, protecting files, and so on, rather than simply writing some database queries.

A flat file solution is a good option if any of these conditions apply:

- Your data format is mature so that you can commit to a specification for a reasonably long period of time.
- Your data processing is I/O bound, so it makes sense to spend development time speeding it up.
- You don't need random access, but can instead read data sequentially.

In this section, we briefly survey some common flat file solutions. You should also remember that you can always create your own file storage format (*https://perma.cc/DU98-FWW9*), although this is fairly complex and is usually worth doing only if you are working in a highly performant language, such as C++ or Java.

If your system is mature and performance-sensitive enough to suit a flat file system, there are several advantages to implementing one, even if it may mean the chore of migrating your data out of a database. These advantages include:

- A flat file format is system agnostic. If you need to share data, you simply make files available in a known shareable format. There is no need to ask a collaborator to remotely access your database or set up their own mirrored database.
- A flat file format's I/O overhead is necessarily less than that for a database, because this amounts to a simple read operation for a flat file rather than a retrieval and a read with a database.
- A flat file format encodes the order in which data should be read, whereas not all databases will do so. This enforces serial reading of the data, which can be desirable, such as for deep learning training regimes.
- Your data will occupy a much smaller amount of memory than it would on a database because you can maximize compression opportunities. Relatedly, you can tune the degree of data compression to explicitly balance your desire to

minimize both the data storage footprint and I/O time in your application. More compression will mean a smaller data footprint but longer I/O waits.

NumPy

If your data is purely numeric, one widely used option for holding that data is a Python NumPy array. NumPy arrays can easily be saved in a variety of formats, and there are many benchmarking publications comparing their relative performance. For example, the `array_storage_benchmark` GitHub repo (*https://perma.cc/ZBS7-PR56*) is designed to test the efficiency and speed of various NumPy file formats.

A downside of NumPy arrays is that they have a single data type, which means you cannot store heterogeneous time series data automatically, but must instead think about whether having just one data type can work for your data in raw or processed form (although there are ways around this restriction). Another downside of NumPy arrays is that it is not natural to add labels to rows or columns, so there is no straightforward way to timestamp each row of an array, for example.

The advantage of using a NumPy array is that there are many options for saving one, including a compressed binary format that takes up less space and has faster I/O than you will see with a database solution. It is also as out-of-the-box as it gets in terms of a performant data structure for analysis as well as storage.

Pandas

If you want easy labeling of data or easy storage of heterogenous time series data (or both), consider the less streamlined but more flexible Pandas data frame. It can be particularly useful when you have a time series that consists of many distinct kinds of data, perhaps including event counts (ints), state measurements (floats), and labels (strings or one-hot encodings). In such cases, you will likely want to stick to Pandas data frames (also remember that the name "pandas" actually comes from an elision of "panel data," so this is a natural format for many use cases).

Pandas data frames are widely used, and there are several resources online to compare the various formats (*https://perma.cc/BNJ5-EDGM*) used to store this data.

Standard R Equivalents

The native formats for storing R objects are `.Rds` and `.Rdata` objects. These are binary file formats, so they will necessarily be more efficient both for compression and I/O than text-based formats, and in this way are akin to Pandas data frames in Python. Relatedly, the `feather` format (*https://perma.cc/4C3J-TBK8*) can be used in both R and Python to save data frames in a language-agnostic file format. For R users, the native binary formats will (of course) be the most performant. Table 5-3 compares the file format options.

Table 5-3. Size and performance comparison of file format options

Format name	Relative size	Relative time to load
.RDS	1x	1x
feather	2x	1x
csv	3x	20x

As we can see, the native format is a clear winner to minimize both storage space and I/O, and those who use a text-based file format rather than binary will pay a heavy price both in storage and I/O slowdowns.

Xarray

When your time series data reaches many dimensions wide, it may be time to think about a more industrial data solution—namely Xarray (*http://xarray.pydata.org/en/stable*), which is useful for a number of reasons:

- Named dimensions
- Vectorized mathematical operations, like NumPy
- Group-by operations, like Pandas
- Database-like functionality that permits indexing based on a time range
- A variety of file storage options

Xarray is a data structure that supports many time series–specific operations, such as indexing and resampling on time, interpolating data, and accessing individual components of a date time. Xarray was built as a high-performance scientific computing instrument, and it is woefully underutilized and underappreciated for time series analysis applications.

Xarray is a data structure in Python, and it comes with a choice of storage options. It implements both pickling and another binary file format called netCDF, a universal scientific data format supported across many platforms and languages. If you are looking to up your time series game in Python, Xarray is a good place to start.

As you've seen, there are many options for flat file storage of time series data, some with associated functionality (Xarray) and some with very streamlined, purely numeric formats (NumPy). When migrating a data set from a database-oriented pipeline to a file-oriented pipeline, there will be some growing pains related to simplifying data and rewriting scripts to move logic out of the database and into the explicit ETL (extract-transform-load) scripts. In cases where performance is paramount, moving your data into files is likely the most important step to reduce latency. Common cases where this is desirable include:

- Latency-sensitive forecasting, such as for user-facing software
- I/O-intensive repetitive data access situations, such as training deep learning models

On the other hand, for many applications, the convenience, scalability, and flexibility of a database is well worth the higher latency. Your optimal storage situation will depend very much on the nature of the data you are storing and what you want to do with it.

More Resources

- On time series database technologies:

Jason Moiron, *"Thoughts on Time-Series Databases," jmoiron blog, June 30, 2015, https://perma.cc/8GDC-6CTX.*
This classic blog post from 2015 provides a glimpse into an earlier time period with time series databases and the cult of recording everything. This high-level overview of options for storing time series databases and typical use cases is highly informative for beginners to database administration and engineering.

Preetam Jinka, *"List of Time Series Databases," Misframe blog, April 9, 2016, https://perma.cc/9SCQ-9G57.*
This lengthy list of time series databases is frequently updated to show databases currently on the market. Each database entry comes with highlights as to how a particular database relates to its competitors and predecessors.

Peter Zaitsev, *"Percona Blog Poll: What Database Engine Are You Using to Store Time Series Data?" Percona blog, February 10, 2017, https://perma.cc/5PXF-BF7L.*
This 2017 poll of database engineers indicated that relational databases (SQL databases) continued to dominate as a group, with 35% of respondents indicating they use these databases to store time series data. ElasticSearch, InfluxDB, MongoDB, and Prometheus were also favorites.

Rachel Stephens, *"The State of the Time Series Database Market," RedMonk, April 3, 2018, https://perma.cc/WLA7-ABRU.*
This recent data-driven write-up by a tech analyst describes the most popular solutions to time series database storage via an empirical investigation of GitHub and Stack Overflow activity. The report also indicates a high degree of fragmentation in the time series storage domain due to a wide variety of

use cases for time series data as well as a larger trend toward segmentation in databases.

Prometheus.io, "Prometheus Documentation: Comparison to Alternatives," n.d., https://perma.cc/M83E-NBHQ.
This extremely detailed and comprehensive list compares Prometheus to a number of other popular time series storage solutions. You can use this model as a quick reference for the dominant data structures and storage structures used by the alternatives to Prometheus. This is a good place to get an overview of what's on offer and what the trade-offs are between various databases designed for temporal data.

- On adapting general database technologies:

Gregory Trubetskoy, "Storing Time Series in PostreSQL Efficiently," Notes to Self blog, September 23, 2015, https://perma.cc/QP2D-YBTS.
This older but still relevant blog post explains how to store time series in Postgres in a performance-focused way. Trubetskoy explains the difficulties of the "naive" approach of a column of values and a column of timestamps, and gives practical advice for approaches based on Postgres arrays.

Josiah Carlson, "Using Redis as a Time Series Database," InfoQ, January 2, 2016, https://perma.cc/RDZ2-YM22.
This article provides detailed advice and examples related to using Redis as a time series database. While Redis has been used for time series data since its creation, there are several gotchas the author points out as well as advantageous data structures that can be applied in many time series use case scenarios. This article is useful for learning more about how to use Redis with time series data, but also as an example of how more general tools can be repurposed for time series–specific uses.

Mike Freedman, "Time Series Data: Why (and How) to Use a Relational Database Instead of NoSQL," Timescale blog, April 20, 2017, https://perma.cc/A6CU-6XTZ.
This blog post by TimescaleDB's founder describes the ways in which his team built a time series database as a relational database with memory layout modifications specific to temporal data. TimescaleDB's contention is that the main problem with traditional SQL databases for time series data is that such databases do not scale, resulting in slow performance as data is shifted into and out of memory to perform time-related queries. TimescaleDB proposes laying out memory and memory mappings to reflect the temporal nature of data and reduce swapping disparate data into and out of memory.

Hilmar Buchta, "Combining Multiple Tables with Valid from/to Date Ranges into a Single Dimension," Oraylis blog, November 17, 2014, https://perma.cc/B8CT-BCEK.

While the title is not very appealing, this is an extremely useful blog post for those of us who are not SQL database experts. It illustrates a good way to deal with the hairy problem of data that is marked as being valid only for a certain range of time and how to combine this data and join with it across multiple tables. This is a surprisingly common and a surprisingly difficult task. The example covered involves a few human resources charts, including an employee's valid dates of employment, valid dates for a given department, valid dates for a specific office location, and valid dates for a company car. A very likely task would be to determine an employee's status on a particular day—was that person employed, and if so, in what department and in what location, and did the employee have a company car? This sounds intuitive, but the SQL to answer this question is difficult and well covered in this post.

## CHAPTER 6

Statistical Models for Time Series

In this chapter, we study some linear statistical models for time series. These models are related to linear regression but account for the correlations that arise between data points in the same time series, in contrast to the standard methods applied to cross-sectional data, in which it is assumed that each data point is independent of the others in the sample.

The specific models we will discuss are:

- Autoregressive (AR) models, moving average (MA) models, and autoregressive integrated moving average (ARIMA) models
- Vector autoregression (VAR)
- Hierarchical models.

These models have traditionally been the workhorses of time series forecasting, and they continue to be applied in a wide range of situations, from academic research to industry modeling.

Why Not Use a Linear Regression?

As a data analyst, you are probably already familiar with linear regressions. If you are not, they can be defined as follows: a linear regression assumes you have *independently and identically distributed* (iid) data. As we have discussed at length in earlier chapters, this is not the case with time series data. In time series data, points near in time tend to be strongly correlated with one another. In fact, when there aren't temporal correlations, time series data is hardly useful for traditional time series tasks, such as predicting the future or understanding temporal dynamics.

Sometimes time series tutorials and textbooks or give an undue impression that linear regression is not useful for time series. They make students think that simple linear regressions simply will not cut it. Luckily this is not the case at all. Ordinary least squares linear regression can be applied to time series data provided the following conditions hold:

Assumptions with respect to the behavior of the time series

- The time series has a linear response to its predictors.
- No input variable is constant over time or perfectly correlated with another input variable. This simply extends the traditional linear regression requirement of independent variables to account for the temporal dimension of the data.

Assumptions with respect to the error

- For each point in time, the expected value of the error, given all explanatory variables for all time periods (forward and backward), is 0.
- The error at any given time period is uncorrelated with the inputs at any time period in the past or future. So a plot of the autocorrelation function of the errors will not indicate any pattern.
- Variance of the error is independent of time.

If these assumptions hold, then ordinary least squares regression is an unbiased estimator of the coefficients given the inputs, even for time series data.[1] In this case, the sample variances of the estimates have the same mathematical form as they do for standard linear regression. So if your data meets the assumptions just listed, you can apply a linear regression, which will no doubt help to offer clear and simple intuitions for your time series behavior.

The data requirements just described are similar to those for standard linear regression applied to cross-sectional data. What we have added is an emphasis on the temporal qualities of the data set.

---

1 Note that ordinary least squares is unbiased even in the case of removing some of these conditions. For example, when errors are correlated and heteroskedastic, ordinary least squares can still provide an unbiased estimate of coefficients, although there are efficiency concerns. For more information about efficiency, start with Wikipedia (*https://perma.cc/4M4H-YKPS*).

Don't force linear regression. Some of the consequences of applying linear regression when your data doesn't meet the required assumptions are:

- Your coefficients will not minimize the error of your model.

- Your *p*-values for determining whether your coefficients are nonzero will be incorrect because they rely on assumptions that are not met. This means your assessments of coefficient significance could be wrong.

Linear regressions can be helpful in offering simplicity and transparency when appropriate, but an incorrect model certainly isn't transparent!

It's fair to question whether time series analysts are overly rigid in applying the assumptions required by standard linear regression so stringently they cannot use the linear regression technique. Real-world analysts take liberties with model assumptions from time to time. This can be productive so long as the potential downsides of doing so are understood.

The importance of adhering to a model's assumptions depends strongly on the domain. Sometimes a model is applied in full knowledge that its baseline assumptions are not met because the consequences are not too serious relative to the payoff. For example, in high-frequency trading, linear models are quite popular for a number of reasons despite no one believing that the data strictly follows all the standard assumptions.[2]

What Is An Unbiased Estimator?

If an estimate is not an overestimate or underestimate, it is using an unbiased estimator. This tends to be a good thing, although you should be aware of the *bias-variance trade-off*, which is a description of the problem for both statistical and machine learning problems wherein models with a lower bias in their parameter estimates tend to have a higher variance of the estimate of the parameter. The variance of the parameter's estimate reflects how variable an estimate will be across different samples of the data.

---

2 There are some mitigating factors to justify the use of standard linear regression in that case. First, some believe that at sufficiently short scales the movements of the financial markets are independent of one another (iid). Second, because linear regressions are so computationally efficient, a fast model, even if inaccurate in its assumptions, is a good model in an industry driven by speed. Third, businesses that use these models manage to make money, so they must be doing something right.

If you find yourself in a situation where a linear regression may be a good fit for your forecasting task, consider taking advantage of `tslm()`, a function in the `forecast` package (*https://perma.cc/TR6C-4BUZ*) designed to provide easy linear regression methodologies for time series data.

Statistical Methods Developed for Time Series

We consider statistical methods developed specifically for time series data. We first study methods developed for univariate time series data, beginning with the very simple case of an autoregressive model, which is a model that says that future values of a time series are a function of its past values. We then work our way up to increasingly complex models, concluding with a discussion of vector autoregression for multivariate time series and of some additional specialized time series methods, such as GARCH models and hierarchical modeling.

Autoregressive Models

The autoregressive (AR) model relies on the intuition that the past predicts the future and so posits a time series process in which the value at a point in time t is a function of the series's values at earlier points in time.

Our discussion of this model will be more detailed to give you a sense of how statisticians think about these models and their properties. For this reason, we start with a fairly lengthy theory overview. You can skim this if you are not interested in the mechanics behind how statistical model properties are derived for time series.

Using algebra to understand constraints on AR processes

Autoregression looks like what many people would use as a first attempt to fit a time series, particularly if they had no information other than the time series itself. It is exactly what its name implies: a regression on past values to predict future values.

The simplest AR model, an AR(1) model, describes a system as follows:

$$y_t = b_0 + b_1 \times y_{t-1} + e_t$$

The value of the series at time t is a function of a constant b_0, its value at the previous time step multiplied by another constant $b_1 \times y_{t-1}$ and an error term that also varies with time e_t. This error term is assumed to have a constant variance and a mean of 0. We denote an autoregressive term that looks back only to the immediately prior time as an AR(1) model because it includes a lookback of one lag.

Incidentally, the AR(1) model has an identical form to a simple linear regression model with only one explanatory variable. That is, it maps to:

$$Y = b_0 + b_1 \times x + e$$

We can calculate both the expected value of y_t and its variance, given y_{t-1}, if we know the value of b_0 and b_1. See Equation 6-1.[3]

Equation 6-1. $E(y_t \mid y_{t-1}) = b_0 + b_1 \times y_{t-1} + e_t$

$$Var(y_t \mid y_{t-1}) = Var(e_t) = Var(e)$$

What Is the | Symbol?

In statistics, you can condition probabilities, expected values, and so on, of one variable on another. Such conditioning is indicated with the | symbol, or pipe, which should be read as "given." So for example, $P(A|B)$ should be read as "the probability of A given B."

Why do we use it in the previous case? In equation (0) we want to determine the expected value of the process at time t, y_t given what the value is at one time step previous. This is a way of expressing that we know information up to time $t - 1$. The expected value of y_t if we do not have any information is the expected value we can calculate more generally from the process. However, if we happen to know the value one time step earlier, we can be more specific about the expected value at time t.

The generalization of this notation allows the present value of an AR process to depend on the p most recent values, producing an AR(p) process.

We now switch to more traditional notation, which uses ϕ to denote the autoregression coefficients:

$$y_t = \phi_0 + \phi_1 \times y_{t-1} + \phi_2 \times y_{t-2} + \dots + \phi_p \times y_{t-p} + e_t$$

As discussed in Chapter 3, stationarity is a key concept in time series analysis because it is required by many time series models, including AR models.

---

3 Only equations referenced subsequently are numbered.

We can determine the conditions for an AR model to be stationary from the definition of stationarity. We continue our focus on the simplest AR model, AR(1) in Equation 6-2.

Equation 6-2. $y_t = \phi_0 + \phi_1 \times y_{t-1} + e_t$

We assume the process is stationary and then work "backward" to see what that implies about the coefficients. First, from the assumption of stationarity, we know that the expected value of the process must be the same at all times. We can rewrite y_t per the equation for an AR(1) process:

$$E(y_t) = \mu = E(y_{t-1})$$

By definition, e_t has an expected value of 0. Additionally the phis are constants, so their expected values are their constant values. Equation 6-2 reduces on the lefthand side to:

$$E(y_t) = E(\phi_0 + \phi_1 \times y_{t-1} + e_t)$$
$$E(y_t) = \mu$$

And on the righthand side to:

$$\phi_0 + \phi_1 \times \mu + 0$$

This simplifies to:

$$\mu = \phi_0 + \phi_1 \times \mu$$

which in turn implies that (Equation 6-3).

Equation 6-3. $\mu = \frac{\phi_0}{1 - \phi_1}$

So we find a relationship between the mean of the process and the underlying AR(1) coefficients.

We can take similar steps to look at how a constant variance and covariance impose conditions on the ϕ coefficients. We begin by substituting the value of ϕ_0, which we can derive from Equation 6-3.

Equation 6-4. $\phi_0 = \mu \times (1 - \phi_1)$

into Equation 6-2:

$$y_t = \phi_0 + \phi_1 \times y_{t\text{-}1} + e_t$$
$$y_t = (\mu - \mu \times \phi_1) + \phi_1 \times y_{t\text{-}1} + e_t$$
$$y_t - \mu = \phi_1(y_{t\text{-}1} - \mu) + e_t$$

If you inspect Equation 6-4, what should jump out at you is that it has very similar expressions on the lefthand and righthand sides, namely $y_t - \mu$ and $y_{t\text{-}1} - \mu$. Given that this time series is stationary, we know that the math at time $t - 1$ should be the same as the math at time t. We rewrite Equation 6-4 in the frame of a time one step earlier as Equation 6-5.

Equation 6-5. $y_{t\text{-}1} - \mu = \phi_1(y_{t\text{-}2} - \mu) + e_{t\text{-}1}$

We can then substitute this into Equation 6-4 as follows:

$$y_t - \mu = \phi_1(\phi_1(y_{t\text{-}2} - \mu) + e_{t\text{-}1}) + e_t$$

We rearrange for clarity in Equation 6-6.

Equation 6-6. $y_t - \mu = e_t + \phi_1(e_{t\text{-}1} + \phi_1(y_{t\text{-}2} - \mu))$

It should catch your eye that another substitution is possible in Equation 6-6 for $y_{t\text{-}2} - \mu$ using the same recursive substitution we used earlier, but instead of working on $y_{t\text{-}1}$ we will work on $y_{t\text{-}2}$. If you do this substitution, the pattern becomes clear:

$$y_t - \mu = e_t + \phi_1(e_{t\text{-}1} + \phi_1(e_{t\text{-}2} + \phi_1(y_{t\text{-}3} - \mu)))$$
$$= e_t + \phi \times e_{t\text{-}1} + \phi^2 \times e_{t\text{-}2} + \phi^3 \times e_{t\text{-}3} + (\text{expressions still to be substituted})$$

So we can conclude more generally that $y_t - \mu = \sum_{i=1}^{\infty} \phi_1^i \times e_{t\text{-}i}$.

In plain English, y_t minus the process mean is a linear function of the error terms.

This result can then be used to compute the expectation of $E[(y_t - \mu) \times e_{t+1}] = 0$ given that the values of e_t at different t values are independent. From this we can conclude that the covariance of y_{t-1} and e_t is 0, as it should be. We can apply similar logic to calculate the variance of y_t by squaring this equation:

$$y_t - \mu = \phi_1(y_{t-1} - \mu) + e_t$$
$$var(y_t) = \phi_1^2 var(y_{t-1}) + var(e_t)$$

Because the variance quantities on each side of the equation must be equal due to stationarity, such that $(var(y_t) = var(y_t - 1))$, this implies that:

$$var(y_t) = \frac{var(e_t)}{1 - \phi_1^2}$$

Given that the variance must be greater than or equal to 0 by definition, we can see that ϕ_1^2 must be less than 1 to ensure a positive value on the righthand side of the preceding equation. This implies that for a stationary process we must have $-1 < \phi_1 < 1$. This is a necessary and sufficient condition for weak stationarity.

Weak Versus Strong Stationarity

When we discuss stationarity in this book, we are always referring to weak stationarity. Weak stationarity requires only that the mean and variance of a process be time invariant.

Strong stationarity requires that the distribution of the random variables output by a process remain the same over time, so, for example, it demands that the statistical distribution of y_1, y_2, y_3 be the same as y_{101}, y_{102}, y_{103} for any measure of that distribution rather than the first and second moments (the mean and variance).

Fun fact: while strong stationarity would seem to fully encompass the case of weak stationarity, this isn't entirely so. Strong stationarity includes weak stationarity where the mean and variance of a process do exist, but if a process does not have a mean or a variance, it may nonetheless be strongly stationary even though it can't be weakly stationary. The classic example of this is the *Cauchy distribution*, which has undefined mean and infinite variance. A Cauchy process cannot meet the requirements of weak stationarity but can meet the requirements of strong stationarity.

We studied the AR(1) process because it is the simplest autoregressive process. In practice, you will fit more complex models all the time. It is possible to derive similar conditions for stationarity for an arbitrary order AR(p), and there are many books where this is demonstrated. If you are interested in seeing this worked out more explicitly, take a look at the resources listed at the end of the chapter. The most important takeaway from this discussion is that time series are quite approachable with some algebra and statistics, and that stationarity is not simply a matter of graphing a model but rather a mathematical concept that can be worked out with respect to the specifics of any given statistical model.

 A *distribution* is a statistical function describing the probability for all possible values that a particular value will be generated by a process. While you might not have encountered this term formally, you have undoubtedly encountered this concept. For example, consider the *bell curve*, which is a probability distribution indicating that most measurements will fall close to, and be distributed equally on both sides of, the mean. It's usually called a *normal* or *Gaussian distribution* in statistics.

Choosing parameters for an AR(p) model

To assess the appropriateness of an AR model for your data, begin by plotting the process and its *partial autocorrelation function* (PACF).The PACF of an AR process should cut off to zero beyond the order p of an AR(p) process, giving a concrete and visual indication of the order of an AR process empirically seen in the data.

On the other hand, an AR process will not have an informative autocorrelation function (ACF), although it will have the characteristic shape of an ACF: exponential tapering with increasing time offset.

Let's take a look at this with some actual data. We use some demand forecasting data published in the UCI Machine Learning Repository (*https://perma.cc/B7EQ-DNLU*).

First we plot the data in chronological order (Figure 6-1). Since we will model this as an AR process, we look to the PACF to set a cutoff on the order of the process (Figure 6-2).

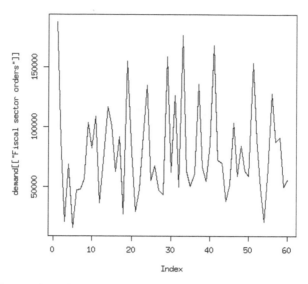

Figure 6-1. Daily number of Banking orders (2).

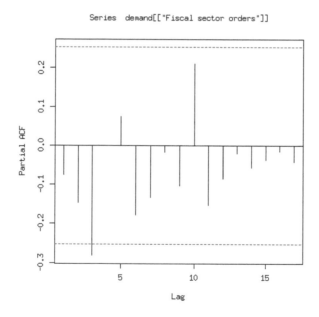

Figure 6-2. PACF of the untransformed orders time series pictured in Figure 6-1.

We can see that the value of the PACF crosses the 5% significance threshold at lag 3. This is consistent with the results from the ar() function available in R's stats package. `ar()` automatically chooses the order of an autoregressive model if one is not specified:

```R
## R
> fit <- ar(demand[["Banking orders (2)"]], method = "mle")
> fit

Call:
ar(x = demand[["Banking orders (2)"]], method = "mle")
Coefficients:
      1           2           3
  -0.1360     -0.2014     -0.3175
```

If we look at the documentation (*https://perma.cc/8H8Z-CX9R*) for the `ar()` function, we can see that the order selected is determined (with the default parameters we left undisturbed) based on the *Akaike information criterion* (AIC). This is helpful to know because it shows that the visual selection we made by examining the PACF is consistent with the selection that would be made by minimizing an information criterion. These are two different ways of selecting the order of the model, but in this case they are consistent.

Akaike Information Criterion

The AIC of a model is equal to AIC = $2k - 2\ln L$ where k is the number of parameters of the model and L is the maximum likelihood value for that function. In general, we want to lessen the complexity of the model (i.e., lessen k) while increasing the likelihood/goodness-of-fit of the model (i.e., L). So we will favor models with smaller AIC values over those with greater AIC values.

A likelihood function is a measure of how likely a particular set of parameters for a function is relative to other parameters for that function given the data. So imagine you're fitting a linear regression of x on y of the following data:

```
x  y
1  1
2  2
3  3
```

If you were fitting this to the model $y = b \times x$, your likelihood function would tell you that an estimate of $b = 1$ was far more likely than an estimate of $b = 0$. You can think of likelihood functions as a tool for helping you identify the most likely true parameters of a model given a set of data.

Notice that the `ar()` function has also provided us with the coefficients for the model. We may, however, want to limit the coefficients. For example, looking at the PACF, we might wonder whether we really want to include a coefficient for the lag – 1 term or whether we should assign that term a mandatory coefficient of 0 given that its PACF value is well below the threshold used for significance. In this case, we can use the `arima()` function also from the `stats` package.

Here, we demonstrate how to call the function to fit an AR(3), by setting the order parameter to `c(3, 0, 0)`, where 3 refers to the order of the AR component (in later examples we will specify other components for the differencing and moving average parameters covered in the next few pages of this chapter):

```R
## R
> est <- arima(x = demand[["Banking orders (2)"]],
>              order = c(3, 0, 0))
> est

Call:
arima(x = demand[["Banking orders (2)"]], order = c(3, 0, 0))

Coefficients:
         ar1 ar2      ar3 intercept
     -0.1358  -0.2013 -0.3176  79075.350
s.e.  0.1299  0.1289 0.1296    2981.125

sigma^2 estimated as 1.414e+09:  log likelihood = -717.42,
                                 aic = 1444.83
```

To inject prior knowledge or opinion into our model, we can constraint a coefficient to be 0. For example, if we want to constrain the lag – 1 term to remain 0 in our model, we use the following call:

```R
## R
> est.1 <- arima(x = demand[["Banking orders (2)"]],
>                order = c(3, 0, 0),
>                fixed = c(0, NA, NA, NA))
> est.1

Call:
arima(x = demand[["Banking orders (2)"]],
      order = c(3, 0, 0),
      fixed = c(0, NA, NA, NA))


Coefficients:
      ar1  ar2      ar3      intercept
        0  -0.1831  -0.3031   79190.705
s.e.    0   0.1289   0.1298    3345.253

sigma^2 estimated as 1.44e+09:  log likelihood = -717.96,
                                aic = 1443.91
```

Setting a value in the vector passed to the fixed parameter of the `arima` function to 0 rather than NA will constraint that value to remain 0:

```R
## R
> fixed <- c(0, NA, NA, NA)
```

We now inspect our model performance on our training data to assess the goodness of fit of our model to this data set. We can do this in two ways. First, we plot the ACF of the residuals (that, is the errors) to see if there is a pattern of self-correlation that our model does not cover.

Plotting the residuals is quite simple thanks to the output of the `arima()` function (see Figure 6-3):

```R
## R
> acf(est.1$residuals)
```

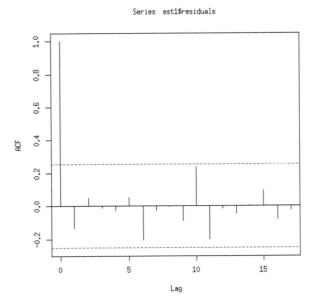

Figure 6-3. ACF of the residuals of the AR(3) model we fit forcing the lag – 1 parameter to 0.

None of the values of the ACF cross the significance threshold. Of course, we should not blindly yield to a significance threshold to assess or reject significance, but this observation is a helpful data point in a model that we already believe reasonable for other reasons.

We do not see a pattern of self-correlation here among the residuals (i.e., the error terms). If we had seen such a pattern, we would likely want to return to our original model and consider including additional terms to add complexity to account for the significant autocorrelation of the residuals.

Another test that is commonly performed is the *Ljung-Box test*, an overall test of the randomness of a time series. More formally it poses the following null and alternate hypotheses:

- H0: The data does not exhibit serial correlation.
- H1: The data does exhibit serial correlation.

This test is commonly applied to AR (and more generally, ARIMA) models, and more specifically to the residuals from the model fit rather than to the model itself:

```R
## R
> Box.test(est.1$residuals, lag = 10, type = "Ljung", fitdf = 3)

    Box-Ljung test

data: est.1$residuals
X-squared = 9.3261, df = 7, p-value = 0.2301
```

We apply the Ljung-Box test to our `est.1` model to assess its goodness of fit. We cannot reject the null hypothesis that the data does not exhibit serial correlation. This is confirmation of what we just found by plotting the ACF of the residuals.

Forecasting with an AR(p) process

In the following sections we will illustrate how to make forecasts with AR processes. We first explore the case of one time step ahead and then discuss how predicting multiple steps ahead differs from the further case. The good news is that from a coding perspective there is not much difference, although the underlying mathematics is more elaborate in the latter case.

Forecasting one time step ahead. We first consider the case that we want to forecast one time step ahead with a known (or estimated) AR model. In such a case, we actually have all the information we need.

We continue working with the model from the demand data, with the lag – 1 coefficient constrained to 0 (fit as `est.1` earlier).

We plot the forecast using the `fitted()` function from the `forecast` package. Here is the code in full; it's quite easy to do:

```R
## R
> require(forecast)
> plot(demand[["Banking orders (2)"]], type = 'l')
> lines(fitted(est.1), col = 3, lwd = 2) ## use the forecast package
```

This results in the plot shown in Figure 6-4.

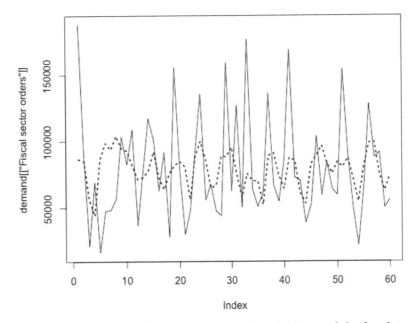

Figure 6-4. Here we see the original time series in the solid line and the fitted time series in the dashed line.

AR(p) Models Are Moving Window Functions

We have an option beyond using the forecast package's fitted functionality to produce a forecast from our AR(3) model: we could handcode a forecast more explicitly by using the zoo package's rollapply() function, which we discussed previously. That function to compute window functions can also compute an AR process. To do this we would take the coefficients from the ar() fit and apply these weights to an input vector representing the values of the different lags to produce a forecast at each point. I leave this as an exercise for you.

Now let's think about the quality of the forecast. If we calculate the correlation between the predicted value and the actual value, we get 0.29. This is not bad in some contexts, but remember that sometimes differencing the data will remove what seemed like a strong relationship and replace it with one that is essentially random.

This will be the case particularly if the data was not truly stationary when we fit it, so that an unidentified trend masquerades as good model performance when it is actually a trait of the data we should have addressed before modeling.

We can difference both the series and the predicted values to see whether the change from one time period to the next is well predicted by the model. Even after differencing, our predictions and the data show similar patterns, suggesting our model is a meaningful one (see Figure 6-5).

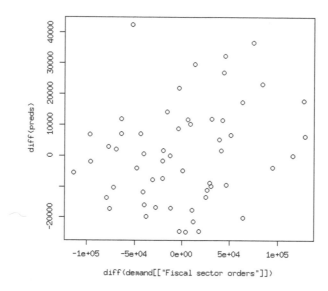

Figure 6-5. The differenced series and the differenced predictions are strongly correlated, suggesting an underlying relationship is indeed identified by the model.

We can also test whether we predict the same moves at the same time, by plotting the differenced series and looking at that correlation. This plot reveals some correlation, which we can confirm by calculating the correlation value. The model works even for predicting the change from one time step to the next.

Looking back at the original plot of the forecast versus actual values, we see that the main difference between the forecast and the data is that the forecast is less variable than the data. It may predict the direction of the future correctly, but not the scale of the change from one time period to another. This is not a problem per se but rather reflects the fact that forecasts are means of the predicted distributions and so necessarily will have lower variability than sampled data.

This property of statistical models is sometimes forgotten in the ease of fast visualizations that tend to suggest a more stable future than will usually be the case. When you present visualizations, make sure to remind your audience of what is being plotted. In this case, our forecasts suggest a much smoother future than is likely to be the case.

Forecasting many steps into the future.   So far we have done a single-step-ahead forecast. However, we may want to predict even further into the future. Let's imagine that we wanted to produce a two-step-ahead forecast instead of a one-step-ahead forecast. What we would do is first produce the one-step-ahead forecast, and then use this to furnish the y_t value we need to predict y_{t+1}.

Notice that in our current model, the shift from a one-step-ahead forecast to a two-step-ahead forecast would actually not require these gymnastics because the y_{t-1} value is not used when predicting y_t. Everything we need to know to make a prediction two time steps ahead is known and there is no need to guesstimate. In fact, we should come up with the same series of values as we did for the one-step-ahead prediction—there will be no new sources of error or variability.

If we want to predict even further, however, we will need to generate forecasted future values as inputs into our prediction. Let's predict y_{t+3}. This will involve a model with coefficients depending on y_{t+1} and y_t. So we will need to predict both of these values—y_{t+1} and y_t—and then in turn use these estimated values in the prediction for y_{t+3}. As before, we can use the fitted() function for the forecast package to do this —it's no more difficult codewise than the one-step-ahead forecast. As mentioned previously, this can also be done with a rollapply() method, but it's more work and more error prone.

We use the fitted() function with an additional parameter now, h for horizon. As a reminder, our object est.1 represents an AR(3) process with the lag – 1 (time minus 1) coefficient constrained to be 0:

```
## R
> fitted(est.1, h = 3)
```

We can use the ease of predicting many time steps into the future to generate many multi-step-ahead forecasts for different horizons. In the next example, we can see the variance for forecasts made increasingly far into the future from the same underlying model. (Note that in the display, rounding and comma delineations modify the original output so it's easier to see what is happening with the variance of the estimation as the forward horizon increases.)

As you can see in Figure 6-6, the variance of the prediction decreases with increasing forward horizon. The reason for this—which highlights an important limitation of the model—is that the further forward in time we go, the less the actual data matters because the coefficients for input data look only at a finite previous set of time points (in this model, going back only to lag – 3; i.e., time – 3). One way of putting this is that forecasts further out in time converge to being the unconditional prediction— that is, unconditioned on data. The future prediction approaches the mean value of the series as the time horizon grows, and hence the variance of both the error term and of the forecast values shrinks to 0 as the forecast values tend toward the unconditional mean value:

```
## R
> var(fitted(est.1, h = 3), na.rm = TRUE)
```

```
[1] 174,870,141
> var(fitted(est.1, h = 5), na.rm = TRUE)
[1] 32,323,722
> var(fitted(est.1, h = 10), na.rm = TRUE)
[1] 1,013,396
> var(fitted(est.1, h = 20), na.rm = TRUE)
[1] 1,176
> var(fitted(est.1, h = 30), na.rm = TRUE)
[1] 3.5
```

Forecasts for sufficiently far out in the future will merely predict the mean of the process, which makes sense. At some point in the distant future, our current data does not give us specific sensible information relevant to the future, and so our forecast increasingly reverts to the baseline properties known about the process, such as its mean.

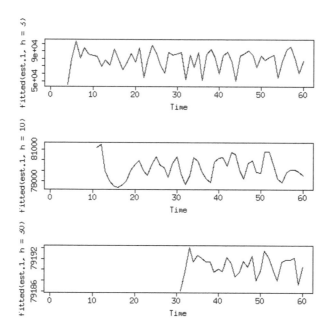

Figure 6-6. Plots of predictions into the future. The values spanned by the y-axis increasingly narrow as we predict further into the future and the model more closely appears to provide a constant prediction of the mean value of the process. The horizons predicted increase from top to bottom and are 3, 10, and 30 time steps, respectively.

The important point to remember from this is that AR (as well as MA, ARMA, and ARIMA, which will be discussed shortly) models are best for making short-term predictions. These models lose predictive power for large future horizons.

For the remaining models, we will go through a similar treatment as we have given to the autoregressive models, albeit with less overall detail. A full discussion is available in all the standard textbooks on time series analysis.

Moving Average Models

A moving average (MA) model relies on a picture of a process in which the value at each point in time is a function of the recent past value "error" terms, each of which is independent from the others. We will review this model in the same series of steps we used to study AR models.

AR MA Equivalence

In many cases, an MA process can be expressed as an infinite order AR process. Likewise, in many cases an AR process can be expressed as an infinite order MA process. To learn more, look into invertibility of an MA process (*https://perma.cc/GJ6B-YASH*), the Wold representation theorem (*https://perma.cc/B3DW-5QGB*), and the duality of MA/AR (*https://perma.cc/K78H-YA6U*) processes generally. The mathematics involved is well beyond the scope of this book!

The model

A moving average model can be expressed similarly to an autoregressive model except that the terms included in the linear equation refer to present and past error terms rather than present and past values of the process itself. So an MA model of order q is expressed as:

$$y_t = \mu + e_t + \theta_1 \times e_{t-1} + \theta_2 \times e_{t-2} \dots + \theta_q \times e_{t-q}$$

Do not confuse the MA model with a moving average. They are not the same thing. Once you know how to fit a moving average process, you can even compare the fit of an MA model to a moving average of the underlying time series. I leave this as an exercise for you.

Economists talk about these error terms as "shocks" to the system, while someone with an electrical engineering background could talk about this as a series of impulses and the model itself as a finite impulse response filter, meaning that the effects of any particular impulse remain only for a finite period of time. The wording is unimportant, but the concept of many independent events at different past times affecting the current value of the process, each making an individual contribution, is the main idea.

The Backshift Operator

The backshift operator, a.k.a. the lag operator, operates on time series points and moves them back by one time step each time it is applied. In general:

$$B^k y_t = y_{t-k}$$

The backshift operator helps to simplify the expression of time series models. For example, an MA model can be rewritten as:

$$y_t = \mu + (1 + \theta_1 \times B + \theta_2 \times B^2 + \dots + \theta_q \times B^q)e_t$$

MA models are by definition weakly stationary without the need to impose any constraints on their parameters. This is because the mean and variance of an MA process are both finite and invariant with time because the error terms are assumed to be iid with mean 0. We can see this as:

$$E(y_t = \mu + e_t + \theta_1 \times e_{t-1} + \theta_2 \times e_{t-2}\dots + \theta_q \times e_{t-q})$$
$$= E(\mu) + \theta_1 \times 0 + \theta_2 \times 0 + \dots = \mu$$

For calculating the variance of the process, we use the fact that the e_t terms are iid and also the general statistical property that the variance of the sum of two random variables is the same as their individual variances plus two times their covariance. For iid variables the covariance is 0. This yields the expression:

$$Var(y_t) = (1 + \theta_1^2 + \theta_2^2 + \dots + \theta_q^2) \times \sigma_e^2$$

So both the mean and variance of an MA process are constant with time regardless of the parameter values.

Selecting parameters for an MA(q) process

We fit an MA model to the same data used fit to AR model, and we can use the ACF to determine the order of the MA process (see Figure 6-7). Before reading on, think about how an MA process works and see if you can reason out why we use the ACF rather than the PACF to determine the order of the process:

```R
## R
> acf(demand[["Banking orders (2)"]])
```

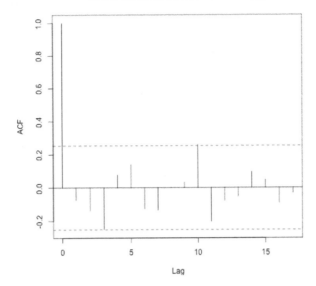

Series demand[["Fiscal sector orders"]]

Figure 6-7. We use the ACF of the demand time series to determine the order of the MA model.

ACF and PACF Patterns Differ from MA and AR Processes

Unlike an autoregressive process, which has a slowly decaying ACF, the definition of the MA process ensures a sharp cutoff of the ACF for any value greater than q, the order of the MA process. This is because an autoregressive process depends on previous terms, and they incorporate previous impulses to the system, whereas an MA model, incorporating the impulses directly through their value, has a mechanism to stop the impulse propagation from progressing indefinitely.

We see significant values at lags 3 and 9, so we fit an MA model with these lags. We need to be careful that we don't accidentally constrain the wrong coefficients to 0 in the model, which we can confirm by printing the display:

```
## R
> ma.est = arima(x = demand[["Banking orders (2)"]],
                order = c(0, 0, 9),
                fixed = c(0, 0, NA, rep(0, 5), NA, NA))
> ma.est
 Call:
 arima(x = demand[["Banking orders (2)"]], order = c(0, 0, 9),
      fixed = c(0, 0, NA, rep(0, 5), NA, NA))
```

```
Coefficients:
       ma1  ma2      ma3  ma4  ma5  ma6  ma7  ma8      ma9  intercept
         0    0  -0.4725    0    0    0    0    0  -0.0120   79689.81
s.e.     0    0   0.1459    0    0    0    0    0   0.1444    2674.60

sigma^2 estimated as 1.4e+09:  log likelihood = -717.31,
                                      aic = 1442.61
```

We should also check our fit, as we did for the AR model, by plotting the ACF of the residuals of the model and, as a second, separate test of model performance, running the Ljung-Box test to check for overall randomness in any fit to the residuals. Note that the Box.test() input requires us to specify the number of degrees of freedom—that is, how many model parameters were free to be estimated rather than being constrained to a specific value. In this case, the free parameters were the intercept as well as the MA3 and MA9 terms:

```
## R
> Box.test(ma.est$residuals, lag = 10, type = "Ljung", fitdf = 3)

Box-Ljung test

data:  ma.est$residuals
X-squared = 7.6516, df = 7, p-value = 0.3643
```

We cannot reject the null hypothesis that there is no temporal correlation between residual points. Likewise, a plot of the ACF of the residuals suggests no temporal correlation (this is left as an exercise for the reader).

Forecasting an MA(q) process

We can generate a forecast again using the techniques shown earlier for an AR process, relying on the fitted() method of the forecast package:

```
## R
> fitted(ma.est, h=1)

Time Series:
Start = 1
End = 60
Frequency = 1
 [1]    90116.64  80626.91 74090.45   38321.61 74734.77 101153.20  65930.90
 [8]   106351.80 104138.05 86938.99  102868.16 80502.02  81466.01  77619.15
[15]   100984.93  81463.10 61622.54   79660.81 88563.91  65370.99 104679.89
[22]    48047.39  73070.29 115034.16  80034.03 70052.29  70728.85  90437.86
[29]    80684.44  91533.59 101668.18  42273.27 93055.40  68187.65  75863.50
[36]    40195.15  82368.91 90605.60   69924.83 54032.55  90866.20  85839.41
[43]    64932.70  43030.64 85575.32   76561.14 82047.95  95683.35  66553.13
[50]    89532.20  85102.64 80937.97   93926.74 47468.84  75223.67 100887.60
[57]    92059.32  84459.85 67112.16   80917.23
```

MA models exhibit strong mean reversion and so forecasts rapidly converge to the mean of the process. This makes sense given that the process is considered to be a function of white noise.

If you forecast beyond the range of the model established by its order, the forecast will necessarily be the mean of the process by definition of the process. Consider an MA(1) model:

$$y_t = \mu + \theta_1 \times e_{t-1} + e_t$$

To predict one time step in the future, our estimate for y_{t+1} is $\mu + \theta_1 \times y_t + e_t$. If we want to predict two time steps in the future, our estimate is:

$$E(y_{t+2} = \mu + e_{t+2} + \theta_1 \times ; e_{t+1}) = \mu + 0 + \theta_1 \times 0 = \mu$$

With an MA(1) process we cannot offer an informed prediction beyond one step ahead, and for an MA(q) process in general we cannot offer a more informed prediction beyond q steps than the mean value emitted by the process. By *informed* prediction, I mean one in which our most recent measurements have an impact on the forecast.

Traditional Notation Is Negative

Note that the MA model is not typically written as it was here. Traditionally, the sign in front of the theta coefficients is negative. This is for reasons of how it can be derived, and one way to think about an MA model is as an AR model with constraints on its parameters. This formulation, with a lot of algebra, leads to negative coefficients on the theta.

We can see this by producing predictions with our MA(9) model that we just fit, and for which we now seek predictions 10 time steps ahead:

```
## R
> fitted(ma.est, h=10)

Time Series:
Start = 1
End = 60
Frequency = 1
 [1]      NA NA      NA NA NA      NA NA NA
 [9]      NA NA 79689.81 79689.81 79689.81 79689.81 79689.81 79689.81
[17] 79689.81 79689.81 79689.81 79689.81 79689.81 79689.81 79689.81 79689.81
[25] 79689.81 79689.81 79689.81 79689.81 79689.81 79689.81 79689.81 79689.81
[33] 79689.81 79689.81 79689.81 79689.81 79689.81 79689.81 79689.81 79689.81
[41] 79689.81 79689.81 79689.81 79689.81 79689.81 79689.81 79689.81 79689.81
```

```
[49] 79689.81 79689.81 79689.81 79689.81 79689.81 79689.81 79689.81 79689.81
[57] 79689.81 79689.81 79689.81 79689.81
```

When we attempt to predict 10 time steps into the future, we predict the mean for every time step. We could have done that without a fancy statistical model!

Common sense is important. If you apply a model without understanding how it works, you can do embarrassing things such as sending your boss the exact same forecast every day after you wasted previous time and computing resources chugging out a model inappropriate to the question you are asking.

Autoregressive Integrated Moving Average Models

Now we that we have examined AR and MA models individually, we look to the Autoregressive Integrated Moving Average (ARIMA) model, which combines these, recognizing that the same time series can have both underlying AR and MA model dynamics. This alone would lead us to an ARMA model, but we extend to the ARIMA model, which accounts for differencing, a way of removing trends and rendering a time series stationary.

What Is Differencing?

As discussed earlier in the book, differencing is converting a time series of values into a time series of changes in values over time. Most often this is done by calculating pairwise differences of adjacent points in time, so that the value of the differenced series at a time t is the value at time t minus the value at time $t - 1$. However, differencing can also be performed on different lag windows, as convenient.

ARIMA models continue to deliver near state-of-the-art performance, particularly in cases of small data sets where more sophisticated machine learning or deep learning models are not at their best. However, even ARIMA models pose the danger of overfitting despite their relative simplicity.

The model

You are probably scratching your head at this point if you've been paying attention because I've just fit the same data to both an AR and an MA process without commenting on it. This is the sort of irritating habit you may sometimes run into in time series analysis textbooks. Some authors will cop to this data laziness, while others will blandly ignore it. We have not investigated in depth whether either of our previous models is a particularly good fit, but it does seem clear from the fitting process we used that there are defensible arguments to describing the data with either an AR or an MA model. This raises the question: might it not be helpful to incorporate both behaviors into the same model?

Table 6-1 can be a handy way of examining a time series process to see whether an AR, MA, or ARMA description of the process is best.

Table 6-1. Determining which model best describes our time series

Kind of plot	AR(p)	MA(q)	ARMA
ACF behavior	Falls off slowly	Sharp drop after lag $= q$	No sharp cutoff
PACF behavior	Sharp drop after lag $= p$	Falls off slowly	No sharp cutoff

This brings us to an autoregressive moving average (ARMA) model, which is applied in the case that neither AR nor MA terms alone sufficiently describe the empirical dynamics. This is a likely case when diagnostics for AR and MA order statistics (PACF and ACF, respectively) point to nonzero values, indicating a term of a given order for either an AR or an MA term. These can be combined with an ARMA model.

Wold's Theorem

Wold's theorem tells us that every covariance-stationary time series can be written as the sum of two time series, one deterministic and one stochastic. Based on this theorem, we can also say that a stationary process can be reasonably approximated by an ARMA model, although it can of course be quite difficult to find the appropriate model.

Here we switch to more traditional statistics notation by applying negative signs to the MA process coefficients:

$$y_t = \phi_0 + \Sigma\,(\phi_i \times\, ;\, r_{t\text{-}i}) + e_t - \Sigma\,(\theta_i \times e_{t\text{-}i})$$

An ARMA Model Is Not Necessarily Unique

Because there can be common factors between the AR and MA portions of the equation it's possible that an ARMA(p, q) model could actually be reduced to another set of parameters. We need to avoid this sort of degenerate situation. In general, you must choose a parsimonious ARMA model. Further discussion is beyond the scope of this book, but additional references are provided at the end of the chapter.

One simple method of identifying cases where this may happen is to use the lag operator (discussed before) formulation of the ARMA model and use good old-fashioned polynomial factorization to look for common factors in your AR and MA models that could divide out to yield an equivalent but simpler model. Here's an example. One way of expressing an ARMA model is to put the AR components on one side and the MA components on another side.

Suppose we have the following ARMA model:

$$y_t = 0.5 \times y_{t-1} + 0.24 \times y_{t-2} + e_t + 0.6 \times e_{t-1} + 0.09 \times e_{t-2}$$

We can rewrite this in a factored form by moving the AR terms (the x coefficients) to one side and leaving the MA terms on the other side:

$$y_t - 0.5 \times y_{t-1} - 0.24 \times y_{t-2} = e_t + 0.6 \times e_{t-1} + 0.09 \times e_{t-2}$$

We then re-express this using the lag operator. So, for example, y_{t-2} is re-expressed as $L^2 y$:

$$y_t - 0.5 \times L \times y_t - 0.24 \times L^2 \times y_t = e_t + 0.6 \times L \times e_t + 0.09 \times L^2 \times e_t$$

Or simplifying by factoring out the common factors of y_t on the lefthand side and e_t on the righthand side, and then factoring the polynomials, we have:

$$(1 + 0.3L) \times (1 - 0.8L) \times y_t = (1 + 0.3L) \times (1 + 0.3L) \times e_t$$

Notice that when factoring we must keep the values operated on by the lag operator to the right of it so that we factor out y_t to the right of the polynomial factors involving the L backshift operator rather than the left as may feel more natural.

Once we have done this factoring, we have a common factor on the lefthand side and righthand side, namely $(1 + 0.3L)$, and we can cancel this out from both sides, leaving us with:

$$(1 - 0.8L) \times y_t = (1 + 0.3L) \times e_t$$

Post factoring, the order of the overall model has been reduced and now involves only values at time $t - 1$ rather than also values at $t - 2$. This should serve as a warning that even a model that doesn't seem too complicated, such as an ARMA(2,2), can sometimes mask the fact that it is actually equivalent to a simpler model form.

It is also good to keep this in mind when computationally fitting models directly. In real-world data sets you may end up with something such that the two sides are quite similar numerically, but not exact. Perhaps on the left side you can find a factor such as $(1 - 0.29L)$ whereas on the right side it is $(1 - 0.33L)$. In such cases it is worth considering that this is close enough to a common factor to drop for model complexity.

The stationarity of the ARMA process comes down to the stationarity of its AR component and is controlled by the same characteristic equation that controls whether an AR model is stationary.

From an ARMA model, it is a simple transition to an ARIMA model. The difference between an ARMA model and an ARIMA model is that the ARIMA model includes the term *integrated*, which refers to how many times the modeled time series must be differenced to produce stationarity.

ARIMA models are far more widely deployed in practice, particularly in academic research and forecasting problems, than are AR, MA, and ARMA models. A quick Google scholar search reveals ARIMA applied to a variety of forecasting problems including:

- Inbound air travel passengers to Taiwan
- Energy demand in Turkey by fuel type
- Daily sales in a wholesale vegetable market in India
- Emergency room demand in the western United States

Importantly, the order of differencing should not be too great. In general, the value of each parameter of an ARIMA(p, d, q) model should be kept as small as possible to avoid unwarranted complexity and overfitting to the sample data. As a not-at-all-universal rule of thumb, you should be quite skeptical of values of d over 2 and values of p and q over 5 or so. Also, you should expect either the p or q term to dominate and the other to be relatively small. These are practitioner notes gathered from analysts and not hard-and-fast mathematical truths.

Selecting parameters

The ARIMA model is specified in terms of the parameters (p, d, q). We select the values of p, d, and q that are appropriate given the data we have.

Here are some well-known examples from the Wikipedia description of ARIMA models:

- ARIMA(0, 0, 0) is a white noise model.
- ARIMA(0, 1, 0) is a random walk, and ARIMA(0, 1, 0) with a nonzero constant is a random walk with drift.
- ARIMA(0, 1, 1) is an exponential smoothing model, and an ARIMA(0, 2, 2) is the same as Holt's linear method, which extends exponential smoothing to data with a trend, so that it can be used to forecast data with an underlying trend.

We choose the order of our model based on a combination of domain knowledge, various fitting evaluation metrics (such as the AIC), and general knowledge of how the PACF and ACF should appear for a given underlying process (as described in Table 6-1). Next we will demonstrate fitting an ARIMA model to a process using both a manual iterative process based on the PACF and ACF and also based on an automated parameter selection tool via the `forecast` package's `auto.arima()` function.

Manually fitting a model.   There are heuristics for choosing the parameters of an ARIMA model, where *parsimony is paramount*. One popular and longstanding method is the Box-Jenkins method, which is an iterative multistep process:

1. Use your data, visualizations, and underlying knowledge to select a class of model appropriate to your data.

2. Estimate the parameters given your training data.

3. Evaluate the performance of your model based on your training data and tweak the parameters of the model to address the weaknesses you see in the performance diagnostics.

Let's work through one example of fitting data. First, we need some data. In this case, for purposes of transparency and knowing the right answer, we generate our data from an ARMA process:

```R
## R
> require(forecast)
> set.seed(1017)
> ## order of arima model hidden on purpose
> y = arima.sim(n = 1000, list(ar = ###, ma = ###))
```

Try not to look at the order of the model created just yet; let's focus on treating this as a mystery. First, we should plot the time series, as we always do, to see whether it appears stationary (Figure 6-8). Then we examine the ACF and PACF of y (Figure 6-9) and compare it with Table 6-1.

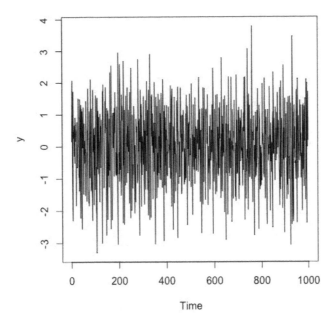

Figure 6-8. Plot of our time series.

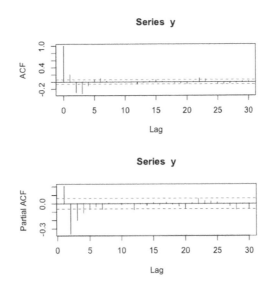

Figure 6-9. ACF and PACF of our time series.

We can see that neither the ACF nor the PACF appears to have a sharp cutoff, suggesting (see Table 6-1) that this is an ARMA process. We start by fitting a relatively simple ARIMA(1, 0, 1) model, as we see no need for differencing and no evidence (Figure 6-10):

```R
## R
> ar1.ma1.model = Arima(y, order = c(1, 0, 1))
> par(mfrow = c(2,1))
> acf(ar1.ma1.model$residuals)
> pacf(ar1.ma1.model$residuals)
```

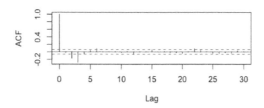

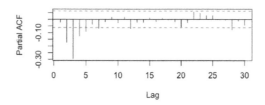

Figure 6-10. ACF and PACF of the residuals of an ARIMA(1, 0, 1) model.

The residuals in Figure 6-10 show particularly large PACF values, suggesting that we have not fully described the autoregressive behavior. For this reason, we build out the model by adding a higher-order AR component, testing an ARIMA(2, 0, 1) model in the following code, and then plotting the ACF and PACF of the residuals of this more complex model (Figure 6-11):

```R
## R
> ar2.ma1.model = Arima(y, order = c(2, 0, 1))
> plot(y, type = 'l')
> lines(ar2.ma1.model$fitted, col = 2)
> plot(y, ar2.ma1.model$fitted)
> par(mfrow = c(2,1))
> acf(ar2.ma1.model$residuals)
> pacf(ar2.ma1.model$residuals)
```

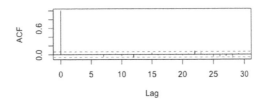

Series ar2.ma1.model$residuals

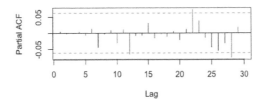

Series ar2.ma1.model$residuals

Figure 6-11. ACF and PACF of the residuals of an ARIMA(2, 0, 1) model.

The residuals in Figure 6-11 no longer show large values for either the ACF or the PACF. Given our desire for a parsimonious model and the dangers of overfitting an ARIMA model, a wise analyst would likely stop here, not seeing any further behavior in the residuals that needs to be fit via an autoregressive, moving average, or differencing component. It is left as an exercise for the reader to consider fitting more complicated models. Though the plots are not shown here, I attempted with the following code to fit more complex models. What I saw was that they did not substantially improve the fit of the model to the data, nor did they contribute to diminishing the ACF or PACF values any further than the previous model. You can verify this for yourself:

```
## R
> ar2.ma2.model = Arima(y, order = c(2, 0, 2))
> plot(y, type = 'l')
> lines(ar2.ma2.model$fitted, col = 2)
> plot(y, ar2.ma2.model$fitted)
> par(mfrow = c(2,1))
> acf(ar2.ma2.model$residuals)
> pacf(ar2.ma2.model$residuals)
>
> ar2.d1.ma2.model = Arima(y, order = c(2, 1, 2))
> plot(y, type = 'l')
> lines(ar2.d1.ma2.model$fitted, col = 2)
> plot(y, ar2.d1.ma2.model$fitted)
> par(mfrow = c(2,1))
> acf(ar2.d1.ma2.model$residuals)
> pacf(ar2.d1.ma2.model$residuals)
```

One quick way of comparing the models is shown here, where I look at how well the forecasts from a fitted model correlate to the actual values:

```R
## R
> cor(y, ar1.ma1.model$fitted)
[1] 0.3018926
> cor(y, ar2.ma1.model$fitted)
[1] 0.4683598
> cor(y, ar2.ma2.model$fitted)
[1] 0.4684905
> cor(y, ar2.d1.ma2.model$fitted)
[1] 0.4688166
```

We see a substantial improvement transitioning from an ARIMA(1, 0, 1) model to an ARIMA(2, 0, 1) model (the first two models), as the correlation goes from 0.3 to 0.47. On the other hand, as we add more complexity, we do not see substantial improvements in the correlation. This offers additional support for the conclusion we drew earlier that the ARIMA(2, 0, 1) model seemed to describe the model behavior well and that there was no need to add further AR or MA components to improve the fit.

Incidentally, we can see how well the fit did below, by comparing the original fit coefficients (shown here, though obscured before) with the fit coefficients:

```R
## R
## original coefficients
> y = arima.sim(n = 1000, list(ar = c(0.8, -0.4), ma = c(-0.7)))
> ar2.ma1.model$coef
        ar1          ar2          ma1     intercept
 0.785028320 -0.462287054 -0.612708282 -0.005227573
```

There is a good match between the fitted coefficients and the actual coefficients used to simulate the data.

There is more to fitting an ARIMA model manually than we have demonstrated here. Over decades, practitioners have developed good rules of thumb for identifying problems such as when there are too many terms of a certain kind, when a model has been overdifferenced, when the pattern in residuals points to certain problems, and so on. One good starting resource is a guide written by a Penn State professor available online (*https://perma.cc/P9BK-764B*).

There are some legitimate criticisms of fitting an ARIMA process "by hand." Fitting by hand can be a somewhat poorly specified process and puts a lot of pressure on the analyst's discretion, and can be time-consuming and path-dependent as to the ultimate outcome. It's a good solution that has worked well for decades to produce actual forecasts used in the real world, but it's not perfect.

Using automated model fitting. Nowadays we can get away from a manually iterative fitting process in favor of an automated model selection in some cases. Our model selection can be driven according to various information loss criteria, such as the AIC we discussed briefly via the `auto.arima()` function in the `forecast` package:

```
## R
> est = auto.arima(demand[["Banking orders (2)"]],
            stepwise = FALSE, ## this goes slower
                            ## but gets us a more complete search
            max.p = 3, max.q = 9)
>  est
Series: demand[["Banking orders (2)"]]
ARIMA(0,0,3) with non-zero mean

Coefficients:
          ma1      ma2      ma3        mean
      -0.0645  -0.1144  -0.4796   79914.783
s.e.   0.1327   0.1150   0.1915    1897.407

sigma^2 estimated as 1.467e+09:  log likelihood=-716.71
AIC=1443.42   AICc=1444.53   BIC=1453.89
```

In this case we put a one-liner with some prior knowledge from our earlier exploration into configuring the inputs. Namely, we specified the maximum order of the AR and MA process that we were prepared to accept, and in fact the model selection chose a much more parsimonious model than we specified, including no AR terms at all. Nonetheless, this model fits well, and we shouldn't data-snoop too much trying to beat it unless we have a principled reason for doing so. Note that according to the AIC criteria, our hand-chosen MA model in the previous section does just slightly better than this model, but the difference does not seem meaningful when we look at plots.

Left for an exercise for the reader is plotting the model against the predictions as well as checking the residuals of this automatically selected model and confirming that the model used does not appear to leave any residual behavior that needs to be addressed by the addition of more terms. The code is no different from what we have used in the previous AR and MA sections. This is also true for the case of making forecasts, and so that is omitted here given that the ARIMA model we selected is not very different from the MA model discussed and used for a forecast earlier.

We can also quickly study how `auto.arima()` would have performed on the model we generated in the previous section when we fit the model by hand:

```
## R
> auto.model = auto.arima(y)
> auto.model
Series: y
ARIMA(2,0,1) with zero mean

Coefficients:
         ar1      ar2      ma1
      0.7847  -0.4622  -0.6123
```

```
s.e.   0.0487    0.0285    0.0522

sigma^2 estimated as 1.019:  log likelihood=-1427.21
AIC=2862.41    AICc=2862.45    BIC=2882.04
```

We did not even use the optional parameters to suggest to `auto.arima()` where it should start its model search, and yet it converged to the same solution as we did. So, as we can see here, in some cases we will find the same solution with different methodologies. We drove our analysis by looking at the ACF and PACF of the residuals of simpler models to build more complex models, whereas `auto.arima()` is largely driven by a grid search to minimize the AIC. Of course, given that we generated the original data from an ARIMA process, this represents a more straightforward case than much real-world data. In the latter case, it won't always be true that our manual fits and the automated model selection come to the same conclusion.

If you are going to make `auto.arima()` or a similar automated model selection tool an important part of your analysis, it's important to read the documentation, experiment with synthetic data, and also read about other analysts' experiences with this function. There are some known scenarios where the function will not perform as naively expected, and there are also known workarounds. Overall this is an excellent solution but not a perfect one.[4] Also, for a good description of how `auto.arima()` works, take a look at the online textbook chapter on this topic (*https://perma.cc/ P92B-6QXR*) written by the author of the function, Professor Rob Hyndman.

We have demonstrated two distinct ways of estimating parameters: either by following a Box-Jenkins approach to fitting the model or using an automated fit from the `forecast` package. In fact, practitioners have very strong opinions on this, with some fiercely advocating only in favor of manual iteration and others equally fiercely in favor of automated selection tools. This remains an ongoing debate in the field. In the long run, as big data takes over more and more of time series analysis, it is likely that automatic exploration and fitting of models will come to dominate large data set time series analysis.

Vector Autoregression

In the real world, we are often lucky enough to have several time series in parallel that are presumably related to one another. We already examined how to clean and align such data, and now we can learn how to make maximal use of it. We can do so by generating an AR(p) model to the case of multiple variables. The beauty of such a model is that it provides for the fact that variables both influence one another and are in turn influenced—that is, there is no privileged y while everything else is designated

---

4 For some sample discussions, see examples such as Stack Overflow (*https://perma.cc/2KM3-Z4R4*) and Rob Hyndman's blog (*https://perma.cc/9DH6-LGNW*).

as x. Instead, the fitting is symmetric with respect to all variables. Note that differencing can be applied as is in other models previously if the series are not stationary.

Exogenous and Endogenous Variables

In statistical speak, once we employ a model where the variables influence one another, we call those variables *endogenous*, meaning their values are explained by what we see within the model. By contrast, *exogenous* variables are those that are not explained within the model—that is, they cannot be explained by assumption—and so we accept their values and do not question the dynamics of how they came to be.

Since every time series putatively predicts every other as well as itself, we will have one equation per variable. Let's say we have three time series: we will denote the value of these time series at time t as $y_{1,t}$ and $y_{2,t}$ and $y_{3,t}$. Then we can write the vector autoregression (VAR) equations of order 2 (factoring in two time lags) as:

$$y_{1,t} = \phi_{01} + \phi_{11,1} \times y_{1,t-1} + \phi_{12,1} \times y_{2,t-1} + \phi_{13,1} \times y_{3,t-1} + \phi_{11,2} \times y_{1,t-2}$$

$$+ \phi_{12,2} \times y_{2,t-2} + \phi_{13,2} \times y_{3,t-2}$$

$$y_{2,t} = \phi_{02} + \phi_{21,1} \times y_{1,t-1} + \phi_{22,1} \times y_{2,t-1} + \phi_{23,1} \times y_{3,t-1} + \phi_{21,2} \times y_{1,t-2}$$

$$+ \phi_{22,2} \times y_{2,t-2} + \phi_{23,2} \times y_{3,t-2}$$

$$y_{3,t} = \phi_{03} + \phi_{31,1} \times y_{1,t-1} + \phi_{32,1} \times y_{2,t-1} + \phi_{33,1} \times y_{3,t-1} + \phi_{31,2} \times y_{1,t-2}$$

$$+ \phi_{32,2} \times y_{2,t-2} + \phi_{33,2} \times y_{3,t-2}$$

Matrix Multiplication

As you will have noticed if you are familiar with linear algebra, expressing the relationships shown in the three previous equations is much simpler when you are using matrix notation. In particular, you can write a VAR in a very similar manner to an AR. In the matrix form, the three equations could be expressed as:

$$y = \phi_0 + \phi_1 \times y_{t-1} + \phi_2 \times y_{t-2}$$

where the y and ϕ_0 are 3×1 matrices, and the other ϕ matrices are 3×3 matrices.

Even for a simple case, you can see that the number of parameters in the model grows very quickly. For example, if we have p lags and N variables, we can see that the predictor equation for each variable is $1 + p \times N$ values. Since we have N values to predict, this translates to $N + p \times N^2$ total variables, meaning that the number of variables grows in $O(N^2)$ proportion to the number of time series studied. Hence, we should not throw in time series gratuitously simply because we have the data but instead reserve this method for when we really expect a relationship.

VAR models are most often used in econometrics. They sometimes come under fire because they do not have any structure beyond the hypothesis that all values influence one another. It is precisely for this reason that the model's goodness of fit can be difficult to assess. However, VAR models are still useful—for example, in testing whether one variable causes another variable. They are also useful for situations where a number of variables need to be forecasted and the analyst does not have domain knowledge to assert any particular kind of relationship. They can also sometimes be helpful in determining how much variance in a forecast of a value is attributable to its underlying "causes."

Here's a quick demonstration. Let's look at the underlying UCI demand information and consider using a second column to predict Banking orders (2) rather than just its own data (note that we will also predict that column due to the symmetric way in which the variables are treated). Let's consider using orders from the traffic control sector. That sounds like it should be quite different, so it may provide a fairly independent source of information relative to the fiscal sector's own past orders. We can also imagine that each column offers underlying information about how the economy is doing and whether demand will increase or decrease in the future.

To determine what parameters to use, we use the `vars` package, which comes with a `VARselect()` method:

```
## R
> VARselect(demand[, 11:12, with = FALSE], lag.max=4,
+                  type="const")
$selection
AIC(n)  HQ(n)  SC(n) FPE(n)
     3      3      1      3

$criteria
                1            2            3            4
AIC(n) 3.975854e+01 3.967373e+01 3.957496e+01 3.968281e+01
HQ(n)  3.984267e+01 3.981395e+01 3.977126e+01 3.993521e+01
SC(n)  3.997554e+01 4.003540e+01 4.008130e+01 4.033382e+01
FPE(n) 1.849280e+17 1.700189e+17 1.542863e+17 1.723729e+17
```

We can see that the function provides a variety of information criteria to choose from. Also note that we indicated we wanted to fit a `"const"` term to accommodate a nonzero mean. We could also elect to fit a drift term or both or neither, but the `"const"` option seems the best fit for our data. Here we will start by looking at the three lags and see how that does:

```
## R
> est.var <- VAR(demand[, 11:12, with = FALSE], p=3, type="const")
```

```
> est.var

> par(mfrow = c(2, 1))
> plot(demand$`Banking orders (2)`, type = "l")
> lines(fitted(est.var)[, 1], col = 2)
> plot(demand$`Banking orders (3)`,
>       type = "l")
> lines(fitted(est.var)[, 2], col = 2)

> par(mfrow = c(2, 1))
> acf(demand$`Banking orders (2)` - fitted(est.var)[, 1])
> acf(demand$`Banking orders (3)` -
>       fitted(est.var)[, 2])
```

This code produces the plots shown in Figures 6-12 and 6-13.

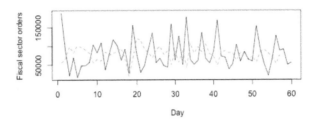

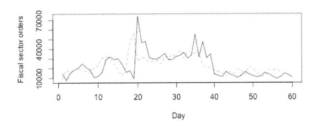

Figure 6-12. At the top we see the actual (solid) versus predicted (dashed) values for Banking orders (2), and at the bottom we see the same for the Banking orders (3). Interestingly, the top plot looks more like a typical forecast, where the forecast is somewhat "slow" in changing relative to the real data, whereas in the bottom plot we see the forecast actually predicted changes somewhat in advance of when they happen. This suggests that Banking orders (2) "lead" Banking orders (3), meaning that Banking orders (2) are helpful in predicting traffic controller orders, but possibly not the inverse, or at least not to such a degree.

Series demand$`Fiscal sector orders` - fitted(est.var)[, 1]

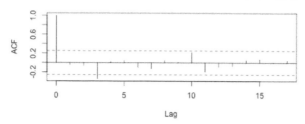

Series demand$`Orders from the traffic controller sector` - fitted(est.v.

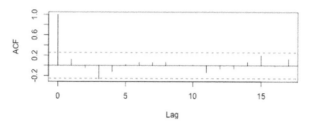

Figure 6-13. We plot the autocorrelation function of the residuals for each time series. Note that for both series there is some borderline significant autocorrelation of the errors at lag 3 that may not be fully accounted for in the model.

The ACF is not as clearly supportive of no autocorrelation in the residuals as we might like, so in this case we can also apply the *Portmanteau test* for serial correlation via the vars package serial.test() method. This test is similar to the serial correlation tests we have looked at in the univariate case:

```
## R
> serial.test(est.var, lags.pt = 8, type="PT.asymptotic")

Portmanteau Test (asymptotic)

data:  Residuals of VAR object est.var
Chi-squared = 20.463, df = 20, p-value = 0.4293
```

Because the *p*-value is so high, we cannot reject the null hypothesis that there is no serial correlation in the residuals. This gives us further evidence that the model does an acceptable job.

Given that for univariate models we examined a variety of models, leading up to the ARMA and ARIMA versions, you may wonder whether there is a VARIMA model. There is indeed, but it is not used very much given that VAR performs relatively well and is already quite complicated. In practice for industrial and academic use cases, you will overwhelmingly see the use of VAR rather than VARIMA.

Another related class of model is the CVAR model, which is the cointegrated vector autoregression. This refers to the case where the individual time series are not stationary, but a linear combination of the time series is stationary without differencing.

Variations on Statistical Models

There are many other kinds of statistical models developed for time series data. Some of these expand the ARIMA model, while others make different underlying assumptions about the temporal dynamics than those used in the ARIMA model. In this section we briefly discuss some of the most commonly used and well-known statistical time series models.

Seasonal ARIMA

A Seasonal ARIMA (SARIMA) model assumes multiplicative seasonality. For this reason, a SARIMA model can be expressed as ARIMA $(p, d, q) \times (P, D, Q)m$. The model postulates that the seasonal behavior itself can be thought of as an ARIMA process, with m specifying the number of time steps per seasonal cycle. What is important in this factoring is that the model recognizes that adjacent points in time can have an influence on one another, either within the same season or in different seasons, but through the usual methods of being temporally proximate.

Identifying a SARIMA is even trickier than identifying an ARIMA model precisely because you need to address seasonal effects. Luckily, `auto.arima()` in the `forecasts` package can handle this just as it handles a standard ARIMA estimation task. As discussed earlier, there are good reasons to go with automated parameter selection unless you have strong knowledge that suggests you override the selected model determined by automated methods.

ARCH, GARCH, and their many brethren

ARCH stands for "Autoregressive Conditional Heteroskedasticity." This model is used almost exclusively in the finance industry. It is often covered in time series courses, so it is worth mentioning here. This class of models is based on the observation that stock prices do not have constant variance, and that in fact the variance itself seems autoregressive conditional on the earlier variances (for example, high-volatility days on the stock exchange come in clusters). In these models, it is the variance of a process that is modeled as an autoregressive process rather than the process itself.

Hierarchical time series models

Hierarchical time series are quite common in the real world, although they are not presented as such. You can easily formulate situations in which they can arise:

- Total dollar monthly demand for a company's products, which can then be broken down by SKU number
- Weekly political polling data for the electorate as a whole, and then that same polling data broken down by demographics (overlapping or nonoverlapping), such as female versus male or Hispanic versus African American
- Total count of tourists landing in the EU daily versus counts of tourists landing in each member nation in particular

One convenient way to handle hierarchical time series is via R's `hts` package. This package can be used both to visualize hierarchical time series data and to generate forecasts.

Forecasts can be generated with a number of different methodologies that have historically been available with this package:

- Generate the lowest-level forecasts (the most individualized) and aggregate these up to produce the higher-level forecasts.
- Generate the highest-level forecasts and then generate lower-level forecasts based on the historical proportions of each component of the aggregate. This methodology tends to be less accurate at making low-level predictions, although this can be somewhat mitigated with a variety of techniques for predicting how proportions of the aggregate will themselves change over time.
- It is possible to try to obtain the best of each methodology by choosing a "middle" approach where a mid-level forecast is generated (assuming you have several layers of hierarchy). This is then propagated up and down in generality to make the other predictions.

Ultimately, the gold standard with the `hts` package is that, for all levels of hierarchy, forecasts can be generated that are independent of other levels of the hierarchy. `hts` then combines these forecasts to assure consistency in a method as per Hyndman et al. (*https://perma.cc/G4EG-6SMP*).

Many of the statistical models we have discussed in this chapter would in turn be applied to the problem of forecasting whatever level of time series was settled upon. The hierarchical portion of the analysis tends to be a wrapper around these fundamental models.

Advantages and Disadvantages of Statistical Methods for Time Series

When you are contemplating whether to apply one of the statistical models described here to a time series problem, it can be good to start with an inventory of advantages and disadvantages. These are laid out in this section.

Advantages
- These models are simple and transparent, so they can be understood clearly in terms of their parameters.
- Because of the simple mathematical expressions that define these models, it is possible to derive their properties of interest in a rigorous statistical way.
- You can apply these models to fairly small data sets and still get good results.
- These simple models and related modifications perform extremely well, even in comparison to very complicated machine learning models. So you get good performance without the danger of overfitting.
- Well-developed automated methodologies for choosing orders of your models and estimating their parameters make it simple to generate these forecasts.

Disadvantages
- Because these models are quite simple, they don't always improve performance when given large data sets. If you are working with extremely large data sets, you may do better with the complex models of machine learning and neural network methodologies.
- These statistical models put the focus on point estimates of the mean value of a distribution rather than on the distribution. True, you can derive sample variances and the like as some proxy for uncertainty in your forecasts, but your fundamental model offers only limited ways to express uncertainty relative to all the choices you make in selecting a model.
- By definition, these models are not built to handle nonlinear dynamics and will do a poor job describing data where nonlinear relationships are dominant.

More Resources

- Classic texts:

Rob J. Hyndman and George Athanasopoulos, Forecasting: Principles and Practice, 2nd ed. (Melbourne: OTexts, 2018), https://perma.cc/9JNK-K6US.
This practical and highly accessible textbook offers an excellent introduction in R, for free, on the internet, of all the basics of preprocessing time series data and using that data to make forecasts. The emphasis is on getting readers up to speed and competent in using practical methods for time series forecasting.

Ruey Tsay, Analysis of Financial Time Series (Hoboken, NJ: John Wiley & Sons, 2001).
This classic textbook introduces a variety of time series models, including a very thorough and accessible chapter on developing AR, MA, and ARIMA models, with applications to historical stock prices. Extensive R examples are also included. This is a middle-of-the-road book in terms of accessibility, presuming some familiarity with statistics and other mathematics but quite readable to anyone with high school calculus and an introductory statistics course under their belt.

Robert H. Shumway, Time Series Analysis and Its Applications (NY, NY: Springer-International, 2017).
This is another classic textbook written at a somewhat more theoretical and less accessible level. It's best to have a look at Tsay's book first, but this book includes additional discussions of the mathematical processes of statistical time series models. It also includes a wider variety of data analysis stemming from the sciences as well as from economic data sources. This book is a more challenging read than the other textbooks listed, but this is because it is dense and informative and not because it requires a much higher level of mathematical proficiency (it does not, although it can feel like it if you read too quickly).

- Heuristical guidance:

Robert Nau, "Summary of Rules for Identifying ARIMA Models," course notes from Fuqua School of Business, Duke University, https://perma.cc/37BY-9RAZ.
These notes give you detailed guidance on how to choose the three parameters of an ARIMA model. You will immediately notice a strong emphasis on parsimonious models when you read this summary.

National Insitute of Standards and Technology (NIST), "Box-Jenkins Models," in NIST/SEMATECH e-Handbook of Statistical Methods (Washington, DC: NIST, US Department of Commerce, 2003), https://perma.cc/3XSC-Y7AG.

This section of the online NIST handbook offers concrete steps to implement the Box-Jenkins method, a commonly used methodology for ARIMA parameter selection. It's also a good example of the resources NIST produces for statistical analysis of time series more generally, as part of the larger meticulously compiled handbook of best practices for time series analysis.

Rob J. Hyndman, "The ARIMAX Model Muddle," Hyndsight blog, October 4, 2010, https://perma.cc/4W44-RQZB.

This concise blog post from noted forecaster Rob Hyndman describes incorporating covariates into ARIMA models, an alternative methodology to VAR for handling multivariate time series.

Richard Hardy, "Cross Validation: Regularization for ARIMA Models," question posted on Cross Validated, StackExchange, May 13, 2015, https://perma.cc/G8NQ-RCCU.

In the scenario of a high order of autoregressive terms or many inputs in the case of VAR, it can make sense to regularize, and in many industries this leads to substantial performance improvements. This post on the Cross Validated Q&A site provides some preliminary discussion as well as links to a computational implementation and related academic research.

## CHAPTER 7

State Space Models for Time Series

State space models are similar to the statistical models we looked at in the previous chapter but with a more "real-world" motivation. They address concerns that emerge in real-world engineering problems, such as how to factor in measurement error when making estimates and how to inject prior knowledge or belief into estimates.

State space models posit a world in which the true state cannot be measured directly but only inferred from what can be measured. State space models also rely on specifying the dynamics of a system, such as how the true state of the world evolves over time, both due to internal dynamics and the external forces that are applied to a system.

While you may not have seen state space models before in a mathematical context, you have likely used them in your day-to-day life. For example, imagine you see a driver weaving in traffic. You try to determine where the driver is going and how you can best defend yourself. If the driver might be intoxicated, you would consider calling the police, whereas if the driver was temporarily distracted for a reason that won't repeat itself, you'd probably mind your own business. In the next few seconds or minutes you would update your own state space model of that driver before deciding what to do.

A classic example of where you would use a state space model is a rocket ship launched into space. We know Newton's laws, so we can write the rules for the dynamics of the system and what the motion should look like over time. We also know that our GPS or sensors or whatever we use to track location will have some measurement error that we can quantify and attempt to factor into the uncertainty about our calculations. Finally, we know that we can't account for all the forces in the world acting on a particular rocket as there are many unknowns in the system, so we want a process that is robust to other unknown sources of noise, perhaps solar wind or earthly wind or both. As it turns out, statistical and engineering advances in the last 50 years have proven quite useful for tackling these sorts of situations.

Two different historical trends led to the development of state space models and interest in the kinds of problems they address. First, around the middle of the 20th century, we entered an age of mechanical automation. There were rockets and spaceships in the sky, navigation systems for submarines, and all sorts of other automated inventions that required estimation of a system state that could not be measured. As researchers thought about how to estimate system state, they began developing state space methods, most importantly to disambiguate measurement errors from other kinds of uncertainty in the system. This led to the first uses of state space methods.

During this time period, too, recordkeeping technology and associated computing were also developing. This led to the creation of much larger data sets for time series, including much longer or more detailed time series data sets. As more time series data became available, more data-intensive methods could be developed for them in conjunction with the new thinking about state space modeling.

In this chapter we will study these commonly used state space methods:

- The Kalman filter applied to a linear Gaussian model
- Hidden Markov Models
- Bayesian structural time series

In each of these cases, the use of such models is quite accessible and well implemented. For each model, we will develop some intuition for the mathematics and discuss what kind of data is appropriate for use with the method. Finally, we will see code examples for each method.

In each case we will distinguish between what we observe and the state that produced our observations. In estimating the underlying state based on observations, we can divide our work into different stages or categories:

Filtering
Using the measurement at time t to update our estimation of the state at time t

Forecasting
Using the measurement at time $t - 1$ to generate a prediction for the expected state at time t (allowing us to infer the expected measurement at time t as well)

Smoothing
Using measurement during a range of time that includes t, both before and after it, to estimate what the true state at time t was

The mechanics of these operations will often be similar, but the distinctions are important. Filtering is a way of deciding how to weigh the most recent information against past information in updating our estimate of state. Forecasting is the prediction of the future state without any information about the future. Smoothing is the use of future and past information in making a best estimate of the state at a given time.

State Space Models: Pluses and Minuses

State space models can be used for both deterministic and stochastic applications, and they can be applied to both continuously sampled data and discretely sampled data.[1]

This alone gives you some inkling of their utility and tremendous flexibility. The flexibility of state space models is what drives both the advantages and disadvantages of this class of models.

There are many strengths of a state space model. A state space model allows for modeling what is often most interesting in a time series: the dynamic process and states producing the noisy data being analyzed, rather than just the noisy data itself. With a state space model, we inject a model of *causality* into the modeling process to explain what is generating a process in the first place. This is useful for cases where we have strong theories or reliable knowledge about how a system works, and where we want our model to help us suss out more details about general dynamics with which we are already familiar.

A state space model allows for changing coefficients and parameters over time, which means that it allows for changing behavior over time. We did not impose a condition of stationarity on our data when using state space models. This is quite different from the models we examined in Chapter 6, in which a stable process is assumed and modeled with only one set of coefficients rather than time-varying coefficients.

Nonetheless, there are also some disadvantages to a state space model, and sometimes the strength of the state space model is also its weakness:

- Because state space models are so flexible, there are many parameters to be set and many forms a state space model can take. This means that the properties of a particular state space model have often not been well studied. When you formulate a state space model tailored to your time series data, you will rarely have statistical textbooks or academic research papers in which others have studied the model too. This leaves you in less certain territory as far as understanding how your model performs or where you may have committed errors.

---

1 In this book we analyze discretely sampled data, which is most common in real-world applications.

- State space models can be very taxing computationally because there are many parameters. Also, the very high number of parameters for some kinds of state space models can leave you vulnerable to overfitting, particularly if you don't have much data.

The Kalman Filter

The Kalman filter is a well-developed and widely deployed method for incorporating new information from a time series and incorporating it in a smart way with previously known information to estimate an underlying state. One of first uses of the Kalman filter was on the Apollo 11 mission—the filter was chosen when NASA engineers realized that the onboard computing elements would not allow other, more memory-intensive techniques of position estimation. As you will see in this section, the benefits of the Kalman filter are that it is relatively easy to compute and does not require storage of past data to make present estimates or future forecasts.

Overview

The mathematics of the Kalman filter can be intimidating to a newcomer, not because it is especially difficult but because there are a fair number of quantities to keep track of, and it's an iterative, somewhat circular process with many related quantities. For this reason, we will not derive the Kalman filter equations here, but instead go through a high-level overview of those equations to get a sense of how they work.[2]

We begin with a linear Gaussian model, positing that our state and our observations have the following dynamics:

$$x_t = F \times x_{t-1} + B \times u_t + w_t$$

$$y_t = A \times x_t + v_t$$

That is, the state at time t is a function of the state at the previous time step ($F \times x_{t-1}$), an external force term ($B \times u_t$), and a stochastic term (w_t). Likewise, the measurement at time t is a function of the state at time t and a stochastic error term, measurement error.

Let's imagine that x_t is the real position of a spaceship, while y_t is the position we measure with a sensor. v_t is the measurement error in our sensor device (or range of

---

2 I would also endorse reading many alternative simple explanations of the Kalman filter on Mathematics StackExchange (*https://perma.cc/27RK-YQ52*).

devices). The fundamental equation applicable to the Kalman filter is this equation illustrating how to update our estimation given new information for time t:

$$\hat{x}_t = K_t \times y_t + (1 - K_t) \times \hat{x}_{t-1}$$

What we see here is a filtering step—that is, a decision about how to use the measurement at time t to update our state estimation at time t. Remember, we have posited a situation in which we can observe only y_t and make inferences about the state but can never be sure of the state exactly. We see above that the quantity K_t establishes a balance in our estimation between old information ($\hat{x}_{t-1}$) and new information (y_t).

To go into the more detailed mechanics, we need to define some terms. We use P_t to denote our estimate of the covariance of our state (this can be a scalar or a matrix, depending on whether the state is univariate or multivariate, the latter being more common). P^-_t is the estimate for t before our measure at time t is taken into account.

We also use R to denote the measurement error variance, that is the variance of v_t, which again can either be a scalar or a covariance matrix depending on the dimensionality of the measurements. R is generally well defined for a system, as it describes well-known physical properties of a particular sensor or measuring device. The apposite value for w_t, Q, is less well defined and subject to tuning during the modeling process.

Then we begin with a process such that we have known or estimated values for x and P at time 0. Then, going forward for times after time 0, we follow an iterative process of a prediction and updating phase, with the prediction phase coming first, followed by the update/filtering phase, and so on:

Prediction:

$$\hat{x}^-_t = F \times \hat{x}_{t-1} + B \times u_t$$
$$P^-_t = F \times P_{t-1} \times F^T + Q$$

Filtering:

$$\hat{x}_t = \hat{x}^-_t + K_t \times (y_t - A \times \hat{x}^-_t)$$
$$P_t = (I - K_t \times A) \times P^-_t$$

where K_t, the Kalman gain, is:

$$K_t = P^-_t \times A^T \times (A \times P^-_t \times A^T + R)^{-1}$$

You will see many visualizations of this recursive process. Some will break this into many steps, perhaps up to four or five, but the simplest way of thinking about it is that there are the computations undertaken to predict the values at time t, without a measurement for y_t (the prediction), and then there are the steps undertaken at time t, after the measurement y_t is known (the filtering).

To start, you need the following values:

- Estimates for R and Q—your covariance matrices for measurement error (easy to know) and state stochasticity (usually estimated), respectively
- Estimates or known values for your state at time 0, $\hat{x}_0$ (estimated based on y_0)
- Advance knowledge of what forces are planned to be applied at time t and how these impact state—that is, the matrix B and the value u_t
- Knowledge of system dynamics that determine the state transition from one time step to another, namely F
- Knowledge of how the measurement depends on the state, namely A

There are many ways to derive the Kalman filter equations, including from a probabilistic perspective in expectation values, a least squares minimization problem, or as a maximum likelihood estimation problem. Derivations of the Kalman filter are widely available, and the interested reader can pursue this topic with a few internet searches.

Code for the Kalman Filter

We consider a classic use case: trying to track an object subject to Newton's mechanics with error-prone sensors. We generate a time series based on Newton's laws of motion, namely that the position of an object is a function of its velocity and acceleration. We imagine taking discrete measurements even though the underlying movement is continuous. We first determine a series of accelerations, then assume that both position and velocity begin at 0. While it is not physically realistic, we assume instant acceleration changes at the beginning of each time step, and a steady acceleration value:

```
## R
## rocket will take 100 time steps
ts.length <- 100

## the acceleration will drive the motion
a <- rep(0.5, ts.length)

## position and velocity start at 0
x  <- rep(0, ts.length)
v  <- rep(0, ts.length)
for (ts in 2:ts.length) {
```

```
      x[ts] <- v[ts - 1] * 2 + x[ts - 1] + 1/2 * a[ts-1] ^ 2
      x[ts] <- x[ts] + rnorm(1, sd = 20) ## stochastic component
      v[ts] <- v[ts - 1] + 2 * a[ts-1]
   }
```

If you don't know or remember Newton's laws of motion, you may wish to familiarize yourself with them, although you can also take these at face value for the current purposes (the computation of x[ts] and v[ts]).

A quick plotting exercise shows us the motion we have created with the acceleration structured as just shown (see Figure 7-1):

```
## R
par(mfrow = c(3, 1))
plot(x,          main = "Position",      type = 'l')
plot(v,          main = "Velocity",      type = 'l')
plot(a,          main = "Acceleration", type = 'l')
```

We posit that these variables would represent a full description of the state, but that the only data available to us is the object's position and that this data is available only from a noisy sensor. This sensor is x in the following code, and we plot how the measured value relates to the actual position in Figure 7-2:

```
## R
z <-  x + rnorm(ts.length, sd = 300)
plot (x, ylim = range(c(x, z)))
lines(z)
```

In Figure 7-1 we see a constant acceleration (bottom plot) driving a linearly increasing velocity (middle plot) to produce a parabolic shape in displacement (top plot). If these mechanical relationships are not familiar, you can take them at face value or look for a quick review of basic mechanics in an introductory physics textbook.

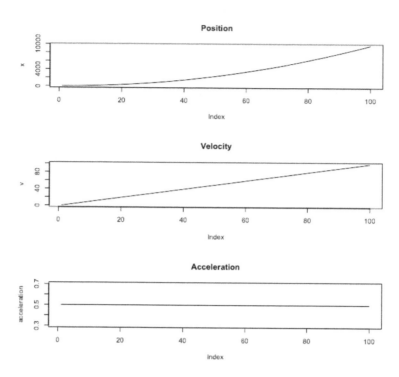

Figure 7-1. The position, velocity, and acceleration of our rocket.

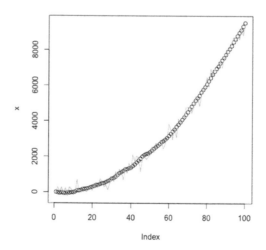

Figure 7-2. The true position (points) versus our noisy measurement (line). Notice that the position x does not reflect a perfect parabola due to the noise we put into the state transition equation.

Now we apply a Kalman filter. First, we code up a general function that reflects our discussion and derivation earlier in this section:

```R
## R
kalman.motion <- function(z, Q, R, A, H) {
  dimState = dim(Q)[1]

  xhatminus <- array(rep(0, ts.length * dimState),
                     c(ts.length, dimState))
  xhat      <- array(rep(0, ts.length * dimState),
                     c(ts.length, dimState))

  Pminus <- array(rep(0, ts.length * dimState * dimState),
                  c(ts.length, dimState, dimState))
  P      <- array(rep(0, ts.length * dimState * dimState),
                  c(ts.length, dimState, dimState))

  K <- array(rep(0, ts.length * dimState),
             c(ts.length, dimState)) # Kalman gain

  # intial guesses = starting at 0 for all metrics
  xhat[1, ] <- rep(0, dimState)
  P[1, , ]  <- diag(dimState)

  for (k in 2:ts.length) {
    # time update
    xhatminus[k, ] <- A %*% matrix(xhat[k-1, ])
    Pminus[k, , ] <- A %*% P[k-1, , ] %*% t(A) + Q

    K[k, ] <- Pminus[k, , ] %*% H %*%
                         solve( t(H) %*% Pminus[k, , ] %*% H + R )
    xhat[k, ] <- xhatminus[k, ] + K[k, ] %*%
                         (z[k]- t(H) %*% xhatminus[k, ])
    P[k, , ] <- (diag(dimState)-K[k,] %*% t(H)) %*% Pminus[k, , ]
  }

  ## we return both the forecast and the smoothed value
  return(list(xhat = xhat, xhatminus = xhatminus))
}
```

We apply this function, making only the rocket's position measurable (so not the acceleration or velocity):

```R
## R
## noise parameters
R <- 10^2 ## measurement variance - this value should be set
          ## according to known physical limits of measuring tool
          ## we set it consistent with the noise we added to x
          ## to produce x in the data generation above
Q <- 10   ## process variance - usually regarded as hyperparameter
          ## to be tuned to maximize performance

## dynamical parameters
```

```
A <- matrix(1) ## x_t = A * x_t-1 (how prior x affects later x)
H <- matrix(1) ## y_t = H * x_t   (translating state to measurement)

## run the data through the Kalman filtering method
xhat <- kalman.motion(z, diag(1) * Q, R, A, H)[[1]]
```

We plot the true position, the measured position, and the estimated position in Figure 7-3.

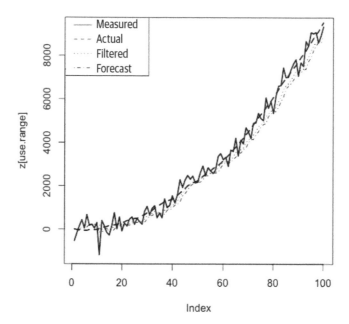

Figure 7-3. Many related quantities: the measured position, the actual/true position, the filtered estimation of the position (that is, best estimate for position at time t incorporating the measurement at time t), and the forecast of the position (that is, the best estimate for position at time t incorporating only the known system dynamics plus measurements up to time t − 1 and not including time t).

The Kalman filter takes out much of the noise from the measurement error. How much it does so will depend on our value for *R*, the measurement noise parameter, which reflects how much the filter should weigh the most recent value relative to earlier values. As we can see, the filter does a satisfactory job of forecasting the data. In particular, note that there is not a lag between the forecast data and the actual data, which would suggest that the method is merely predicting the current value based on the last value.

Here we have worked through a simple example of a Kalman filter. The Kalman filter is widely studied because it is so useful in a variety of applications, particularly those where the internal dynamics of the system are very well understood. This makes it an ideal tool in cases such as the simple rocket example, where we understand the dynamics that are driving the system.

Note that in this simple example, the whole power and utility of the Kalman filter is not fully realized. It is particularly useful when we have multiple kinds of measurements, say measuring different quantities or measuring the same thing simultaneously with many devices. Also there are many extensions to the Kalman filter that are worth studying if this is a promising area for your domain of interest.

One of the great benefits of the Kalman filter, as we've illustrated, is that it is recursive. This means that it is not necessary to look at all data points on each iteration of the process. Rather, at each time step, all the information from prior time steps is already incorporated in the best possible way in the few estimated parameters, namely the most recent state and covariance estimate. The beauty of this method is that we can update in a smart way just using these "summary statistics" like measures, and we already know how to weigh them intelligently relative to the most recent data. This makes the Kalman filter very useful for real-world applications where computational time or resources are at a premium. In many cases this also coincides with the dynamics of real systems, in that the processes are relatively Markovian (memoryless other than the immediately prior state), and a function of an underlying state that can only be measured with some error.

There are many useful extensions to the Kalman filter that we have not discussed here. One principal use of the Kalman filter is to adapt it to *smoothing*, which means using data both before and after time t to make the best estimate of the true state at time t. The mathematics and code are similar to what has been presented already. Also similar is the Extended Kalman Filter (EKF), which adapts the Kalman filter to data with nonlinear dynamics. This is also relatively straightforward to implement and is widely available in a variety of R and Python packages.

The Kalman filter is $O(T)$ with respect to the length of the time series but $O(d^2)$ with respect to d, the dimension of the state. For this reason, it is important not to overspecify the state when a more streamlined specification will do just as well. However, it is this linearity with respect to the length of the time series that makes the Kalman filter commonly used in real production scenarios and much more popular than other filters developed for the purpose of state space modeling of time series.

Hidden Markov Models

Hidden Markov Models (HMMs) are a particularly useful and interesting way of modeling a time series because it is a rare instance of unsupervised learning in time series analysis, meaning there is no labeled correct answer against which to train. An HMM is motivated by an intuition similar to what we used when experimenting with the Kalman filter earlier in this chapter, namely the idea that the variables we are able to observe may not be the most descriptive variables of the system. As with the Kalman filter applied to a linear Gaussian model, we posit the idea that the process has a state, and our observations give information about this state. And again, as before, we need to have some opinion as to how the state variables influence what we can observe. In the case of an HMM, what we posit is that the process is a nonlinear one characterized by jumps between discrete states.

How the Model Works

An HMM posits a system in which there are states that are not directly observable. The system is a Markov process, which means that it is "memoryless" in the sense that the probabilities of future events can be fully calculated given only the system's current state. That is, knowing the system's current state and its previous states is no more useful than simply knowing the system's current state.

Markov processes are often described in terms of matrices. For example, suppose there was a system that fluctuated between state A and state B. When in either state, the system was statistically more likely to remain in the same state than to flip to the other state at any distinct time step. One such system would be described by the following matrix:

```
    A    B
A  0.7  0.3
B  0.2  0.8
```

Let's imagine that our system is in state A, namely (1, 0). (State B would be (0, 1).) In such a case, the probability that the system would remain in state A is 0.7, whereas the probability that the system would flip is 0.3. We don't need to know what states the system was in before its most recent moment in time. This is what it means to be a Markov process.

A Hidden Markov Model represents the same kind of system, except that we are not able to directly infer the state of the system from our observations. Instead, our observations offer clues as to the state of the system (see Figure 7-4).

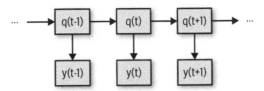

Figure 7-4. The process for an HMM. The actual states of the system at a given time are represented by x(t), whereas the observable data at a given time is represented by the y(t). Only x(t) is relevant to thinking about y(t). In other words, x(t – 1) offers no additional information to predict y(t) if we know x(t). Likewise, only x(t) has any bearing for predicting x(t + 1) and there is no additional information from knowing x(t – 1). This is the Markov aspect of the system.

Note that for realistic applications, the states will usually produce outputs that overlap, so that it is not 100% clear what state is producing output. For example, we are going to apply an HMM to data that looks like Figure 7-5. This is data simulated from four states, but it is not obvious from a simple visual inspection of the time series how many states there are, what their divisions are, or where the system transitions between states.

Time series with four underlying states

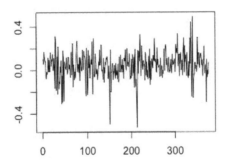

Figure 7-5. This time series was simulated with four states, but it's not clear from a visual observation that there are four states, nor is it clear where one state ends and another begins.

Some examples of real-life use cases of HMMs are:

- Identifying regime shifts in the financial markets (*https://perma.cc/JRT2-ZDVJ*)
- Classifying, predicting, and correcting DNA sequencing information (*https://perma.cc/4V4A-53TZ*)

- Recognizing stages of sleep as reflected in ECG data (*https://perma.cc/G37Y-XBQH*)

How We Fit the Model

We are positing that there is a state we cannot measure directly, and there is no way to get a demonstrably correct answer in many of the data sets where we could apply this technique. So how can this algorithm identify hidden states without knowing anything about them a priori? The answer is: iteratively. There is no magic bullet for deriving the most probable sequence of hidden states to explain the observations, but it is possible to edge our way toward an estimation once we have fully specified the system.

In an HMM, we posit that the system is fully described with the following information:

- The *transition probability* of going from $x(t)$ to $x(t) + 1$. This is equivalent to specifying a matrix like the one just mentioned: transitioning between states A and B. The size of the matrix would depend on the number of hypothesized states.
- The *emission probability*, or the probability of observation $y(t)$ given $x(t)$.
- The initial state of the system.

More concretely, here is a listing of the variables needed to characterize and fit an HMM process:

- $Q = q_1, q_2, \ldots q_N$ the distinct states of the system
- $A = a_{i,j} = a_{1,1}, a_{1,2}, \ldots a_{N,N}$ the transition probability matrix indicating the transition at any given time step of changing from state i to state j
- $O = o_1, o_2, \ldots o_T$ a sequence of observations sampled from this process in order, that is a time series of observations
- $b_{i(ot)}$ indicating the emission probabilities, that is the probabilities of seeing a given observation value, o_t if the state is q_i
- $p = p_1, p_2, \ldots p_N$, the initial probability distributions, namely the probability that the system will start in state $q_1, q_2, \ldots q_N$, respectively

However, with real data generally none of these variables are known. All that is known is the actual sequence of observables $y_1, y_2, \ldots y_t$.

Baum Welch algorithm

In estimating the parameters of a Hidden Markov Model, we use the *Baum Welch algorithm*. This guides us in our complex task to estimate the values of all the parameters, as they were detailed in the previous section. Our task is multifold. We are seeking to:

- Identify the distinct emission probabilities for each possible hidden state and identify the transition probabilities from each possible hidden state to each other possible hidden state. For this we use the Baum-Welch algorithm.

- Identify the most likely hidden state for each time step given the full history of observations. For this we use the Viterbi algorithm (described shortly).

These are two related tasks, and each one is difficult and computationally taxing. What's more, they are dependent. In the case of two interrelated tasks, parameter estimation and likelihood maximization, we can use the *expectation maximization* algorithm to iterate between these two steps until an acceptable solution is found.

To apply the Baum Welch algorithm, the first step is to specify the likelihood function, which is the probability of observing a given sequence given the hypothesized parameters. In our case, the hypothesized parameters would be the mathematical parameters per postulated state.

For example, if we assumed the states produced Gaussian outputs with distinct means and standard deviations of the observed values that depended on the state, and if we assumed a two-state model, we might describe the model in terms of μ_1, σ_1, μ_2, and σ_2, with $\mu_{u=i}$ indicating the ith state's mean and σ_i indicating the ith state's standard deviation. These could describe the emission probabilities, and we collectively denote them as θ. We could also posit the state sequence as x_1, x_2, ...x_t (I'll summarize this as X_t for shorthand), which we were not able to observe, but imagine for the moment that we could.

Then the likelihood function would describe the likelihood of observing the sequence we had observed given the parameters of the emission probability (that is, the probability of a given observation given a specific state) and the sequence of hidden states as the integral over all possible X_t of $p(y_1, y_2, ...y_t | \mu_1, \sigma_1, \mu_2, \sigma_2, ...\mu_N, \sigma_{Nt}) = p(y_1, y_2, ... y_t | \mu_1, \sigma_1, \mu_2, \sigma_2, ...\mu_N, \sigma_N)$.

However, this is a difficult problem for several reasons, including the fact that the complexity grows exponentially with the number of time steps, meaning that an exhaustive grid search is not realistic. So, we simplify the task by applying the EM algorithm as follows:

1. Randomly initialize the emission probability variables.

2. Compute the probability of each possible X_t, assuming the emission probability values.

3. Use these values of X_t to generate a better estimate of the emission probability variables.

4. Repeat steps 2 and 3 until convergence is reached.

What this means, more informally, is that in the event of positing (at random) two distributions, we would then look at each time step and determine the probability that for each time step a particular state was occupied (e.g., at time step t what was the probability of state A? of state B?) Once we had then assigned a putative state to each time step, we would use those labels to re-estimate the emission probabilities (zeroing in on a better mean and standard deviation for the state). Then we would repeat the process again, using the newly updated emission probability variables to improve our estimate of the X_t trajectory.

One important point to remember about such uses of the EM algorithm is that you are not guaranteed to find a globally optimal set of parameters. For this reason it's worth running many fits to see what the global consensus is across numerous initializations. It's also important to remember that a fit will require a burn-in period, the appropriate length of which will depend on the details of your data and model.

Viterbi algorithm

Once the parameters of an HMM process have been estimated, as via the Baum Welch algorithm, the next question of interest is what the most likely series of underlying states was given the measured time series of observable values.

Unlike the Baum Welch algorithm, the Viterbi algorithm is guaranteed to provide you with the best solution to the question you are asking. This is because it is a dynamic programming algorithm designed to fully and efficiently explore the range of possible fits by saving the solutions to portions of a path so that as the path is lengthened there's no need to recompute all possible paths for all path lengths (see Figure 7-6).

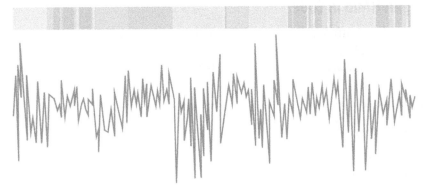

Figure 7-6. The Viterbi algorithm searches all possible paths that could explain a given observed time series, where a path indicates which state was occupied at each time step.

Dynamic Programming

Data scientists do not always have a standard algorithms course in their toolkit, but basic algorithms can be particularly helpful for time series analysis, where inference and recurrence are helpful concepts in thinking about temporally ordered data.

The easiest way to explain what dynamic programming is tends to be with an example, and the classic one is the Fibonacci sequence. Imagine you were asked to calculate the eighth Fibonacci number. The easiest way to do this would be if you already knew the sixth and seventh Fibonacci numbers. To find these out, you need the fourth and fifth Fibonacci numbers and so on. Since you know that solutions to later Fibonacci problems are determined from earlier ones, as you compute Fibonacci numbers, you should store them somewhere so that they can be used for later, more complicated problems. For this reason, dynamic programming is also described as a technique of *memoization*.

Hallmarks of a situation where dynamic programming could apply are:

- You can calculate a solution for a problem of size N if you know the solution to a problem of size $N - 1$. For this reason, remembering, or *memoizing*, the solutions to earlier smaller problems will help you solve later problems more efficiently.

- The problem has a clearly defined ordering to scale from a smaller problem to a larger problem.

- You can identify a single base case that is directly (and easily) calculable.

Fitting an HMM in Code

While the HMM fitting process is very complicated, it is implemented in a number of packages in R. Here we will work with the depmixS4 package. First, we need to formulate an appropriate time series. We do that with the following code:

```R
## R
## notice in this case we have chosen to set a seed
## if you set the same seed, our numbers should match
set.seed(123)

## set parameters for the distribution of each of the four
## market states we want to represent
bull_mu    <-  0.1
bull_sd    <-  0.1

neutral_mu <-  0.02
neutral_sd <-  0.08

bear_mu    <- -0.03
bear_sd    <-  0.2

panic_mu   <- -0.1
panic_sd   <-  0.3

## collect these parameters in vectors for easy indexing
mus <- c(bull_mu, neutral_mu, bear_mu, panic_mu)
sds <- c(bull_sd, neutral_sd, bear_sd, panic_sd)

## set some constants describing the time series we will generate
NUM.PERIODS     <- 10
SMALLEST.PERIOD <- 20
LONGEST.PERIOD  <- 40

## stochastically determine a series of day counts, with
## each day count indicating one 'run' or one state of the market
days <- sample(SMALLEST.PERIOD:LONGEST.PERIOD, NUM.PERIODS,
               replace = TRUE)

## for each number of days in the vector of days
## we generate a time series for that run of days in a particular
## state of the market and add this to our overall time series
returns   <- numeric()
true.mean <- numeric()
for (d in days) {
  idx = sample(1:4, 1, prob = c(0.2, 0.6, 0.18, 0.02))
  returns <- c(returns, rnorm(d, mean = mus[idx], sd = sds[idx]))
  true.mean <- c(true.mean, rep(mus[idx], d))
}
```

In the preceding code we use a stock-market-inspired example, with bull, bear, neutral, and panic modes. A random number of days for a state to persist is selected, as are variables describing the emission probability distribution for each state (that is, the _mu and _sd variables, which indicate what kind of values we expect to see measured given a particular state).

We can get a sense of what our generated time series looks like, and the frequency of each state, by seeing how many days in the sample correspond to each `true.mean`, which is the variable we are using to track state:

```
## R
> table(true.mean)
true.mean
-0.03   0.02    0.1
  155    103     58
```

Catastrophe! Although we intended to include four states in the simulated series, only three are included. This is likely because the fourth state had a very low probability of inclusion (0.02). We see that the least probable state wasn't even selected for inclusion in the series. We couldn't always know that for a given time series—that not all possible states were actually included in a time series—which highlights part of why it's difficult to fit HMMs and why things can be somewhat unfair to the algorithm. Nonetheless, we will move forward with the analysis specifying four groups, to see what this gives.[3]

We still need to fit an HMM. The resulting HMM will provide a time series of posterior probabilities for each state for as many states as we indicate are desired. Consistent with the earlier description of the EM algorithm, nothing more than the number of putative states needs to be specified. The rest will be gradually determined via an iterative back-and-forth process.

As is often the case, with the right package, the difficult bit of the analysis is actually quite easy in terms of the amount of code that needs to be written. In this case we use the depmixS4[4] package in R. The model is fit in two steps. First, it is specified with the depmix() function, which indicates the expected distribution, the number of states, and the input data to be used in fitting. Then the model is fit via the fit function, which takes the model specification as its input. Finally, to generate the posterior distribution of the state labels given the fit to the data, we use the posterior() function. At this point the model itself has already been fit, so this is a separate task to label the

---

3 Notice that another problem with the simulated data is that we didn't established a transition probability matrix to control the flow of the hidden state from one state to another. Essentially we built in the assumption that a state is much more likely to stay as is for many days in a row and then jump to any of the other states with equal probability. We left out the formal specification and use of a transition matrix simply to keep the code more simple.

4 The name of this package reflects another name used for HMM, *dependent mixture models*.

data now that the parameters describing the state distributions and transition likelihoods have been estimated:

```R
## R
require(depmixS4)
hmm.model  <- depmix(returns ~ 1, family = gaussian(),
                     nstates = 4, data=data.frame(returns=returns))
model.fit  <- fit(hmm.model)
post_probs <- posterior(model.fit)
```

Here we produce an `hmm.model` where we indicate the observable is the `returns` vector. We also indicate the number of states (4) and that the emission probabilities come in a Gaussian distribution via our specification for the `family` parameter. We fit the model via a `fit()` function and then calculate the posterior probabilities with the `posterior()` function. The posterior probabilities give us the probability of a given state at a given time for the model parameters we have determined with the fitting process.

Now we can visualize the states along with the measured values as follows (see Figure 7-7):

```R
## R
plot(returns, type = 'l', lwd = 3, col = 1,
     yaxt = "n", xaxt = "n", xlab = "", ylab = "",
     ylim = c(-0.6, 0.6))

lapply(0:(length(returns) - 1, function (i) {
  ## add a background rectangle of the appropriate color
  ## to indicate the state during a given time step
  rect(i,-0.6,(i + 1),0.6,
       col = rgb(0.0,0.0,0.0,alpha=(0.2 * post_probs$state[i + 1])),
       border = NA)
}
```

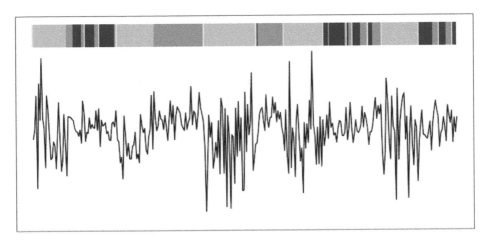

Figure 7-7. In this plot the background indicates a distinct state, and the solid black line depicts the actual values. The vertical white line sort of objects are actually very narrow slices that represent the times when the process is estimated to be in what turns out to be the rarest of the four states.

We can access information about the hypothesized distributional parameters the model determined via its attributes. As we review this, let's keep in mind our original data generation parameters:

```
bull_mu     <-  0.1
bull_sd     <-  0.1

neutral_mu <-  0.02
neutral_sd <-  0.08

bear_mu     <- -0.03
bear_sd     <-  0.2

panic_mu    <- -0.1
panic_sd    <-  0.3
```

If we were going to match (remembering that the panic regime didn't make it into the data), it would look like, of the states actually present in the data, the following groups roughly correlated:

```
> attr(model.fit, "response")
[[1]]
[[1]][[1]] <- coincidentally has a mean near the panic regime
              but that regime did not actually produce any data
              in the sample that was fit. instead the algorithm
              allotted this fourth state to more negative values
Model of type gaussian (identity), formula: returns ~ 1
Coefficients:
(Intercept)
```

```
-0.09190191
sd  0.03165587


[[2]]
[[2]][[1]] <- could be consistent with the bear market regime
Model of type gaussian (identity), formula: returns ~ 1
Coefficients:
(Intercept)
-0.05140387
sd  0.2002024


[[3]]
[[3]][[1]]  <- could be consistent with the bull market regime
Model of type gaussian (identity), formula: returns ~ 1
Coefficients:
(Intercept)
  0.0853683
sd  0.07115133


[[4]]
[[4]][[1]] <- could be somewhat consistent with the neutral market regime
Model of type gaussian (identity), formula: returns ~ 1
Coefficients:
  (Intercept)
-0.0006163519
sd  0.0496334
```

One possible reason the fit doesn't line up better with our baseline hidden states is that we didn't use a proper transition matrix, whereas the model used one for a fit. Our transitions between states were not properly Markovian, and this could impact the fit. Additionally, we fit to a relatively short time series with few transitions between states, whereas HMMs will perform better on longer time series with more opportunities to observe/infer state transitions. I'd recommend coming up with more realistic synthetic data to test a proposed HMM when you are experimenting with this technique and preparing to fit it to real data. Remember, in most cases of real data you are positing an unobservable state, so you want to think about the limits of the model performance in a more controlled setting (with synthetic data) before setting yourself up for more ambitious projects.

HMMs are apt for analyzing many kinds of data. HMMs have been used to model whether a financial market is in a growth or recession phase, determine what stage folding a protein within a cell is in, and describe human motion (before the arrival of deep learning). These models continue to be useful, more often for understanding the dynamics of a system than for making predictions. Also, HMMs provide more than just a point estimate or forecast. Finally, we can inject prior knowledge or beliefs into our model, such as by specifying the number of states used to fit our HMM. In this way, we reap the benefits of statistical methods but also have parameters to parameterize our prior knowledge about the system.

The mathematics and computation of HMMs are quite interesting and approachable. You can learn a lot of accessible programming techniques and numerical optimization algorithms by looking into the most common ways that HMMs are fit to data. You'll also learn about dynamic programming techniques, which are helpful to a data scientist or a software engineer.

As with a Kalman filter, HMMs can be used for a variety of tasks. In fact, the variety of inference problems related to HMM systems is even more complex due to the increased complexity of discrete states, each with its own emission probability. Some of the inference tasks you may face when using an HMM include:

- Determining the most likely description of the states producing a series of observations. This involves estimating the emission probabilities of these states as well as the transmission matrix that describes how likely one state is to lead to another. We did this, although we did not look explicitly at the transition probabilities.

- Determining the most likely sequence of states given a series of observations and a description of the states and their emission and transition probabilities. We did this as well as part of the previous task. This is sometimes referred to as the "most likely explanation" and is commonly calculated with the Viterbi algorithm.

- Filtering and smoothing. In this situation, filtering would correspond to estimating the hidden state of the most recent time step given the most recent observation. Smoothing would correspond to determining the most likely distribution of the hidden state in a particular time step given observations before, during, and after that time step.

Bayesian Structural Time Series

Bayesian structural time series (BSTS) are related to the linear Gaussian model we worked through with Kalman filtering earlier. The main difference is that Bayesian structural time series provide a way to use preexisting components to build more complex models that can reflect known facts or interesting hypotheses about a system. We can then design the structure, use robust fitting techniques to estimate parameters for the built model in the case of our data, and see whether the model does a good job of describing and predicting a system's behavior.

The mathematics of BSTS models is fairly complicated and computationally taxing compared to the linear Gaussian model we covered in our discussion of the Kalman filter. For this reason, we'll stick to a general overview and then apply the code.

There are four steps to fitting a BSTS model, carried out in the following order:

1. A structural model is defined, including specification of priors.
2. A Kalman filter is applied to update estimates of state based on observed data.
3. The spike-and-slab method is used to perform variable selection within the structural model.[5]
4. Bayesian model averaging is applied to combine results for the purpose of generating a forecast.

In the next example we will focus only on steps 1 and 2, wherein we define a flexible model out of preexisting modular components and then use this to fit our data with a Bayesian method that updates parameter estimation as time passes.

Code for bsts

Here we use a popular and powerful BSTS package from Google, bsts, and an open data set from OpenEI.org (*https://openei.org*).

We first plot the data to get a sense of what we are trying to model (Figure 7-8):

```R
## R
elec = fread("electric.csv")

require(bsts)
n = colnames(elec)[9]
par(mfrow = c(2, 1))
plot(elec[[n]][1:4000])
plot(elec[[n]][1:96])
## as discussed earlier in the book
## appropriate temporal scale is key to understanding ts data
```

---

5 To read more about the spike-and-slab method, start with Wikipedia (*https://perma.cc/4GNC-VDQY*). The math is fairly complicated, so we will not discuss it further here. The spike-and-slab method is most useful in cases where you have many inputs and need variable selection to simplify your model.

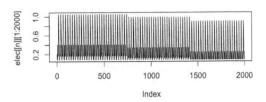

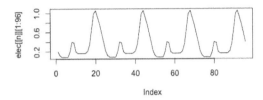

Figure 7-8. The full series we fit (two thousand consecutive hourly measurements) in the upper plot and a shorter, more understandable subset of the data in the lower plot. The plot makes more sense once we look at it in a way that makes the daily patterns evident.

Looking at our data gives us a sense of how to model. We can see that there is certainly a daily pattern, and there may even be a day-of-week pattern. These reflect the seasonal behavior we will describe in our model. We also want to allow for a trend in the data given the nonstationary behavior we see when we plot the whole data in the upper panel of Figure 7-8:

```R
## R
ss <- AddLocalLinearTrend(list(), elec[[n]])
ss <- AddSeasonal(ss, elec[[n]], nseasons = 24, season.duration = 1 )
ss <- AddSeasonal(ss, elec[[n]], nseasons = 7,  season.duration = 24)
```

The local linear trend in this model assumes that both the mean and slope of a trend in the data follow a random walk.[6]

The seasonal component of the model takes two arguments, one indicating the number of distinct seasons and one indicating the duration of the season. In the first seasonal component we add, which reflects a daily cycle, we want one season for each hour of the day, and each season lasts only the one hour. In the second seasonal component we add, which reflects a weekly cycle, we want one season for each day of the week and we want each season to last 24 hours.

While you might wonder whether we are truly starting at 12:01 am on Monday (or however we want to define the week), being consistent is more important than is

---

6 More info in the docs (*https://perma.cc/2N77-ALJ4*).

whether a seasonal label of day 1 exactly corresponds with a Monday. In the recurrent pattern we see here, it seems like any way of chopping the data into 24 hours would likely be acceptable for seasonality analysis.

The most taxing part of the code, computationally, is shown next. The beauty of the bsts package is that we are able to run a number of Markov Chain Monte Carlo (MCMC) computations of the posterior:

```R
## R
model1 <- bsts(elec[[n]],
               state.specification = ss,
               niter = 100)
plot(model1, xlim = c(1800, 1900))
```

We can also inspect the seasonal components. For example, we inspect the day of the week seasonal component like so (see Figures 7-9 and 7-10):

```R
## R
plot(model1, "seasonal", nseasons = 7, season.duration = 24)
```

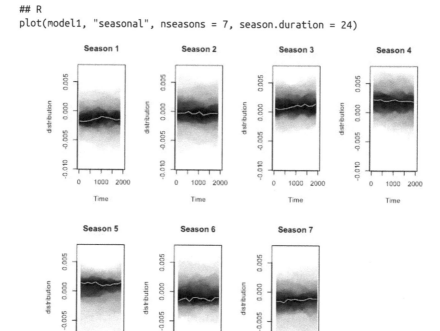

Figure 7-9. The day of the week seasons, which show that there is a difference in different days of the week. It also shows that the day of the week parameter distributions are stable over time for each day of the week.

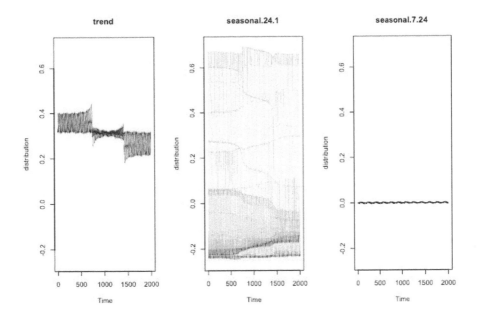

Figure 7-10. Distributions of contributions from the trend of the data as well as the daily and day of week seasonal components. If you sum these three components, you will get the value for the prediction.

The day of the week seasonal component shows great stability, whereas the hour of the day seasonality shown in the middle plot in Figure 7-10 tends to show trends over time, likely relating to changing daylight hours. In Figure 7-10 we also see the parameter fit for our local linear trend, which shows a decreasing pattern in electricity demand overall.

Finally, we forecast, complete with a full graph of the posterior distribution of the predictions into the future (see Figure 7-11). Notice that we have the flexibility up until the end of this modeling process to indicate how many time horizons we want to predict forward. Remember this is hourly data, so predicting 24 time horizons forward may seem quite ambitious but amounts to only one day. We also indicate that we want to see the 72 time periods prior to the prediction for context:

```
pred <- predict(model1, horizon = 24, quantiles = c(0.05, 0.95))
plot(pred, plot.original = 72)
```

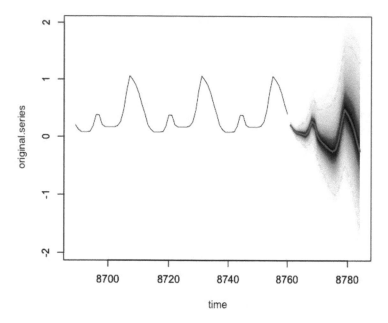

Figure 7-11. The last 72 hours in our data combined with a forecast for the next 24 hours as well as 5% and 95% quantile boundaries on the forecasts. Notice that the distribution of the forecast grows more spread out as we forecast further into the future.

There is a fair amount of optionality in the `bsts` package and in Bayesian structural time series modeling that we haven't even taken advantage of yet, namely:

- We did not specify nonstandard priors.
- We did not use the spike-and-slab method to select out regressors.
- We did not use Bayesian model averaging.

These are all possible to do with the `bsts` package, and you can readily find guidance in the documentation.

We have only scratched the surface in our example of what can be done with BSTS. Here are a few other important available options with BSTS:

- BSTS allow you to inject any kind of prior into your modeling. The standard linear Gaussian model we covered when discussing the Kalman filter is just one way of injecting a fairly vanilla prior, while BSTS offers many options (for example, nonsymmetric priors).
- BSTS models allow you to perform variable selection.

- BSTS models can be combined via Bayesian model averaging, which can help account for the uncertainty associated with selecting a model in the first place.

While we did not make use of these options in our current modeling case, they are not difficult to incorporate when using the `bsts` package, and you can find many examples online.

More Resources

- On Kalman filters and linear Gaussian state space models:

Greg Welch and Gary Bishop, "An Introduction to the Kalman Filter," technical report, University of North Carolina at Chapel Hill, 1995, https://perma.cc/ZCU8-MXEF.
> This introductory overview of the Kalman filter offers a derivation and matrix-formulation of the filter. The introduction also discusses the Extended Kalman Filter, a more common filtering in real-life scenarios that can be used for nonlinear processes or nonlinear measurement errors.

R.E. Kalman, "A New Approach to Linear Filtering and Prediction Problems," Transactions of the ASME—Journal of Basic Engineering 82, Series D (1960): 35–45, https://perma.cc/GNC4-YLEC.
> This 1960 research article is the original introduction of the Kalman filter. The mathematics is quite approachable with a basic background in statistics and calculus, and this is an interesting read to learn the original motivations for the Kalman filter and the intellectual context for its creator.

R. Labbe, "Kalman and Bayesian Filters in Python," GitHub repository, https://perma.cc/CMU5-Y94A.
> This epic GitHub repository includes dozens of examples of using Kalman filters and related techniques of "filtering" more generally. The repository is textbook-like, featuring worked examples, a related PDF book, and exercises with solutions.

Marie Auger-Méthé et al., "State-Space Models' Dirty Little Secrets: Even Simple Linear Gaussian Models Can Have Estimation Problems," Scientific Reports 6, no. 26677 (2016), https://perma.cc/9D8V-Z7KJ.
> This article highlights a case where even simple linear Gaussian models, such as those we used in our discussion of the Kalman filter, are easily prone to extreme misspecification, particularly in the case of a measurement error that is large relative to the values of a time series. The authors focus on this problem from the perspective of ecologists, but the general concern remains valid for a variety of data-driven disciplines and offers a counterbalancing view compared to the many advantages that are emphasized for this method.

- On Hidden Markov Models:

Andrew Moore, "Hidden Markov Models," lecture notes, School of Computer Science, Carnegie Mellon University, https://perma.cc/K3HP-28T8.
These comprehensive lecture notes offer an overview of HMMs, complete with illustrations of estimation algorithms and examples from robotics of how HMMs are used in real applications.

Dan Klein, "Artificial Intelligence: Hidden Markov Model," lecture notes, University of California Berkeley, https://perma.cc/V7U4-WPUA.
This is another set of accessible reference notes. These give examples of the usefulness of HMMs for digitizing speech and also for developing AI to play strategic games.

user34790, "What Is The Difference Between the Forward-Backward and Viterbi Algorithms?" question posted on Cross Validated, StackExchange, July 6, 2012, https://perma.cc/QNZ5-U3CN.
This StackExchange post offers an interesting discussion and outline relating to the many estimation algorithms deployed for specific HMM use cases. This post will help you get a sense of the ways HMMs can be used to understand time series data even if you are not interested in the details of the related modeling algorithms.

- On Bayesian structural time series:

Mark Steel, "Bayesian Time Series Analysis," in Macroeconometrics and Time Series Analysis, ed. Steven N. Durlauf and Lawrence E. Blume (Basingstoke, UK: Palgrave Macmillan, 2010), 35–45, https://perma.cc/578D-XCVH.
This brief read offers a comprehensive overview of different techniques associated with Bayesian analysis of time series as well as concise commentary as to the strengths and weaknesses of each method.

Steven Scott and Hal Varian, "Predicting the Present with Bayesian Structural Time Series," unpublished paper, June 28, 2013, https://perma.cc/4EJX-6WGA.
This Google paper based on economic time series presents an example of applying a time series forecasting problem to data about a given time that becomes available with different lags. In particular, the authors use current Google searches to predict unemployment rates where the latter rates are only published periodically, whereas the Google search counts were continuously available. This is an example of what is often called "nowcasting" to indicate a forecast that is actually being made about the present due to a reporting lag. The paper uses a combination of Bayesian structural time series and ensemble techniques.

Jennifer Hoeting et al., "Bayesian Model Averaging: A Tutorial," Statistical Science 14, no. 4 (1999): 382–401, https://perma.cc/BRP8-Y33X.

This article provides a comprehensive overview of how Bayesian model averaging works via several different methods. As this article describes, the purpose of Bayesian model averaging is to account for uncertainty in the modeling process due to model selection. Through worked examples, the authors provide a way to better estimate the uncertainty in predictions. A related, simpler summary is provided in an overview of BMA (*https://perma.cc/U7M4-PRMW*), an R package for Bayesian Model Averaging.

## CHAPTER 8

Generating and Selecting Features for a Time Series

In the previous two chapters we examined methods of time series analysis that rely on using all the data points in a time series to fit a model. However, in preparation for the next chapter's discussion of the application of machine learning to time series analysis, in this chapter we will study feature generation and selection for time series. If you are unfamiliar with the concept of feature generation, you will not remain so for long. It's an intuitive process and one that enables a creative side to data analysis.

Feature generation is the process of finding a quantitative way to encapsulate the most important traits of time series data into just a few numeric values and categorical labels. You are compressing the raw times series data into a shorter representation via a set of features to describe that time series (we'll work through a quick example momentarily). For example, a very simple feature generation could describe every time series with its mean value and the number of time steps in the series. This would be one way of describing that time series without going through all the raw data step by step.

The purpose of feature generation is to compress as much information about the full time series as possible into a few metrics or, alternately, to use those metrics to identify the most important information about the time series and discard the rest. This is important for machine learning methods, most of which were developed on nontemporal data but which can be fruitfully applied to time series problems, provided we can digest a time series into a properly formatted input. In this chapter we will focus particularly on packages that allow us to automatically generate commonly used time series features so there will be no need for us to reinvent or handcode them.

Once we have generated some putatively useful features, we must ensure that they are indeed useful. While you are unlikely to craft too many unhelpful features by hand, this is a problem you will run into when you use code that automatically generates a large number of features of a time series for downstream use in machine learning. For this reason, we must inspect the features, once generated, to see which can be discarded in subsequent analyses.

Traditional machine learning models were not originally developed with time series in mind, and so they do not automatically lend themselves to time series analytical applications. However, one way to make these models work with temporal data is feature generation. For example, by describing a univariate time series not with a series of numbers detailing the step-by-step outputs of a process but rather by describing it with a set of features, we can access methods designed for cross-sectional data.

In this chapter we will first work through a very simple example of feature generation for a short time series. We will then review feature generation packages for time series, both in R and Python. Finally, we'll work through an example of automated feature generation and feature selection. After reading this chapter you will have all the skills needed to preprocess a time series data set for downstream machine learning applications in Chapter 9.

Introductory Example

Imagine the past week's morning, midday, and evening temperatures were as shown in Table 8-1.

Table 8-1. Temperatures for the past week

Time	Temperature (°F)
Monday morning	35
Monday midday	52
Monday evening	15
Tuesday morning	37
Tuesday midday	52
Tuesday evening	15
Wednesday morning	37
Wednesday midday	54
Wednesday evening	16
Thursday morning	39
Thursday midday	51
Thursday evening	12
Friday morning	41
Friday midday	55
Friday evening	20
Saturday morning	43
Saturday midday	58

Time	Temperature (°F)
Saturday evening	22
Sunday morning	46
Sunday midday	61
Sunday evening	35

You could plot this data and you'd see elements of periodicity (a daily cycle) and also a trend of overall increasing temperatures. But we can't store an image of a plot in a database, and most methods that accept a picture as an input are data-intensive and seek to strip down the picture into summary metrics. So we should do the summary metrics ourselves. Instead of describing the 21 numbers in Table 8-1 as a time series, we could describe the series with a few words and numbers:

- Daily/periodic
- Increasing trend; we could make this more quantitative by computing a slope
- Mean values for each of morning, midday, and evening

By doing so, we'd summarize the 21-point time series with 2 to 5 numbers—quite a bit of data compression without losing too much detail. This is a simple case of feature generation. Then, feature *selection* would entail paring away any features that were not descriptive enough to justify inclusion. What justifies inclusion will depend on our downstream use of the features.

General Considerations When Computing Features

As with any aspect of analysis, when you are computing time series features for a time series data set, you will want to think through whether your analysis makes sense and whether the effort you put into generating features is more likely to lead to overfitting from a surfeit of features than it is to lead to meaningful insights.

The best approach is to develop a set of potentially useful features as you run through time series exploration and cleaning. As you visualize data and think about what distinguishes different time series in the same data set or different time periods in the same time series, you will develop ideas about what kinds of measurements would be useful for labeling or predicting a time series. You can also draw useful assistance from any background knowledge you have about a system or even a working hypotheses you'd like to test with subsequent analysis.

Next we discuss a few distinct concerns you should keep in mind when generating time series features.

The Nature of the Time Series

As you decide what time series features to generate, you need to keep in mind the underlying attributes of your time series, which you determined during data exploration and cleaning.

Stationarity

Stationarity is one consideration. Many time series features assume stationarity and are useless unless the underlying data is stationary or at least ergodic. For example, using the mean of a time series as a feature is practical only where the time series is stationary so that the idea of a mean makes sense. This value is not very meaningful where we have a nonstationary time series, as the value measured as the mean in that case is more or less an accident, a result of too many entangled processes, such as a trend or a seasonal cycle.

Stationary Versus Ergodic Time Series

An *ergodic* time series is one in which every (reasonably large) subsample is equally representative of the series. This is a weaker label than stationarity, which requires that these subsamples have equal mean and variance. Ergodicity requires that each slice in time is "equal" in containing information about the time series, but not necessarily equal in its statistical measurements (mean and variance). You can find a helpful discussion on StackExchange (*https://perma.cc/5GW4-ZENE*).

Length of time series

Another consideration for feature generation is the length of the time series. Some features may be sensible for a stationary time series but become unstable as the length of the time series increases, such as the minimum and maximum value of the series. For the same underlying process, a longer time series will likely measure more extreme maximum and minimum values than a shorter time series produced by the same process, simply because there were more opportunities for data collection.

Domain Knowledge

Domain knowledge should be key for time series feature generation where you are lucky enough to have some insights. Some examples of how domain knowledge is applied to generate specific time series features are provided later in this chapter, but for now we'll focus on the more general point.

For example, if you are working with a physics time series, you should quantify features that make sense on the timescale of the system you are studying, as well as make sure that the features you select would not be unduly influenced by the characteristics of, say, the error of a sensor rather than the characters of an underlying system.

As another example, imagine you are working with data from a specific financial market. To ensure financial stability, this market imposes maximum price changes in a given day. If the price changes too much, the market shuts down. You might consider whether, in this context, to generate a feature indicating the maximum price seen on a given day.

External Considerations

The extent of your computational resources and associated storage resources is also important. Likewise, your motivation for generating features matters. Are you generating features that will be stored so that you can throw out voluminous raw data? Or are you merely computing the features for a single analysis and planning to keep only the raw data?

The purpose of your time series feature generation may influence how many features you decide to compute and whether you should contemplate particularly computationally demanding features. This may also depend on the overall size of the data set you are analyzing. For a small data set all these decisions will be low stakes, but in the case of extremely large time series data sets, you may risk embarking on a feature generation task that will be left half-done, wasting computational energy and coding.

After considering all these factors, try putting together a list of features and running them on a small data set to get an idea of how fast or slow they run. If the small set runs too slowly, you should pare down your time series substantially before continuing your analysis. Likewise, you might consider exploring the usefulness of computationally taxing features on a subset of your data before undertaking an analysis with the full data set.

A Catalog of Places to Find Features for Inspiration

Time series feature generation is limited only by your data, your imagination, your coding skill, and your domain knowledge. So long as you can think of a reasonably general and well-defined way to quantify the behavior of a time series, you can generate a feature. Some simple and oft-used time series features amount to the same summary statistical functions you will have used in other applications, such as:

- Mean and variance
- Maximum and minimum
- Difference between last and first values

You will also visually identify other features that are more computationally challenging to compute but are often useful. Some examples include:

- Number of local maxima and minima
- Smoothness of the time series
- Periodicity and autocorrelation of the time series

In such cases, you will need to make some implementation definitions, as there are different ways to identify these commonly used features. It will help to keep your own personal library of feature generation code available, but you may also want to look into feature generation libraries for time series data, particularly as you become interested in the more computationally demanding features. In such cases, you should look for an excellent implementation such that the code is both reliable and efficient.

Now we'll turn to the use of time series feature generation libraries, paying particular attention to the wide range of features you can benefit from via automatic feature generation.

Open Source Time Series Feature Generation Libraries

There have been many efforts to automate the creation of time series features because they tend to be interesting, descriptive, and even predictive across domains.

The tsfresh Python module

One particularly compelling example of automatic feature generation in Python is the `tsfresh` module, which implements a large and general set of features. We can get a sense of the breadth of implemented features (*https://perma.cc/2RCC-DJLR*) by considering some of the general categories of features that are available. These include:

Descriptive statistics
These are driven by the traditional statistical time series methodologies we studied in Chapter 6, including:

- An Augmented Dickey–Fuller test value
- An AR(k) coefficient
- The autocorrelation for a lag, k

Physics-inspired indicators of nonlinearity and complexity
This category includes:

- The function `c3()`, which is a proxy for calculating the expected value of $L^2(X^2) \times L(X) \times X$ (L is the lag operator). This has been proposed as a measure of *nonlinearity* in a time series.

- The function `cid_ce()`, which calculates the square root of the sum from 0 to n − 2 × lag of $(x_i − x_i + 1)^2$. This has been proposed as a measure of the *complexity* of a time series.

- The function `friedrich_coefficients()`, returns coefficients of a model fitted to describe complex nonlinear motion.

History-compressing counts

This category comprises features such as:

- The sum of the values in a time series that occur more than once

- The length of the longest consecutive subsequence that is above or below the mean

- The earliest occurrence within the time series of the minimum or maximum value

A module like `tsfresh` can help you save time and choose efficient implementations for feature selection. It can also educate you about ways of describing data that could be relevant but that you might not have come across in your own research. There are numerous other benefits to using a module, as there always are when you combine your analysis with open source, well-vetted tools, including:

- There is no need to reinvent the wheel when you are computing standard features. By using a shared library, you have the assurance of some accuracy checking by other users, rather than doubting your own code and having to verify it.

- A library such as this one provides a framework for calculating features, not just a laundry list of features. For example, `tsfresh` has a feature calculator class, which you can use to extend this library for your own purposes but with the benefits of a systematic framework.

- This library is designed to hook up with downstream consumers of features, most importantly with `sklearn`, so that your features can easily be passed to machine learning models.

The `tsfresh` library has a particularly technical flavor in that many of the features are derived from ideas about analyzing scientific experimental data.

The Cesium time series analysis platform

A more approachable but equally extensive catalog of generated features is the list implemented in the Cesium (*http://cesium-ml.org/docs/index.html*) library. The current list is available in its documentation (*http://cesium-ml.org/docs/feature_table.html*), and next we pick out a few interesting features for discussion and inspection. The general categories are broken down in the source code (*https://perma.cc/8HX4-MXBU*), but we break them down further here:

- Features that describe the overall distribution of the data values, without regard to its temporal relationships. This category can include a diverse set of features that are all nonetheless time-agnostic:
  - How many local peaks there are in a histogram of the data?
  - What percentage of the data points are within a fixed window of values close to the median of the data?
- Features that describe the distribution of the timing of data:
  - Features that take the distribution of time between measurements as their own distribution and compute similar statistics to those just described, now on the distribution of time differences rather than data values
  - Features that compute the probability that the next observation will occur within *n* time steps given the observed distribution
- Features that describe measures of the periodicity of the behavior within the time series. Often these features are associated with the *Lomb-Scargle periodogram*.

Periodograms

A periodogram of a time series is an estimate of how much different underlying frequencies contribute to the time series. A perfect periodogram could translate a time series from the time-value phase space, which we most commonly work with, to a power-frequency space whereby a time series would be described in terms of how much of it was shaped by recurring processes at various timestamps. If you are unfamiliar with this concept, Figure 8-1 shows it visually.

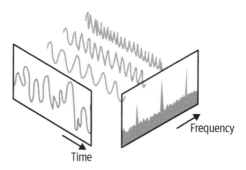

Figure 8-1. The same data represented in the time domain, where it appears similar to a sum of sine time series. In the frequency domain, we see all the spectral mass loading onto the three frequencies that characterize the sines that were summed to produce the series.

There are many ways of computing periodograms. One definition and computational method is the *Fourier transform* of the autocorrelation function. A more intuitive way of thinking about creating a periodogram is as a least squares fit of the original data to a set of sine curves of varying frequencies.

This latter is what the much-vaunted Lomb-Scargle method does. The Lomb-Scargle method constructs a periodogram for irregularly sampled time series data, unlike a traditional periodogram, which assumes a regularly sampled time series. Lomb, and then Scargle, developed techniques to study irregular time series with this method and established that the statistical properties of the periodogram constructed from such data had the same statistical properties as a traditional periodogram sampled from regularly sampled time series data. Lomb's and Scargle's advancements enabled the study of many natural science problems for which irregularly sampled times series are inevitable, such as climate data, astrophysical data, and geological data.

The features just described can be computed either over the entire time series or as rolling or expanding window functions. Given what we learned in earlier chapters about the mechanics of coding up rolling and expanding window functions, we could certainly implement these features ourselves, and we are in a good position to understand the documentation and meaning of what this library does. In this case, we would be applying rolling window functions to summarize data rather than to clean it. The same techniques in time series analysis come into use in many different but equally useful cases.

The `cesium` library provides supplementary functionality in addition to feature generation. For example, it includes a web-based GUI to perform feature generation and also integrates with `sklearn`.

If you try these libraries on your data, you will notice that time series generation is extremely time consuming. For this reason, you should think carefully about how many features you need to generate for your data and when it truly makes sense to automatically generate features rather than thoughtfully developing your own.

Many of the features generated by these libraries are computationally taxing and—given how extensive the lists of features are—will often not address points of interest for the question you are trying to answer. With some domain knowledge, you may even recognize that a particular kind of feature is irrelevant, noisy, or not predictive. Do not calculate these useless features unnecessarily. It will slow down your entire analysis without adding any clarity. Automatic feature generation libraries are useful but should be used judiciously rather than indiscriminately.

R's tsfeatures package

tsfeatures, developed by Rob Hyndman et al., is a convenient R package for generating a variety of commonly used and useful time series features. The documentation includes a listing of the features, which include helpful functions such as:

- acf_features() and pacf_features(), which each compute a number of related values given a general sense of how important autocorrelation is in the behavior of a series. For the acf_features() function, the documentation describes the following return values: "A vector of 6 values: first autocorrelation coefficient and sum of squared of first 10 autocorrelation coefficients of original series, first-differenced series, and twice-differenced series. For seasonal data, the autocorrelation coefficient at the first seasonal lag is also returned."

- lumpiness() and stability(), which are tiled window–driven functions, and max_level_shift() and max_var_shift(), which are rolling window–driven functions. In each of these cases differences and variety-measuring statistics are applied to values measured on either overlapping (rolling) or nonoverlapping (tiled) windows of the time series.

- unitroot_kpss() (*https://perma.cc/WF3Y-7MDJ*) and unitroot_pp() (*https://perma.cc/54XY-4HWJ*).

The tsfeatures package usefully consolidates and includes work from a variety of academic projects related to studying time series features, as well as from other ongoing efforts to improve the process of creating time series features that are useful in a variety of domains. These include:

- compengine() computes the same time series features developed by the comp-engine.org project, which have been found to be helpful across a wide variety of time series data from many domains.[1]

- A number of features borrowed from the hctsa package (*https://github.com/benfulcher/hctsa*), which is intended to run highly comparative time series analysis in Matlab. Some of these features are in: autocorr_features(), first min_ac(), pred_features(), and trev_num(). You will also find others from reading the documentation.

---

1 Interested readers who want to explore the details are referred to the Catch22 set of time series features (*https://perma.cc/57AG-V8NP*), which have proven useful across a wide variety of time series data sets. Cutting down the set of features from over 4,000 to just 22 reduced computation time by a factor of 1,000 with a reduction in accuracy on a classification task of only 7%. It is also quite educational to read about the pipeline the researchers used to select features and ensure a set of relatively independent but still accurate features from their starting set.

The `tsfeatures` documentation (*https://perma.cc/Y8E9-9XCK*) also includes helpful illustrations of the uses and outputs of each of the feature-generating functions, and an extensive bibliography linking to related statistical and machine learning work on the time series feature set available in the package.

Domain-Specific Feature Examples

Another source of inspiration can come from domain-specific features that have been developed for a variety of time series data. Often these features have been developed over decades, either from heuristics that empirically work even if they are not well understood or from scientific knowledge about how a system's underlying mechanics work.

Next we review some domain-specific features for two kinds of time series data: financial and healthcare.

Technical stock market indicators

Technical stock market indicators are probably the most widely documented and formalized sets of indicators used for a domain-specific time series application. Economists studying time series data over the last century have an extensive list of features they commonly use to quantify time series in the financial markets and make predictions. Even if you have no interest in the financial markets, this list is inspirational in showing how a domain-specific list of features can also be quite extensive, descriptive, and creative.

A nonexhaustive list of these features is included here so that you can get an idea of how complex and highly specific to the financial markets these indicators are. Given the complexity, it's no surprise that people can spend their entire career attempting to understand how these "signals" can forecast the rise and fall of the financial markets.

Relative Strength Index (RSI)
> This measure is equal to $100 - 100 / (1 + RS)$ where RS is the ratio of the mean gain during an "up" period (rising prices) divided by the mean loss during a "down" period (decreasing prices), and the lookback period for these up and down periods is an input parameter, so that you can have different RSI values for different lookback periods. Traders have developed rules of thumb about what cutoff RSI values indicate that an asset is undervalued or overvalued relative to its true worth. The RSI is known as a "momentum indicator" because it relies on measures of an asset's movement.

Moving Average Convergence/Divergence (MACD)
> This indicator is itself composed of three time series:

- The MACD time series is a time series of the difference between a short-term exponential moving average of the asset ("fast") and a long-term exponential moving average of the asset ("slow").

- The "average" time series is an exponential moving average of the MACD time series.

- The "divergence" time series is the difference between the MACD time series and the "average" time series. This is the value that is usually used for making financial forecasts. The other input series (MACD and average) are usually only prepared to create the "divergence" series.

Chaikin Money Flow (CMF)
This indicator measures the direction in which spending is trending. To calculate it:

- Calculate the money flow multiplier: ((*Close* – *Low*) – (*High* – *Close*))/(*High* – *Low*).

- Calculate the money flow volume, which is the day's trading volume multiplied by the money flow multiplier.

- Sum this money flow volume for a particular period of days and divide it by the volume for that same period of days. This indicator is an "oscillator" that ranges between –1 and 1. It indicates "buying pressure" or "selling pressure," a measure of the direction of the market.

As you can see from this small taste of the many technical features that can be constructed from simple financial time series, there are many ways to describe a time series. The financial markets are a particularly rich and much-studied domain for feature generation.[2]

Healthcare time series

Healthcare is another area where time series features have domain-specific meanings and even names. As we discussed in Chapter 1, health data offers quite a wide array of time series data. One example is EKG data (see Figure 8-2). Reading EKGs is both a science and an art, and various features are manually identified by physicians and used to read a time series. If you were going to be selecting features for a machine learning study on EKG data, you would certainly want to start by studying these features and speaking to a knowledgeable physician to understand the purpose of the features and what they indicate.

---

2 For an extensive catalog of features that can be used in machine learning for financial markets time series, see the Kaggle blog post (*https://perma.cc/Q84C-44XD*). Unfortunately, it does not seem as though that code met with great success, but it is an excellent example of exhaustive preparation of potentially useful features based on their domain relevance.

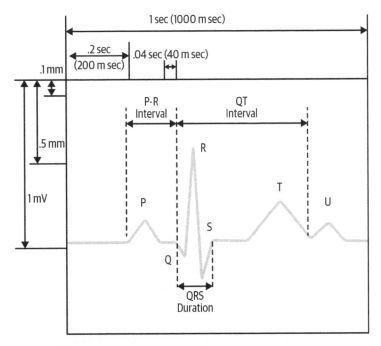

Figure 8-2. Examples of time series features medical professionals use to read EKG time series data.

Similarly, if you are analyzing high-resolution blood glucose time series data, it is also helpful to understand the kind of patterns that tend to affect daily data as well as how a healthcare professional would understand and label this data (see Figure 8-3).

The two time series illustrated in Figures 8-2 and 8-3 provide good examples of when we would be interested in finding local maxima or distances between them. Hence, we can easily envision specific features from the `tsfresh` or `cesium` libraries that are relevant to healthcare time series given domain-specific knowledge.

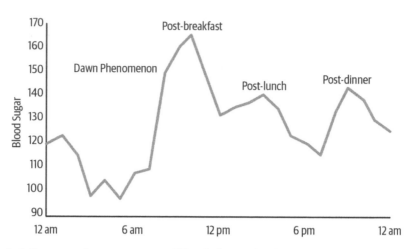

Figure 8-3. In a one-day time series of blood glucose levels, we can see there are four features most healthcare professionals would identify in a typical day, one of which is not related to food but is known as the "dawn phenomenon." This is something we would want to consider identifying in one or several features.

How to Select Features Once You Have Generated Them

Suppose you have generated many features automatically to describe your large time series data set. You may not be able to look at all proposed features on a first pass through your data, so it can be helpful to complement the use of automatic feature *generation* with the use of automatic feature *selection*. One helpful feature selection algorithm is the FRESH algorithm, which is implemented in the `tsfresh` package described earlier. FRESH stands for feature extraction based on scalable hypothesis tests.

The FRESH algorithm is motivated by the ever-increasing amount of time series data that is available, often stored in a distributed manner, which facilitates parallelization of computation. The algorithm evaluates the significance of each input feature with respect to a target variable via the computation of a *p*-value for each feature. Once computed, the per-feature *p*-values are evaluated together via the Benjamini-Yekutieli procedure, which determines which features to keep based on input parameters about acceptable error rates and the like. The *Benjamini-Yekutieli procedure* is a method of limiting the number of false positives discovered during hypothesis testing used to produce the *p*-values in the initial step of the FRESH algorithm.

To implement these steps on our own would be quite a daunting task, but we can accomplish this in a few lines of code via `tsfresh`. Here we follow the code used in an illustration in the module's documentation. First, we download time series data relating to robot execution failures:

```
## python
>> from tsfresh.examples.robot_execution_failures import
                      download_robot_execution_failures,
                      load_robot_execution_failures
>> download_robot_execution_failures()
>> timeseries, y = load_robot_execution_failures()
```

Then we extract the features without needing to specify them, because the package automatically calculates all features. In this sense, it goes against the advice given in this chapter in being extremely inclusive without concern for computational resources. In this test data set there are not too many data points, but you would likely not want to blindly deploy this on your data set without shrinking it down to a reasonable number of points:

```
## python
>> from tsfresh import extract_features
>> extracted_features = extract_features(timeseries,
                                 column_id   = "id",
                                 column_sort = "time")
```

While tsfresh does provide a way for you to specify which features you would like to calculate, in this example we simply opt to include all features. You can also manually set the parameters for those features that take parameters into account for their calculation rather than using the defaults. This is all described and illustrated in the documentation (*https://perma.cc/D5RS-BJ6T*).

If you do perform a full extraction, as we have with the example data provided by tsfresh, you can see that there are numerous features calculated:

```
## python
>> extracted_features.columns
Index(['F_x__abs_energy', 'F_x__absolute_sum_of_changes',
       'F_x__agg_autocorrelation__f_agg_"mean"',
       'F_x__agg_autocorrelation__f_agg_"median"',
       'F_x__agg_autocorrelation__f_agg_"var"',
       'F_x__agg_linear_trend__f_agg_"max"__chunk_len_10__attr_
                                         "intercept"',
       'F_x__agg_linear_trend__f_agg_"max"__chunk_len_10__attr_
                                         "rvalue"',
       'F_x__agg_linear_trend__f_agg_"max"__chunk_len_10__attr_
                                         "slope"',
       'F_x__agg_linear_trend__f_agg_"max"__chunk_len_10__attr_
                                         "stderr"',
       'F_x__agg_linear_trend__f_agg_"max"__chunk_len_50__attr_
                                         "intercept"',

       ...
       'T_z__time_reversal_asymmetry_statistic__lag_1',
       'T_z__time_reversal_asymmetry_statistic__lag_2',
       'T_z__time_reversal_asymmetry_statistic__lag_3',
       'T_z__value_count__value_-inf', 'T_z__value_count__value_0',
       'T_z__value_count__value_1', 'T_z__value_count__value_inf',
```

```
        'T_z__value_count__value_nan', 'T_z__variance',
        'T_z__variance_larger_than_standard_deviation'],
       dtype='object', name='variable', length=4764)
```

There are 4,764 columns. This is a far greater number of features than we could have calculated by hand, but this is also very time-consuming to run on a realistic data set. As you decide how and when to deploy such an outsize set of features, try to be realistic about both your computing power and your ability to carefully review the results. Remember that for time series data, outliers can be particularly influential in a nasty and unhelpful way for subsequent analysis. You will want to ensure that the features you choose are resistant to outliers.

While the FRESH algorithm is helpful in accounting for dependence among the features, it is somewhat difficult to reason about. We can also use a more traditional and transparent feature selection technique, *recursive feature elimination* (RFE). We can use RFE to complement our use of the FRESH algorithm and enhance our understanding of the degree of difference between those features selected by the FRESH algorithm and those not selected.

RFE describes an incremental approach to feature selection whereby features are gradually eliminated from a more inclusive model, creating a less inclusive model down to the minimum number of features to be included, which is set at the start of the selection procedure.

This technique is known as *backward selection* because you start with the most inclusive model and move "backward" to a simpler model. In contrast, in *forward selection* features are incrementally added until the maximum number of specified features, or some other stopping criterion, is reached.

We can use RFE both for feature selection and also as a way of ranking feature importance. To run an experiment, we combine 10 randomly selected features from the list of features kept by the FRESH algorithm with 10 randomly selected features from the list of features rejected by the FRESH algorithm:

```
## R
>> x_idx = random.sample(range(len(features_filtered.columns)), 10)
>> selX = features_filtered.iloc[:, x_idx].values
>> unselected_features = list(set(extracted_features.columns)
                        .difference(set(features_filtered.columns)))
>> unselected_features = random.sample(unselected_features, 10)
>> unsel_x_idx = [idx for (idx, val) in enumerate(
          extracted_features.columns) if val in unselected_features]
>> unselX = extracted_features.iloc[:, unsel_x_idx].values
>> mixed_X = np.hstack([selX, unselX])
```

With this set of 20 features, we can then perform RFE to get a sense of the ranked importance of these features for the data set and for the model we use within the RFE:

```
>> svc = SVC(kernel="linear", C=1)
>> rfe = RFE(estimator=svc, n_features_to_select=1, step=1)
>> rfe.fit(mixed_X, y)
>> rfe.ranking_
array([ 9, 12,  8,  1,  2,  3,  6,  4, 10, 11,
       16,  5, 15, 14,  7, 13, 17, 18, 19, 20])
```

Here we can see the relative rankings of the 20 features we fed into the RFE algorithm. We would hope that the first 10 features—those selected, among others, by the FRESH algorithm—would be ranked higher than those not selected by the FRESH algorithm. This is largely but not exclusively the case. For example, we can see that in the second half of the array, representing the ranks of the unselected features, we actually have the 5th and 7th most important features out of all 20. However, we would not expect a perfect match-up, and the results are largely consistent.

We can use RFE on the selected features as a way of further culling our features. We can also use it as a sanity check if we attempt to fine-tune the input parameters of the FRESH algorithm, or the number of features we are generating as input to the FRESH algorithm in the first place.

Notice that the FRESH algorithm itself is essentially parameter free, so the number and quality of the features we input are our best way of affecting its output. The other parameter we set for the FRESH algorithm is the fdr_level, which is the percentage of irrelevant features we expect after generating the features. This parameter defaults to .05, but you might decide to set this value much higher to enhance the selectivity of the feature filtering, particularly when you are generating a large number of features without any consideration of whether they are appropriate for your domain of interest.

Concluding Thoughts

In this chapter we have discussed the motivation for feature selection as well as a simple example of how feature generation can work to convert even a short time series to a more compressed set of numbers that is nearly as informative as the original. We also looked at examples of two Python modules designed to implement automated feature generation and selection on time series data, which can handily generate thousands of features of a time series. Because there is a danger that many of the features generated this way will not be especially useful, we also looked at methods for selecting the most useful features to pass on further down our analytical pipeline so that feature generation does not produce noisy or uninformative features.

Feature generation is useful for a number of purposes:

- Producing downstream data about time series in a format that is conducive to use in machine learning algorithms, most of which are designed to accept sets of features per data point rather than a time series.

- Summarizing time series data in a way that compresses temporal observations into the shorthand of a few numbers and qualitative indicators. This can be useful not just for analysis but also for storing time series data in a more succinct and readable format in cases where we do not need to keep the full time series.

- Providing a common set of metrics to describe, and identify similarities across, data that may have been measured under many different conditions. By summarizing our data more broadly, we can make data comparable that otherwise may not seem easy to compare.

In Chapter 9 we will use feature generation to prepare data input for a number of machine learning algorithms that rely on input of time series features rather than raw time series data for classification and forecasting purposes.

More Resources

- On feature-based time series analysis:

  Ben D. Fulcher, "Feature-Based Time-Series Analysis," eprint arXiv:1709.08055, 2017, https://perma.cc/6LZ6-S3NC.
    This approachable review paper was written by the same person who implemented the tsfresh algorithm in Matlab. It offers an extensive taxonomy of different kinds of features that can be implemented on time series data. It also includes helpful illustrations of the categories of features discussed. There is also an emphasis on the use of feature generation as a form of data exploration to understand the appropriate analysis for a given time series data set.

- On feature selection:

  Maximilian Christ et al., "Time Series FeatuRe Extraction on basis of Scalable Hypothesis tests (tsfresh—A Python package)," Neurocomputing 307 (2018): 72–7, https://oreil.ly/YDBM8.
    This article introduces the tsfresh Python package and included FRESH algorithm, which were discussed in this chapter. Information about the computational efficiency of specific features to compute and sample usage patterns are also presented.

  Maximilian Christ et al., "Distributed and Parallel Time Series Feature Extraction for Industrial Big Data Applications," paper presented at ACML Workshop on

*Learning on Big Data (WLBD), Hamilton, NZ, November 16 2016, https://
arxiv.org/pdf/1610.07717.pdf.*

This more technical explanation and testing of the FRESH algorithm provides more details about how the algorithm works and why it is appropriate
for industrial time series application, and an extensive list of background
reading in the references section.

- On domain-specific features:

*Wikipedia, "Technical Analysis for Financial Markets," https://perma.cc/8533-
XFSZ.*

This Wikipedia article gives a history of how technical analysis was developed for the financial markets as well as an extensive listing of many popu-
larly used indicators.

*Amjed S. Al-Fahoum and Ausilah A. Al-Fraihat, "Methods of EEG Signal Features
Extraction Using Linear Analysis in Frequency and Time-Frequency Domains,"
ISRN Neuroscience 2014, no. 730218 (2014), https://perma.cc/465U-QT53.*

This article provides an example of testing standard time series features in
the domain of EEG signals. There is an entire academic industry of generating features for medical time series, and this paper provides an accessible
example of how it's done.

*Juan Bautista Cabral et al., "From FATS to Feets: Further Improvements to an
Astronomical Feature Extraction Tool Based on Machine Learning, Astronomy
and Computing 25 (2018), https://perma.cc/8ZEM-Y892.*

This article discusses a recent redesign of a Python package designed for the
extraction of features from astronomical time series data. While this document assumes some familiarity with astronomy and the specific packages, it's
still a useful read for getting an idea of how software is designed for feature
extraction and for understanding some specific challenges and changes that
have been occurring in astronomical time series data in particular.

*Alvin Rajkomar et al., "Supplementary Information for Scalable and Accurate
Deep Learning for Electronic Health Records," npj Digital Medicine 1, no. 18
(2018), https://perma.cc/2LKM-326C.*

This supplementary information relates to an interesting paper published by
Google illustrating a highly successful use of deep learning to understanding
electronic health records. The supplementary information provides detailed
information about how healthcare records were turned into quantitative
input features that could be accessed by a neural network. If you are not
familiar with deep learning, it may be worth returning to this resource only
after reading Chapter 10. Additionally, it would be a good idea to first get a
basic understanding of embeddings for deep learning (*https://perma.cc/
3KAZ-9A3Y*).

- On periodograms:

"The Periodogram," lecture notes, Eberly College of Science, Pennsylvania State University, https://perma.cc/5DRZ-VPR9.
> These lecture notes from Penn State's course on Applied Time Series Analysis offer an introduction to what a periodogram is, how to interpret one, and how to calculate one using R.

Jacob T. VanderPlas, "Understanding the Lomb-Scargle Periodogram," Astrophysical Journal Supplement Series 236, no. 1 (2018), https://arxiv.org/pdf/1703.09824.pdf.
> This expansive article provides an intuitive understanding of periodicity estimators generally and the Lomb-Scargle method in particular.

## CHAPTER 9

Machine Learning for Time Series

In this chapter, we will look at a few examples of applying machine learning methods to time series analysis. This is a relatively young area of time series analysis but one that has shown promise. The machine learning methods we will study were not originally developed for time series–specific data—unlike the statistical models we studied in the past two chapters—but they have proven useful for it.

This turn to machine learning is a shift from our previous work in forecasting in earlier chapters of this book. Up to this point we have focused on statistical models for time series forecasts. In developing such models, we formulated an underlying theory about the dynamics of a time series and the statistics describing the noise and uncertainty in its behavior. We then used the hypothesized dynamics of the process to make predictions and also to estimate our degree of uncertainty about the predictions. With such methods, both model identification and parameter estimation required that we think carefully about the best way to describe the dynamics of our data.

We now turn to methodologies in which we do not posit an underlying process or any rules about that underlying process. We instead focus on identifying patterns that describe the process's behavior in ways relevant to predicting the outcome of interest, such as the appropriate classification label for a time series. We will also consider unsupervised learning for time series, in the form of time series clustering.

We cover prediction and classification with tree-based methodologies as well as clustering as a form of classification. In the case of tree-based methodologies, formulating features of our time series is a necessary step along the way of using the methodology, as trees are not a "time-aware" methodology, unlike, say, an ARIMA model.

In the case of clustering and distance-based classification, we will see that we have the option to use features or to use the original time series as an input. To use the time series itself as an input, we study a distance metric known as *dynamic time warping*, which can be applied directly to time series, preserving the full chronological set of information in our data rather than collapsing it into a necessarily limited set of features.

Time Series Classification

In this section we work through an example of converting raw electroencephalogram (EEG) time series data into features, which can in turn be used for machine learning algorithms. We then use decision tree methods to classify EEG data after we have extracted features from the EEG time series.

Selecting and Generating Features

In the previous chapter, we went through a general discussion of the purposes of time series feature generation. We also worked through a brief example of generating features for a time series data set via `tsfresh`. Now we will generate features with another time series feature package that was discussed: `cesium`.

One of the very handy attributes of the `cesium` package is that it comes with a variety of helpful time series data sets, including an EEG data set originally derived from a 2001 research paper (*https://perma.cc/YZD5-CTJF*). In this paper, you can read more on the data preparation details. For our purposes it suffices to know that the five categories of EEG time series present in the data set all represent equal-length segments cut out of continuous time readings of EEG samples from the following:

- EEG recordings of healthy people with eyes both opened and closed (two separate categories)
- EEG recordings of epilepsy patients during seizure-free times from two non-seizure-related areas of the brain (two separate categories)
- An intracranial recording of EEG during a seizure (one category).

We download that data set via a convenience function supplied by `cesium`:

```python
## python
>>> from cesium import datasets
>>> eeg = datasets.fetch_andrzejak()
```

It can be helpful to first view a few examples of the data we are analyzing to get an idea of how we would like to classify these time series:

```python
## python
>>> plt.subplot(3, 1, 1)
>>> plt.plot(eeg["measurements"][0])
>>> plt.legend(eeg['classes'][0])
>>> plt.subplot(3, 1, 2)
```

```
>>> plt.plot(eeg["measurements"][300])
>>> plt.legend(eeg['classes'][300])
>>> plt.subplot(3, 1, 3)
>>> plt.plot(eeg["measurements"][450])
>>> plt.legend(eeg['classes'][450])
```

These plots show some differences between the classes of EEG measurement (see Figure 9-1). It is not surprising that the EEG plots are markedly different: they are measuring activity in different parts of the brain during different activities in both healthy subjects and epilepsy patients.

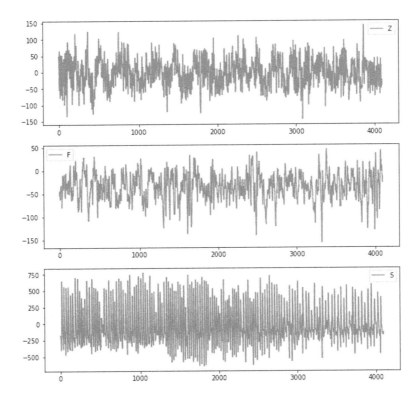

Figure 9-1. We plot three randomly selected samples from the EEG data set. These are independent samples and not contemporaneous measurements from different parts of the same brain. Each is an independent time series measurement taken of a different patient at a different time.

These visualizations provide guidance for feature generation. For example, class Z and G seem to have less skewed data than class S. Additionally, each class has quite a different range of values, as we can see by inspecting the y-axis. This suggests that an amplitude feature could be useful. Also, it is not just the overall amplitude but the overall distribution of the points that seems characteristically different in the three classes. We will use these features as well as a few others in our analysis, and next we show the code to generate such features.

Visualization Meets Machine Learning

This EEG data would be a great opportunity to make use of a 2D histogram. In particular, a 2D histogram would form a handy composite image of each kind of time series and could confirm (or negate) the likely representativeness of the features we've just discussed, such as establishing whether amplitude or skew is a consistent difference between classes.

Also useful would be univariate histograms, forming one histogram per class. You could also consider going a step further to compute and plot a *kernel density estimate* for each class.

A kernel density estimate is like a histogram, but it can be shaped to have properties such as continuity and smoothness, effectively transforming raw histogram data to a smooth but nonparametric estimate of a possible distribution for the underlying variable. Wikipedia (*https://perma.cc/P3UT-UBCL*) offers a good overview of kernel density estimation, which can be accomplished by a variety of means, such as via scikit-learn (*https://perma.cc/DN7N-YD5R*) in Python or the density() function in the R stats package.

Here, we generate the features with `cesium`:

```python
## python
>>> from cesium import featurize.featurize_time_series as ft
>>> features_to_use = ["amplitude",
>>>                    "percent_beyond_1_std",
>>>                    "percent_close_to_median",
>>>                    "skew",
>>>                    "max_slope"]
>>> fset_cesium = ft(times          = eeg["times"],
>>>                  values         = eeg["measurements"],
>>>                  errors         = None,
>>>                  features_to_use = features_to_use,
>>>                  scheduler      = None)
```

This produces our features as depicted in Figure 9-2, a screenshot from a Jupyter notebook.

```
fset_cesium.head()
```

feature	amplitude	percent_beyond_1_std	percent_close_to_median	skew	max_slope
channel	0	0	0	0	0
0	143.5	0.327313	0.505004	0.032805	11107.796610
1	211.5	0.290212	0.640469	-0.092715	20653.559322
2	165.0	0.302660	0.515987	-0.004100	13537.627119
3	171.5	0.300952	0.541128	0.063678	17008.813559
4	170.0	0.305101	0.566268	0.142753	13016.949153

Figure 9-2. Numerical values of features we generated for the first few samples in our data set.

Note that many of these values are not normalized, so that would be something we'd want to keep in mind were we using a technique that assumed normalized inputs.

We should also confirm that we understand what our features are indicating and that our understanding matches what cesium computes. As an illustration of error checking and common sense affirmation, we can verify the percent_beyond_1_std for one time series sample:

```python
## python
>>> np.std(eeg_small["measurements"][0])
40.411
>>> np.mean(eeg_small["measurements"][0])
-4.132
>>> sample_ts = eeg_small["measurements"][0]
>>> sz = len(sample_ts)
>>> ll = -4.13 - 40.4
>>> ul = -4.13 + 40.4
>>> quals = [i for i in range(sz) if sample_ts[i] < ll or
                                     sample_ts[i] > ul  ]
>>> len(quals)/len(ser)
0.327 ## this checks out with feature generated in Figure 9-2
```

Features Should Be Ergodic

In choosing features to generate for a time series, be sure to select features that are *ergodic*, meaning that the values measured will each converge to a stable value as more data from the same process is collected. An example where this is not the case is a random walk, for which the measurement of the mean of the process is meaningless and will not be ergodic. The mean of a random walk will not converge to a specific value.

In the EEG data we've plotted, different subsamples from any given time series are clearly comparable and the series itself is weakly stationary, so the features we generated make sense.

You should be able to verify any of the features you are using. This is a simple matter of responsible analysis. You should not be providing information to your algorithm that you cannot understand, explain, and verify.

Don't overuse feature generation libraries. It's not difficult to write your own code to generate features. If you work in a domain where a particular set of features is generated often and in the same combinations repeatedly, you should write your own code even if you have initially used a package.

You will be able to optimize your code in ways that authors of a general exploratory package cannot. For example, if you have several features that rely on the calculation of the mean of a time series, you can create code to calculate that mean only once rather than once per separate feature calculated.

Decision Tree Methods

Tree-based methods mirror the way humans make decisions: one step at a time, and in a highly nonlinear fashion. They mirror the way we make complicated decisions: one step at a time, thinking about how one variable should affect our decision, and then another, much like a flow chart.

I assume you have already worked with decision trees or can quickly intuit what decision trees are. If you could use more support, pause here and check out some background reading (*https://perma.cc/G9AA-ANEN*).

Figure 9-3 shows a simple example of a decision tree that might be used to make an estimate of someone's body weight.

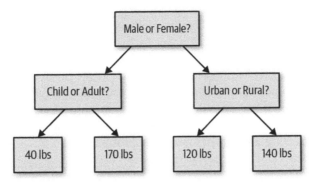

Figure 9-3. In this simple regression tree, a series of logical branches are used to arrive at a predicted weight for a human. This is a crude model, but it illustrates the nonlinear and variable approach that even a simple tree model can apply to a regression problem.

There are abundant examples of humans behaving like a decision tree when analyzing time series data. For example, a discretionary stock market trader may use technical indicators, but they will likely use them in a serial hierarchical fashion, just like a tree—first asking, for example, which direction the momentum is trending according to one technical indicator before then asking how volatility is evolving over time, with the answer to this second question interacting with the first in a tree-like nonlinear fashion. Likely they have something like a decision tree structure in their brain that they use to make forecasts about how the market will move.

Similarly, when a medical professional reads an EEG or ECG, it is not uncommon to first look for the presence of one feature before considering another, sequentially working through a series of factors. If one feature is present and another isn't, it will lead to a different diagnosis than in the converse case and hence a different forecast regarding a patient's prognosis.

We will use the features we generated from the EEG data as inputs to two different decision tree methods, random forests and gradient boosted trees, each of which can be used for classification tasks. Our task will be classifying the EEG data we've discussed solely on the basis of the features we generated from the raw data.

Random forest

A random forest is a model in which we use not one decision tree but many. Our classification or regression is the result of averaging the outputs of these trees. Random forests look to the "wisdom of the crowd" where the crowd is composed of many simple models, none of which may itself be particularly good, but all of which together often outperform a highly refined but single decision tree.

The idea of assembling a collection of models to produce a forecast rather than merely striving to find the single "best" model was articulated as early as 1969 in *The Combination of Forecasts*, a research paper by two venerable statisticians, J.M. Bates and C.W.J. Granger. That paper showed that combining two separate forecasts of airline passenger data could lead to a model that had lower mean squared error, a surprising and, at the time, unintuitive result. The younger generation of analysts, who often have worked their way into data analysis via machine learning rather than statistics, finds such an idea intuitive rather than unsettling, and the random forest has become a workhorse for all sorts of forecasting problems.

A random forest is constructed according to parameters specifying the number of trees to train, as well as the maximum allowed depth of those trees. Then, for each individual tree, a random sample of the data and of its features is used to train that tree. The trees are generally parameterized to be quite simple so that overfitting can be avoided and the model can average over many general models, none of which is especially good but all of which are sufficiently general to avoid "traps" in the data.

As mentioned earlier, we will input the features computed for each time series sample into the model as our training outputs. In theory, we could think of ways to input our raw time series data, rather than the computed features, but there are a number of problems with this:

- It would be complicated to deal with time series of unequal length.

- Such a large number of inputs (equal to or close to the number of time steps) would result in computationally expensive models and training.

- Assuming no particular time step is very important (since any given time step would correlate to one feature), there would be a great deal of noise and very little signal to train on from the perspective of the tree, which would be seeing each time step as an input.

So random forests are not a good tool for working with time series data in its raw form, but they can be useful for working with it once it has been compressed into summary features. Here are a few specific reasons:

- From an efficiency/computational resources perspective, it is wonderful to think that we can distill extremely long time series into a handful of features and find a model with reasonable accuracy.

- It is helpful that a random forest reduces the risk of overfitting. As we have discussed previously, overfitting is particularly a problem for time series analysis because of unfortunate synergies between overfitting and lookahead. Having a deliberately dumb/simple methodology combats some of this concern.

- Random forests may be particularly apt for time series data for which we do not have a working model or hypothesis regarding the underlying mechanics of the process.

As a general rule, analysts have had more success deploying random forests for cases of time series classification rather than for cases of time series forecasting. In contrast, the next method we discuss, gradient boosted trees, has been largely successful at both tasks.

Gradient boosted trees

Boosting is another way of building an ensemble of predictors. Boosting creates models sequentially with the idea that later models should correct the mistakes of earlier models and that data misfit by earlier models should be more heavily weighted by later models.

Gradient boosted trees have become the go-to boosting methodology, and a particularly successful one for time series as evidenced by their success in some data science competitions in recent years.

XGBoost works by building trees sequentially, with each tree seeking to predict the residuals of the combination of prior trees. So, for example, the first tree built by XGBoost will attempt to match the data directly (a category or a numeric value). The second tree will attempt to predict the true value minus the predicted value. The third tree will attempt to predict the true value minus the first tree's predicted value minus the second tree's prediction of the first tree's residuals.

However, XGBoost does not simply build models infinitely, attempting to minimize the residuals of the predicted residuals of the predicted residuals ad infinitum. The XGBoost algorithm minimizes a loss function that also includes a penalty term for model complexity, and this penalty term limits the number of trees that will be produced. It is also possible to directly limit the number of trees produced.

Over the last few years many have reported much greater success with XGBoost than with machine learning methods for time series, such as in Kaggle competitions or at industry machine learning conferences.

Bagging (or, more formally, *bootstrap aggregating*) refers to a technique for training models wherein randomly generated training sets are created for each different model in an ensemble. Random forests generally use a bagging methodology in model training.

Boosting, as noted, refers to a technique for training models in which an ensemble is composed of sequentially trained models, each of which focuses on correcting the error made by its predecessor. Boosting is at the core of how gradient boosted tree models are trained.

Code example

In the case of random forests and XGBoost, it can be easier to code a machine learning model than it is to understand how that model works. In this example, we will train both a random forest and a gradient boosted tree model to classify our EEG data based on the features we generated.

We use sklearn to divide our data into training and testing data sets:

```python
## python
>>> from sklearn.model_selection import train_test_split
>>> X_train, X_test, y_train, y_test = train_test_split(
        fset_cesium.values, eeg["classes"], random_state=21)
```

We begin with a random forest classifier. Here we see the ease with which we can create a model to classify our EEG data:

```python
## python
>>> from sklearn.ensemble import RandomForestClassifier
>>> rf_clf = RandomForestClassifier(n_estimators = 10,
>>>                                 max_depth    = 3,
>>>                                 random_state = 21)
>>> rf_clf.fit(X_train, y_train)
```

We are then able to determine the out-of-sample accuracy of our data with a method call on the Classifier object:

```python
## python
>>> rf_clf.score(X_test, y_test)
0.616
```

In just a few lines of code, we have a model that can do better than I would be able to do as a human classifier (one without a medical education). Remember, too, that thanks to feature selection this model sees only summary statistics rather than an entire EEG.

The code for the XGBoost classifier is similarly straightforward and succinct:

```python
## python
>>> import xgboost as xgb
>>> xgb_clf = xgb.XGBClassifier(n_estimators    = 10,
```

```
>>>                                max_depth    = 3,
>>>                                random_state = 21)
>>> xgb_clf.fit(X_train, y_train)
>>> xgb_clf.score(X_test, y_test)
0.648
```

We can see that the XGBoost classifier model does slightly better than the random forest model. It also trains slightly faster, as we can see with this quick experiment to calculate the training time for each model:

```
## python
>>> start = time.time()
>>> xgb_clf.fit(X_train, y_train)
>>> end = time.time()
>>> end - start
0.0189

## Random Forest
>>> start = time.time()
>>> rf_clf.fit(X_train, y_train)
>>> end = time.time()
>>> end - start
0.027
```

This execution speed is a substantial improvement, with the random forest taking 50% more time than XGBoost. While this is not a definitive test, it does point to an advantage of XGBoost, particularly if you are dealing with large data sets. You'd want to make sure that this advantage scaled up when you used larger data sets with more examples and more features.

This Is Not the Way to Do Performance Testing!

There's a lot to criticize about the preceding code. For starters, there are more compact ways to write a method to test performance. Also, you'd want to average the training time and run it on realistic data in realistic training conditions, not a toy example in a Jupyter notebook.

Most importantly, just because a commonly available algorithm runs slowly doesn't mean the algorithm is intrinsically slow. Sometimes algorithms in modules are written more for code transparency than for performance, and this can mean they run slower than is necessary. Other times the algorithms are implemented for the most general case possible, whereas you may find that your data needs are highly specialized in ways that mean you can modify standard implementations of algorithms to suit your purposes.

We will discuss performance considerations specific to time series in Chapter 12.

We could fairly ask whether there is something about our particular set of hyperparameters that gave the advantage to XGBoost over a random forest. For example, what if we use less complex trees by setting a lower depth? Or what if we allow fewer total decision trees to exist in the model? It's quite easy to test these possibilities, and again we see that XGBoost tends to maintain its edge.

For example, if we allow the same number of decision trees in the ensemble but lessen the complexity by decreasing the depth of the trees, we see that the gradient boosted model maintains a higher accuracy than does the random forest model:

```python
## python
>>> ## Test the same number of trees (10) but with less complexity
>>> ## (max_depth = 2)
>>>
>>> ## XGBoost
>>> xgb_clf = xgb.XGBClassifier(n_estimators = 10,
                                max_depth     = 2,
                                random_state = 21)
>>> xgb_clf.fit(X_train, y_train)
>>> xgb_clf.score(X_test, y_test)
0.616

>>> ## Random Forest
>>> rf_clf = RandomForestClassifier(n_estimators = 10,
                                    max_depth     = 2,
                                    random_state = 21)
>>> rf_clf.fit(X_train, y_train)
>>> rf_clf.score(X_test, y_test)
0.544
```

This is true even when we reduce the tree complexity further:

```
>>> ## Test the same number of trees (10) but with less complexity
>>> ## (max_depth = 1)

>>> ## XGBoost
>>> xgb_clf = xgb.XGBClassifier(n_estimators = 10,
                                max_depth     = 1,
                                random_state = 21)
>>> xgb_clf.fit(X_train, y_train)
>>> xgb_clf.score(X_test, y_test)
0.632

>>> ## Random Forest
>>> rf_clf = RandomForestClassifier(n_estimators = 10,
                                    max_depth     = 1,
                                    random_state = 21)
>>> rf_clf.fit(X_train, y_train)
>>> rf_clf.score(X_test, y_test)
0.376
```

There are several possible reasons to explain the performance and the putative advantage of gradient boosted trees over random forests. One important consideration is that we do not know for certain that all the features we selected for our classification are particularly useful. This highlights an example of when boosting (gradient boosted trees) could be preferable to bagging (random forest). Boosting will be more likely to ignore useless features because it will always make use of the full set of features and privilege the relevant ones, whereas some trees that result from bagging will be forced to use less meaningful features.

This also suggests the helpfulness of boosting when paired with the supercharged feature generation libraries we discussed in the last chapter. If you do take the approach of generating hundreds of time series features—far more than you can reasonably inspect—boosting may be a safeguard from truly disastrous results.

 Gradient boosted trees are particularly useful for large data sets, including large time series data sets. The implementation you choose will likely affect your accuracy slightly and your training speed. In addition to XGBoost you should also consider LightGBM and CatBoost. These latter packages have sometimes been reported to perform quite a bit faster than XGBoost, though sometimes at a slight decrease in out-of-sample test accuracy.

Classification versus regression

In the previous examples we considered random forests and gradient boosted tree methodologies for time series classification. These same methodologies can also be used for time series predictions.

Many statisticians argue that machine learning has been less successful—or no more successful—than traditional time series statistical analysis in the domain of forecasting. However, in the last several years, gradient boosted trees for prediction have taken off and are indeed often outperforming traditional statistical models when given sufficiently large data sets, in both forecasting competitions and in industry applications. In such cases, however, extensive time is taken to tune the models' parameters as well as to prepare time series features.

One of the strengths of gradient boosted tree models is that they approach "autopilot" in their ability to weed out irrelevant or noisy features and focus on the most important ones. This tendency alone, however, will not be enough to get state-of-the-art performance from a model. Even for a seemingly automatic method, such as gradient boosted trees, the outputs can only be as good as the inputs. The most important way to improve your model will still be to provide high-quality and well-tested input features.

There are many options for improving upon the current model. We could learn from it by using XGBoost's option to produce feature importance metrics. This could help us identify traits of useful features and nonuseful features, and we could then expand the data set by adding similar features to those that are already judged useful. We could also do a hyperparameter grid search to tune our model parameterization. Finally, we could look at our mislabeled data's raw time series to see whether there are traits of the mislabeled data that may not be represented by our current set of features. We could consider adding features that might better describe the mislabeled data, augmenting our inputs further.

Clustering

The general idea of clustering is that data points that are similar to one another constitute meaningful groups for purposes of analysis. This idea holds just as true for time series data as for other kinds of data.

As with our earlier discussion, I assume that you have some familiarity with the relevant machine learning concepts in a non-time-series context. If you are not familiar with clustering techniques, I recommend that you pursue some short background reading (*https://perma.cc/36EX-3QJU*) on clustering before continuing this section.

Clustering for time series can be used both for classification and for forecasting. In the case of classification, we can use clustering algorithms to identify the desired number of clusters during the training phase. We can then use these clusters to establish types of time series and recognize when new samples belong to a particular group.

In the case of forecasting, the application can be pure clustering or can be inspired by clustering in the form of using relevant distance metrics (more on distance metrics soon). There are a few options for generating forecasts at a horizon, h, in the future based on clustering and related techniques. Remember that in this case, we will not have fully observed a time series but only its first N steps, from which we want to forecast its value at time step $N + h$. In such a case there are a few options.

One option is to use class membership to generate a forecast based on typical behavior in that class. To do this, first determine which cluster a time series sample belongs to based on its first N time steps and then infer likely future behavior based on cluster membership. Specifically, look at how values of time series in this cluster tend to change their values between time step N and time step $N + h$. Note that you would want to perform the original clustering for all the time series based on their first N steps rather than all parts of the time series to avoid lookahead.

Another option is to predict future behavior of a sample time series based on the behavior of its nearest neighbor (or neighbors) in the sample space. In this scenario, based on metrics from the first N time steps, find a time series sample's nearest neighbor(s), for which the full trajectory is known. Then average the $N + h$ behavior of these nearest neighbors, and this is your forecast for the current sample.

In the case of both classification and forecasting, the most important consideration is how to assess similarity between time series. Clustering can be done with a variety of distance metrics, and much research has been devoted to thinking about how to measure distance in high-dimensional problems. For example, what is the "distance" between two job applicants? What is the "distance" between two blood samples? These challenges are already present in cross-sectional data, and they persist in time series data.

We have two broad classes of distance-metric options when applying clustering techniques to time series data:

Distance based on features
> Generate features for the time series and treat these as the coordinates for which to calculate data. This does not fully solve the problem of choosing a distance metric, but it reduces the problem to the same distance metric problem posed by any cross-sectional data set.

Distance based on the raw time series data
> Find a way to determine how "close" different time series are, preferably in a way that can handle different temporal scales, a different number of measurements, and other likely disparities between time series samples.

We will apply both of these distance metrics to a time series data set in which each sample represents the projection of a handwritten word from a 2D image to a 1D time series.

Generating Features from the Data

We have already discussed ways of generating and selecting features. Here we consider how to assess the distance between time series data sets based on the similarity of their features.

In an ideal scenario, we would already have culled unimportant or uninteresting time series, perhaps by using a tree to assess feature importance. We do not want to include such features in a distance calculation since they may falsely indicate dissimilarities between two time series, when in fact they simply do not indicate a relevant similarity relative to the classes in our classification task or the outcome in our forecasting task.

As we did with the EEG data set, we start our analysis by taking a look at some class examples and noticing what obvious differences there are over time and in the structure of the time series.

Our data is a subset of the FiftyWords (*https://oreil.ly/yadNp*) data set available from the UEA and UCR Time Series Classification Repository. This data set was released by the authors of a 2003 paper (*https://oreil.ly/01UJ8*) on clustering handwritten words in historical documents. In that paper, the authors developed "word profiles" as a way of mapping the 2D image of a handwritten word into a 1D curve, consisting of the same number of measurements regardless of word length. The repository data set is not identical to that in the paper, but the same principles apply. The purpose of the original paper was to develop a method of tagging all similar or identical words in a document with a single label so that humans could go back and label these words digitally (a feat that nowadays might very well be directly accomplished by a neural network, with 20 years of technological improvement to help with the task).

So in this example we see sample *projection profiles* of the words, where "projection" refers to the fact that they converted an image from a 2D space to a 1D space, and this latter space, since ordering mattered, is amenable to time series analysis. Note that the "time" axis is not actually time but rather left-to-right progressions of written words. Nonetheless the concept is the same—ordered and equally spaced data—so for simplicity I will use the words *time* and *temporal* in the analysis even though this is not strictly true. For our use case, there isn't a distinction.

In Figure 9-4 we see examples of a few distinct words.[1]

When we examine this data, and particularly when we consider the patterns obvious to the human eye in the plots, just as with the EEG we can formulate a starting point of some features to consider for our analysis—for example, the heights and locations of the peaks as well as their characteristics, such as how sharply they rise and what shape there is at the top of the peak.

Many of the features I am describing begin to sound more like image recognition features than like a time series, and this is a helpful perspective to have for feature generation. After all, visual data is often data we can easily process and have more intuition about than time series data. It could be helpful to think in terms of images when thinking about features. It is also a perspective that illustrates why generating features can be surprisingly difficult. In some cases it may be obvious from examining a time series how to distinguish two classes, but we may find that how to write the code is not so obvious. Or we may find that we can write the code, but it is extremely taxing.

---

[1] Note that information about the actual content of each "word" is not available and was not especially relevant when the original data set was put together. The idea was to recognize that all words of one label were the same so as to cut down on the human effort needed to label documents.

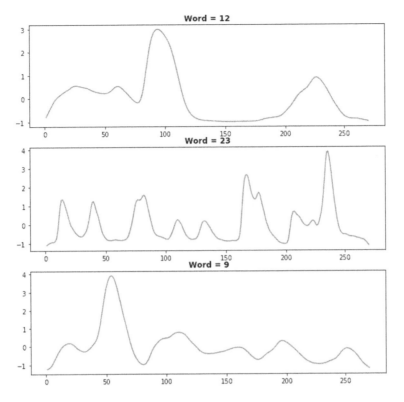

Figure 9-4. The projection profiles of three different words (12, 23, and 9) are quite different from one another. We also already see some features that could distinguish these words from one another: the temporal location (x-axis) of the largest peak or even of the second-largest peak, how many local peaks there are, the overall range of values, and the average convexity of the curves.

In the last decade, deep learning has emerged as the strongest performer for image classification, and with sufficient data, we could train a deep learning classifier on the images of these plots (more on this in Chapter 10). For now, it will be helpful for us to think of ways around the difficulty of programming. For example, it would be difficult to generate features locating each peak, as peak finding is programmatically demanding and something of an art form.

We can also use a 1D histogram, either of all the class examples or of an individual example. This may suggest computationally less taxing ways either of identifying peaks or finding other proxy values that will map onto the overall shapes we see in the time series. Here we plot the same individual class members plotted before, now accompanied by their 1D histograms (see Figure 9-5):

```python
## python
>>> plt.subplot(3, 2, 1)
>>> plt.plot(words.iloc[1, 1:-1])
>>> plt.title("Word = " + str(words.word[1]), fontweight = 'bold')
>>> plt.subplot(3, 2, 2)
>>> plt.hist(words.iloc[1, 1:-1], 10)
>>> plt.subplot(3, 2, 3)
>>> plt.plot(words.iloc[3, 1:-1])
>>> plt.title("Word = " + str(words.word[3]), fontweight = 'bold')
>>> plt.subplot(3, 2, 4)
>>> plt.hist(words.iloc[3, 1:-1], 10)
>>> plt.subplot(3, 2, 5)
>>> plt.plot(words.iloc[5, 1:-1])
>>> plt.title("Word = " + str(words.word[11]), fontweight = 'bold')
>>> plt.subplot(3, 2, 6)
>>> plt.hist(words.iloc[5, 1:-1], 10)
```

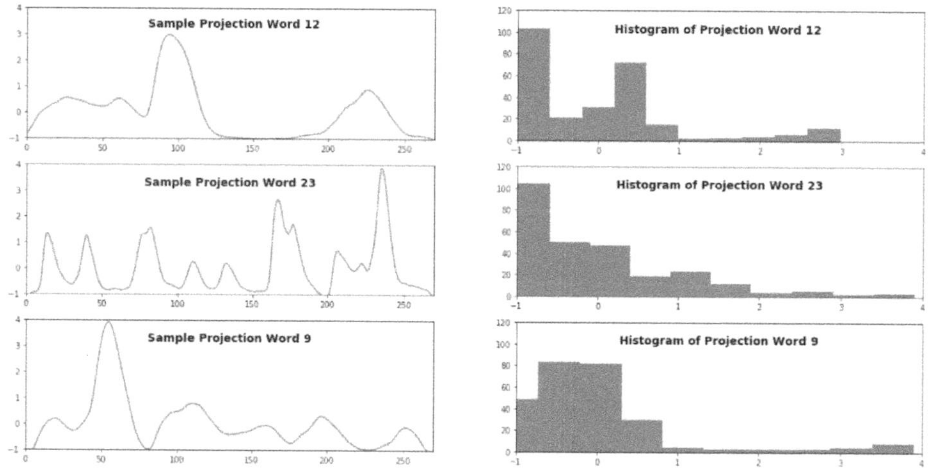

Figure 9-5. Another way of measuring classes to brainstorm useful features. In particular, the histogram of individual class examples indicates that attributes of the histogram, such as the number of local peaks, the skewness, and the kurtosis would be helpful and possibly good proxies for some of the attributes of the time series curves themselves that are obvious to the human eye but not easy to identify with code.

We also want to make sure that the examples we are looking at are not outliers compared to the other examples for these words. For this reason, we construct the 2D histogram for two words to get an idea of the individual variation (see Figure 9-6):

```python
## python
>>> x = np.array([])
>>> y = np.array([])
>>>
>>> w = 12
>>> selected_words = words[words.word == w]
```

```
>>> selected_words.shape
>>>
>>> for idx, row in selected_words.iterrows():
>>>     y = np.hstack([y, row[1:271]])
>>>     x = np.hstack([x, np.array(range(270))])
>>>
>>> fig, ax = plt.subplots()
```

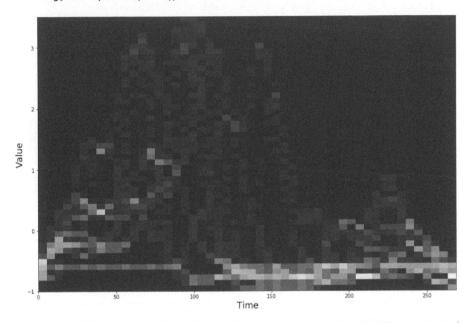

Figure 9-6. A 2D histogram of the 1D word projections for word = 12. The y-axis is the value at a given time step, and the x-axis represents the 270 time steps for each time series sample/word projection.

Figure 9-6 shows the 2D histogram for all the word = 12 members of the data set. While the individual curve in Figure 9-5 suggested that we focus on finding the two large peaks that seemed to dominate the time series, here we see that what most of the members of this class have in common is likely the flat span between these peaks, which appears to go from around time steps 120 through 200, based on the intensity of the points in that region.

We can also use this 2D histogram to establish a cutoff point for the maximum peak for this class, which seems to vary in location between time steps 50 and 150. We may even want to code up a feature that is as specific as "is the maximum value reached between points 50 and 150?"

We plot another 2D histogram for the same reason, this time choosing word class 23, which has many small bumps in the example we plotted in Figure 9-5, a difficult-to-quantize feature (see Figure 9-7).

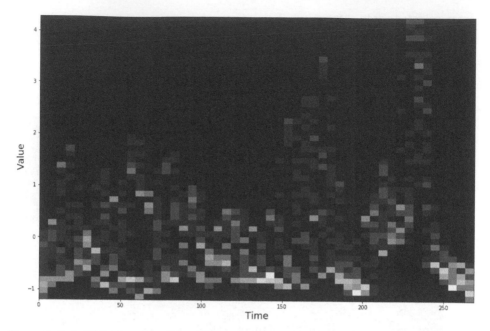

Figure 9-7. A 2D histogram of the 1D word projections for word = 23. The y-axis is the value at a given time step, and the x-axis represents the 270 time steps for each time series sample/word projection.

It is not surprising to see a particularly "smeared" histogram for word class 23 in Figure 9-7, given that even the example we plotted in Figure 9-5 showed so many features that we would expect to see a lot of smearing in the 2D histogram if the features did not match up exactly between samples. However, we also see that the maximum point value in this class comes at a nonoverlapping range of time steps here, as compared to word class 12. For this class, the maximum value comes after 150, which makes sense given that the two largest peaks we saw in the word class 23 example were in this range. The 2D histogram tends to substantiate that the earlier peak is not as tall as the later peak, suggesting other ways to quantize the shape of this time series to distinguish it from other classes.

The 2D histograms are helpful in letting us know the variability of a feature within an individual class, so that we do not unduly rely on a single class example when thinking about how to form our features.

In this case we choose to generate a set of features derived from the shape of the word projection, and an additional set of features derived from the shape of the word projection's histogram (projecting a 1D summary into a different 1D summary, from which we generate features). This is in response to the large "smears" we see in the 2D histograms, indicating that there are peaks but that their location is not particularly stable. Using a histogram to generate a second characteristic shape for each word projection may prove more reliable and characteristic than the word projection itself. Histograms characterize what kinds of values appear in the series without characterizing their location in the series, which is what is important to us given that the peaks in the projections don't have especially stable temporal locations within the time series.

First, we generate features for the times series, which are 270 time steps in length apiece. In this case, we shorten the name of the function used to generate features for code readability:

```python
from cesium import featurize.featurize_time as ft

## python
>>> word_vals     = words.iloc[:, 1:271]
>>> times         = []
>>> word_values   = []
>>> for idx, row in word_vals.iterrows():
>>>     word_values.append(row.values)
>>>     times.append(np.array([i for i in range(row.values.shape[0])]))
>>>
>>> features_to_use = ['amplitude',
>>>                    'percent_beyond_1_std',
>>>                    'percent_close_to_median']
>>> featurized_words = ft(times          = times,
>>>                       values         = word_values,
>>>                       errors         = None,
>>>                       features_to_use = features_to_use,
>>>                       scheduler      = None)
```

Next we generate histograms and use these as another time series for which to generate features:[2]

```python
## python
>>> ## create some features derived from histogram
>>> times = []
>>> hist_values = []
>>> for idx, row in words_features.iterrows():
>>>     hist_values.append(np.histogram(row.values,
>>>                                     bins=10,
>>>                                     range=(-2.5, 5.0))[0] + .0001)
>>>                                     ## 0s cause downstream problems
>>>     times.append(np.array([i for i in range(9)]))
>>>
>>> features_to_use = ["amplitude",
>>>                    "percent_close_to_median",
```

---

2 Just like the word projections themselves, for which the x-axis is not actually time but another ordered, evenly spaced axis that may as well be time, the same is true for the histograms. We can think of their x-axis as time for purposes of our analysis, such as generating features.

```
>>>                    "skew"
>>>                ]
>>>
>>> featurized_hists = ft(times          = times,
>>>                       values         = hist_values,
>>>                       errors         = None,
>>>                       features_to_use = features_to_use,
>>>                       scheduler      = None)
```

We made sure that all histograms use the same number of bins and the same range of values as the basis for the bins, via the parameters we pass to `np.histogram()`. This ensures that all the histograms are directly comparable, having the same range of bin values, which will be the "temporal" axis when these histograms are run through the time series feature generation. If we did not enforce this consistency, the features generated would not necessarily be meaningful in comparing one histogram to another.

Finally, we combine these two sources of features:

```
## python
>>> features = pd.concat([featurized_words.reset_index(drop=True),
>>>                       featurized_hists],
>>>                       axis=1)
```

Temporally Aware Distance Metrics

When running clustering analysis we have to choose a distance metric. With time series features, as we just did, we can apply a variety of standard distance metrics to them, as is done in standard clustering analysis on cross-sectional data. If you are not familiar with the process of choosing a distance metric in such cases, I recommend a short detour to do some background reading (*https://perma.cc/MHL9-2Y8A*).

In this section we will focus on the problem of measuring similarity between time series by defining a distance metric between them. One of the most well-known examples of such metrics is dynamic time warping (DTW). DTW is apt for clustering a time series whose most salient feature is its overall shape, as is the case with our word projection data.

The technique's name is inspired by the methodology, which relies on temporal "warping" to align time series along their temporal axis so as to compare their shapes. A picture is far more valuable than words for conveying the concept of dynamic time warping, so take a look at Figure 9-8. The temporal (x) axis is warped—that is, expanded or contracted as convenient—to find the best alignment of points between the two curves depicted (i.e., two time series), in order to compare their shape.

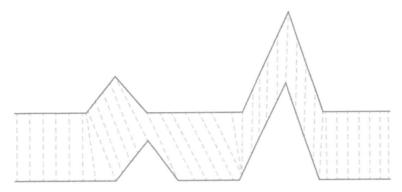

Figure 9-8. How dynamic time warping works. Each point on one time series is mapped to a point on the opposite time series, but there is no requirement that there must be a one-to-one mapping of points. This has a few implications: (1) Time series need not be of the same length or even of the same timescale. What matters is the shape. (2) Time does not always move forward during the fitting process, and may not move at the same pace for each time series. By time moving, I mean progressing along the curve in the x-axis direction. Source: Wikipedia (https://perma.cc/F9ER-RTDS).

Note that the actual time values on the time axis of one curve as compared to another are not relevant in the standard form of this algorithm. We could be comparing one time series that is measured in nanoseconds to another that is measured in millennia (although this would likely not be a sensible exercise). The purpose of the algorithm is akin to comparing the visual "shape" of that algorithm rather than thinking about how much time is passing. Indeed, "time" is really only meant in the more general sense of an ordered evenly spaced set of points along the x-axis rather than time proper.

The rules of DTW are as follows:

- Every point in one time series must be matched with at least one point of the other time series.

- The first and last indices of each time series must be matched with their counterparts in the other time series.

- The mapping of points must be such that time moves forward and not backward. There is no going back in time by matching a point in one time series with a point in the other time series that has already been passed on the time axis. However, time does not need to move forward constantly. For example, two consecutive time steps in the original series could be warped by being condensed to the same place on the x-axis during the fit, as is illustrated in Figure 9-8 at the first "kink" in the upper curve/solid line.

There are many ways that the temporal alignment can be adjusted to follow these rules, but the selected match is the one that minimizes the distance between the curves. This distance, or cost function, is often measured as the sum of absolute differences between matched points, where the absolute difference is the difference between the points' values.

Now that we have an idea of how DTW works intuitively, we can take a look at the code:

```python
## python
>>> def distDTW(ts1, ts2):
>>>     ## this is setup
>>>     DTW={}
>>>     for i in range(len(ts1)):
>>>         DTW[(i, -1)] = np.inf
>>>     for i in range(len(ts2)):
>>>         DTW[(-1, i)] = np.inf
>>>     DTW[(-1, -1)] = 0
>>>
>>>     ## this is where we actually calculate the optimum
>>>     ## one step at at time
>>>     for i in range(len(ts1)):
>>>         for j in range(len(ts2)):
>>>             dist = (ts1[i] - ts2[j])**2
>>>             DTW[(i, j)] = dist + min(DTW[(i-1, j)],
>>>                                      DTW[(i, j-1)],
>>>                                      DTW[(i-1, j-1)])
>>>             ## this is an example of dynamic programming
>>>
>>>     ## once we have found the complete path, we return
>>>     ## the associated distance
>>>     return sqrt(DTW[len(ts1)-1, len(ts2)-1])
```

As indicated in the comments, the solution to this problem is an example of dynamic programming, and DTW distance is a classic dynamic programming problem. We can take one step at a time on the path from the beginning of each time series to the end, and we know that we can build on the solution one step at a time and refer back to our earlier knowledge to make later decisions.

There are a number of different DTW implementations, with a variety of ideas for making the search for the optimum solution, or a near-optimum solution, more efficient. These should be explored, particularly if you are working with a larger data set.

There are also other ways of measuring distances between time series. Here are a few:

Fréchet distance

This is the maximum distance between two curves during a time-warping-like traversal of the curves that always seeks to minimize the distance between two curves. This distance metric is often explained by the analogy of a dog and its human companion walking on the two curves with a leash between them. They each need to traverse a separate curve from beginning to end, and they can go at different speeds and vary their speeds along the curve so long as they always move in the same direction. The Fréchet distance is the shortest length of leash necessary for them to complete the task following the optimal trajectory (assuming they can find it!).

Pearson correlation

The correlation between two time series can be a way of measuring the distance between them. Unlike with other distance metrics, you will minimize the distance between the time series by maximizing the metric of the correlation. Correlation is relatively easy to compute. However, this method requires that the time series have the same number of data points or that one be downsampled to match the fewer data points of the other. The time complexity of computing correlation is $O(n)$, which makes it a particularly efficient metric from a computational resources perspective.

Longest common subsequence

This distance measure is appropriate for time series that represent a sequence of categorical or integral values. In such cases, to consider the similarity of two time series, we can determine the length of the longest common subsequence, meaning the longest length of consecutive values that are exactly identical, although their exact location in the time series is not required to match. As with DTW, this means that we are more concerned with finding a shape of commonality rather than where in time the common shape is occurring. Also note that like DTW, but unlike Pearson correlation, this does not require that the time series have the same length. A related measure is *edit distance*, whereby we find the number of changes we would need to make to one time series to make it identical to another and use this value to define a distance metric.

Distance Versus Similarity

The literature on measuring distances between time series also uses the term similarity to describe these metrics. In most cases, you can treat these terms interchangeably, namely as a way to establish which time series are more or less like one another. That said, some metrics will be proper distances, such as the Fréchet distance, which can be computed with proper units (such as "feet" or "kg/dollar" or whatever metric a time series is measuring). Other measures are unit-free, such as a correlation.

Sometimes a little creativity can go a long way toward finding a simple but apt solution, so it is always a good idea to consider exactly what your needs are and define them as specifically as possible. Consider a Stack Overflow post (*https://perma.cc/389W-68AH*) seeking a distance metric for a specific application, namely classifying time series to match them to one of three centroids from a prior clustering analysis. The three classes were:

- A flat line.
- A peak at the beginning of the time series, and otherwise a flat line.
- A peak at the end of the time series, and otherwise a flat line. The user found that several standard distance metrics, including Euclidean distance and DTW, failed to do the trick. In this case, DTW was too generous and rated any time series with a peak as equally close to the time series with a peak at the end and a peak at the beginning (so DTW is not a panacea, despite being computationally taxing!).

In this case, a clever commenter suggested a transform that would make distance metrics work better, namely to compare the cumulative summed time series rather than the original time series. After this transform, both the Euclidean and DTW distances gave the correct ordering such that a time series with a peak at the beginning showed the least distance to the prototype of that class, rather than an equal distance to the prototype with a peak at the beginning of the time series and the prototype with a peak at the end of the time series. This should remind us of earlier analyses we have studied, in which transforming a time series can make ARIMA appropriate even if the raw data does not meet the necessary conditions.

Avoid Euclidean Distance Measures for Time Series

You may have noticed that Euclidean distance was not mentioned in our discussion of time series similarity measures. Euclidean distances tend to perform poorly in assessing the similarity between time series in a way that we care about. To see why, let's compare two sine curves and a flat line. If the sine curves have the same period but have a different phase (displacement along the time/x-axis) or if the sine curves have sufficiently different amplitudes, the straight line will have a shorter Euclidean distance to the sine curve than will the other sine curve. This is usually not the desired outcome but shows a few advantages offered by some of the previous distance metrics:

- Ability to compensate for phase—that is, displacement along the time axis that isn't really important for comparison
- Ability to recognize similarity of shapes rather than similarity of magnitudes

This is an easy exercise to code up, so give it a try.

Should you decide that a Euclidean distance measure is appropriate, consider using a Fourier transform distance rather than a Euclidean distance. In this case you can use dimension reduction by having a frequency ceiling, which is appropriate for most time series data where higher frequencies tend to be less important to the overall

---

shape and dynamics of a time series. Another option is to use the Symbolic Aggregate Approximation (SAX) technique (first developed in 2007) to reduce the dimensionality of the time series data before computing a lower bound on the Euclidean distance.

Unfortunately, there is no "autopilot" for choosing a distance metric. You will need to use your best judgment to find a balance of:

- Minimizing the use of computational resources.
- Choosing a metric that emphasizes the features of a time series most relevant to your ultimate goal.
- Making sure that your distance metric reflects the assumptions and strengths/ weaknesses of the analytical methods you are pairing it with. For example, *k*-means clustering does not use pairwise distances but rather minimizes variances, such that only Euclidean-like distances make sense for this technique.

Clustering Code

Now that we have discussed how to generate features for clustering analysis and how to measure distance directly between time series as a distance metric for clustering, we will perform the clustering with our selected features and with our pairwise DTW distance matrix to compare the results.

Hierarchical clustering of normalized features

We computed features for our words-as-time-series for both the time series of the original recording and the histogram of the time series. These features may occur on quite different scales, so if we want to apply a single distance metric to them, we normalize them as is standard operating procedure for feature-based clustering generally:

```python
## python
>>> from sklearn import preprocessing
>>> feature_values = preprocessing.scale(features.values)
```

We select a hierarchical clustering algorithm and perform a fit for 50 clusters, since we are seeking to match these clusters to the 50 words in our data set:

```python
## python
>>> from sklearn.cluster import AgglomerativeClustering
>>> feature_clustering = AgglomerativeClustering(n_clusters = 50,
>>>                                               linkage   = 'ward')
>>> feature_clustering.fit(feature_values)
>>> words['feature_labels'] = feature_clustering.fit_predict(p)
```

Then we want to see whether the clusters (whose labels are arbitrary with respect to the original word labels) show useful correspondences to the word labels:

```python
## python
>>> from sklearn.metrics.cluster import homogeneity_score
>>> homogeneity_score(words.word, words.feature_labels)
0.508
```

We are lucky that we are working with labeled data; we might otherwise draw incorrect conclusions based on the clusters we formed. In this case fewer than half of the clusters relate strongly to a single word. If we go back and think about how to improve this result, we have a number of options:

- We only used six features. This is not a large number of features, so we could add more.

- We could look for features that would be relatively uncorrelated, which we did not do here.

- We are still missing obviously useful features. There were some features we noticed from the visual exploration of the data that we did not include, such as the number and location of distinctive peaks. We should probably rework this analysis to include that information or some better proxy for it.

- We should explore using other distance metrics, perhaps some that will weight certain features more strongly than others, privileging the features the human eye finds useful.

Hierarchical clustering with the DTW distance matrix

We have already completed the difficult portion of direct clustering based on time series clustering by calculating the pairwise distance matrix via DTW. This is computationally taxing, which is why we are careful to save the results in case we want to revisit the analysis:

```python
## python
>>> p = pairwise_distances(X, metric = distDTW)
>>> ## this takes some time to calculate so worth saving for reuse
>>> with open("pairwise_word_distances.npy", "wb") as f:
        np.save(f, p)
```

Now that we have them, we can use a hierarchical clustering algorithm:

```python
## python
>>> from sklearn.cluster import AgglomerativeClustering
>>> dtw_clustering = AgglomerativeClustering(linkage   = 'average',
>>>                                          n_clusters = 50,
>>>                                          affinity  = 'precomputed')
>>> words['dtw_labels'] = dtw_clustering.fit_predict(p)
```

And finally, as before, we compare the correspondence between the fitted clusters and the known labels:

```python
## python
>>> from sklearn.metrics.cluster import homogeneity_score,
>>>                                     completeness_score
>>> homogeneity_score(words.word, words.dtw_labels)
0.828
>>> completeness_score(words.word, words.dtw_labels)
0.923
```

We see that this DTW-based clustering does substantially better than our feature-based clustering. However, if you run the DTW distance computation code on your own computer—particularly if it's a standard-issue laptop—you will see just how much longer the DTW takes to compute compared to the features we chose. It's likely that we can improve our feature-based clustering, whereas the DTW distance clustering, now that it's computed, does not leave any straightforward means for improvement. Our alternatives to improving this would be:

- Include features as well as the DTW distance. This is tricky both from a coding perspective and also from the conceptual perspective of deciding how to combine the features with the DTW distance.

- Try other distance metrics. As discussed earlier, the appropriate distance metric will depend on your data, your goals, and your downstream analysis. We would need to define our goal for this word analysis more narrowly and geometrically, and then we could think about whether DTW is really the best metric for what we want to accomplish.

More Resources

- On time series distance and similarity measures:

Meinard Müller, "Dynamic Time Warping," in Information Retrieval for Music and Motion (Berlin: Springer, 2007), 69–84, https://perma.cc/R24Q-UR84.
    This chapter of Müller's book offers an extensive overview of dynamic time warping, including a discussion of common approximations made to reduce the computational complexity of computing DTW.

Stéphane Pelletier, "Computing the Fréchet Distance Between Two Polygonal Curves," (lecture notes, Computational Geometry, McGill University, 2002), https://perma.cc/5QER-Z89V.
    This set of lecture notes from McGill University offers an intuitive visual and algorithmic explanation of what the Fréchet distance is and how it can be calculated.

Pjotr Roelofsen, "Time Series Clustering," master's thesis, Business Analytics, Vrije Universiteit Amsterdam, 2018, https://perma.cc/K8HJ-7FFE.

This master's thesis on time series clustering begins with a helpful and very thorough discussion of the mainstream techniques for calculating distances between time series, including information about the computational complexity of the distance calculation and helpful illustrations that assist with building intuition.

Joan Serrà and Josep Ll. Arcos, "An Empirical Evaluation of Similarity Measures for Time Series Classification," Knowledge-Based Systems 67 (2014): 305–14, https://perma.cc/G2J4-TNMX.

This article offers empirical analysis of out-of-sample testing accuracy for classification models built with seven different measures of time series similarity: Euclidean distance, Fourier coefficients, AR models, DTW, edit distance, time-warped edit distance, and minimum jump costs dissimilarity. The authors tested these measures on 45 publicly available data sets from the UCR time series repository.

- On machine learning for time series:

Keogh Eamonn, "Introduction to Time Series Data Mining," slideshow tutorial, n.d., https://perma.cc/ZM9L-NW7J.

This series of slides gives an overview to preprocessing time series data for machine learning purposes, measuring distance between time series, and identifying "motifs" that can be used for analysis and comparison.

Spyros Makridakis, Evangelos Spiliotis, and Vassilios Assimakopoulos, "The M4 Competition: Results, Findings, Conclusion and Way Forward," International Journal of Forecasting 34, no. 4 (2018): 802–8, https://perma.cc/42HZ-YVUU.

This article summarizes results of the M4 competition in 2018, which compared a variety of time series forecasting techniques, including many ensemble techniques, on a randomly selected set of 100,000 time series, including data collected at various frequencies (yearly, hourly, etc.). In this overview of the competition results, the authors indicate that a few "hybrid" approaches, relying heavily on statistics but also with some machine learning components, took both first and second place in the competition. Results such as these point to the importance of understanding and deploying both statistical and machine learning approaches to forecasting.

## CHAPTER 10

Deep Learning for Time Series

Deep learning for time series is a relatively new endeavor, but it's a promising one. Because deep learning is a highly flexible technique, it can be advantageous for time series analysis. Most promisingly, it offers the possibility of modeling highly complex and nonlinear temporal behavior without having to guess at functional forms—which could potentially be a game changer for nonstatistical forecasting techniques.

If you aren't familiar with deep learning, here's a one-paragraph summary (we'll go into more details later). Deep learning describes a branch of machine learning in which a "graph" is built that connects input nodes to a complicated structure of nodes and edges. In passing from one node to another via an edge, a value is multiplied by that edge's weight and then, usually, passed through some kind of nonlinear activation function. It is this nonlinear activation function that makes deep learning so interesting: it enables us to fit highly complex, nonlinear data, something that had not been very successfully done previously.

Deep learning has come into its own primarily within the past 10 years, as improvements in commercially available hardware have been coupled with massive amounts of data to enable this kind of heavy-duty model fitting. Deep learning models can have millions of parameters, so one way of understanding them is to dream up just about any graph you can think of, with all sorts of matrix multiplications and nonlinear transforms, and then imagine setting loose a smart optimizer that optimizes your model one small group of data at a time, continuously adjusting the weights of this large model so they give increasingly good outputs. This is deep learning in a nutshell.

Deep learning has not yet delivered the amazing results for forecasting that it has for other areas, such as image processing and natural language processing. However, there is good reason to be optimistic that deep learning will eventually improve the art of forecasting while also lessening the brittle and highly uniform nature of assumptions and technical requirements common for traditional forecasting models.

Many of the headaches of preprocessing data to fit a model's assumptions are gone when deep learning models are used:

- There is no requirement of stationarity.
- There is no need to develop the art and skill of picking parameters, such as assessing seasonality and order of a seasonal ARIMA model.
- There is no need to develop a hypothesis about the underlying dynamics of a system, as is helpful with state space modeling.

These advantages should sound familiar after Chapter 9, where we discussed many of the same advantages with respect to applying machine learning to time series. Deep learning is even more flexible for a number of reasons:

- Many machine learning algorithms tend to be fairly brittle in terms of the dimensionality and kinds of input data required for a training algorithm to work. In contrast, deep learning is highly flexible as to the model and the nature of the inputs.
- Heterogenous data can be challenging with many commonly applied machine learning techniques, whereas it is quite commonly used for deep learning models.
- Machine learning models rarely are developed for time series problems, whereas deep learning offers a great deal of flexibility to develop architectures specific to temporal data.

However, deep learning is not a magic bullet. Although there is no requirement of stationarity for deep learning applied to time series, in practice, deep learning does not do a good job of fitting data with a trend unless standard architectures are modified to fit the trend. So we still need to preprocess our data or our technique.

Also, deep learning does best with numerical inputs in different channels all scaled to similar values between −1 and 1. This means that you will need to preprocess your data even though it's not theoretically required. Also, you need to preprocess it in a way that avoids lookahead, which is not something the deep learning community as a whole has spent a lot of time perfecting.

Finally, deep learning optimization techniques and modeling for time-oriented neural networks (the largest class of which is recurrent neural networks, or RNNs) are not as well developed as those for image processing (the largest class of which is convolutional neural networks, or CNNs). This means you will find less guidance on best practices and rules of thumb for selecting and training an architecture than you would for nontemporal tasks.

In addition to these difficulties of applying deep learning to time series, you will find the rewards for doing so are a mixed bag. For starters, deep learning's performance for time series does not consistently outperform more traditional methods for time series forecasting and classification. Indeed, forecasting is an area ripe for improvements from deep learning, but as of this writing these improvements have not yet materialized.

Nonetheless, there is reason to expect both immediate and long-term benefits to adding deep learning to your time series analysis toolkit. First, large tech companies have begun launching deep learning for time series services with custom architectures they have developed in-house, often with industry-specific modeling tasks in mind. You can use these services or combine them with your own analyses to get good performance.

Second, you may have a data set that does amazingly well with deep learning for time series. In general, the stronger your signal-to-noise ratio is, the better you will do. I have had more than one fairly new programmer tell me that they succeeded to a surprising degree with some simple deep learning.

For example, a college found that a simple LSTM (more on this later) did as well as overworked guidance counselors at predicting which students were likely to fail or drop out soon, so that the college could reach out to those students and provide more resources. While the best outcome would be to have more guidance counselors, it is heartwarming to know that a simple LSTM applied to the heterogeneous time series data of a student's grades and attendance record could help change vulnerable students by flagging them for outreach and enhanced support. In the future, I believe we can expect more of such innovative and assistive uses of deep learning for time series.

Remeber that every model has assumptions. Machine learning models, including neural networks, invariably have assumptions built in, both in the architecture and in training methodologies. Even the fact that most neural networks work best with inputs scaled to [–1, 1] suggests there are strong assumptions built into the model, even if these have not been well identified yet.

It may even be that neural network forecasts haven't reached their optimum performance yet, and that better understanding the theoretical underpinnings of and requirements for forecasting neural networks will lead to performance enhancements.

If you are new to deep learning, this chapter will not provide you with all the conceptual and programming equipment you need to get a competent start. However, it will give you a point of departure, and from here, there are many good tutorials, books, and even online courses that you can use to learn more in depth. The good news about deep learning is that there need be no mathematics involved to get a general flavor of how it works and how it is programmed. Additionally, APIs are available at various levels of specificity. This means that a beginner can get by using fairly high-level APIs to try out some introductory techniques, and frankly even experts will use these high-level APIs to save time. Later, when some specific architectural innovation is necessary, and as your understanding develops, you can use lower-level APIs where more is left for you to decide and specify concretely.

This chapter will provide a brief review of the concepts that inspired and supported deep learning as a mathematical and computer science pursuit, as well as concrete examples of the code you can use to apply a deep learning model to data.

Deep Learning Concepts

Deep learning has roots in a number of areas. There was a biological inspiration, with computer scientists and quantitative analysts wondering if the way to build intelligent machines was, ultimately, to mimic brains, with their networks of neurons firing off in response to certain triggers. There was mathematical inspiration in the form of the universal approximation theorem being proven for various activations, with many of these proofs originating in the late 1980s and early 1990s. Finally, there was the growth of computing capabilities and availability coupled with a burgeoning field of machine learning, whose success showed that, with enough data and parameters, complicated systems could be modeled and predicted. Deep learning took all these ideas even further by creating networks described by millions of parameters, trained on large data sets, and with a theoretical grounding that a neural network should be able to represent an arbitrary and nonlinear function to a large degree of accuracy. Figure 10-1 shows a simple neural network, a multilevel *perceptron* (or fully connected network).

In Figure 10-1, we can see how multichannel inputs are served to the model in the form of a vector of dimension d. The nodes represent the input values, while the edges represent the multipliers. All the edges entering into a node represent a previous value multiplied by the value of the edge it traversed. These values from all the inputs at a single node are summed and then, usually, passed through a nonlinear activation function, creating a nonlinearity.

We can see that the input consists of three channels, or a vector of length 3. There are four hidden units. We multiply each of the three inputs by a different weight for each of the four hidden units for which it is destined, meaning that we need $3 \times 4 = 12$ weights to fully describe the problem. Also, since we will then sum the results of these various multiplications, matrix multiplication is not just *analogous* to what we are doing but *exactly* what we are doing.

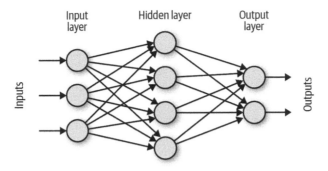

Figure 10-1. A simple feed forward network.

If we wanted to write out the steps for what we are doing, they would go something like this:

1. Input vector X has three elements. Layer 1 weights are designated by W_1, a 4×3 matrix, such that we compute the hidden layer values as $W_1 \times X_1$. This results in a 4×1 matrix, but this is not actually the output of the hidden layer:

    $$W_1 \times X_1$$

2. We need to apply a nonlinearity, which we can do with various "activation functions," such as hyperbolic tan (tanh) or the sigmoid function (σ). We will usually also apply a bias, B1, inside the activation function, such that the output of the hidden layer is really:

    $$H = a(W_1 \times X_1 + B_1)$$

3. In the neural network depicted in Figure 10-1, we have two outputs to predict. For this reason we need to convert the four-dimensional hidden state output to two outputs. Traditionally, the last layer does not include a nonlinear activation function, unless we count applying a softmax in the case of a categorization problem. Let's assume we are just trying to predict two numbers, not two probabilities or categories, so we simply apply a last "dense layer" to combine the four outputs of the hidden layer per ultimate output. This dense layer will combine four inputs into two outputs, so we need a 2×4 matrix, W_2:

    $$Y = W_2 \times H$$

Hopefully this gave you some sense of how these models work. The overall idea is to have lots of parameters and opportunities for nonlinearities.

It's something of an art form to choose the right number and form of parameters, the right training hyperparameters, and, of course, a reasonably accessible problem. The magic comes in learning how to train these parameters, initialize them in a smart way from the start, and make sure the model goes in the right direction toward a reasonably good solution. This is a nonconvex class of models, and the intention is not about finding the global optimum. Instead, the thinking goes, so long as you find clever ways to regularize the model, you will find a "good enough" local optimum to suit your needs.

Programming a Neural Network

Understanding how neural networks work in principle can feel quite a bit easier than understanding how the associated programming frameworks applied to this problem actually operate. As we'll discuss here, however, there are a few broad themes that these frameworks tend to have in common.

Symbolic and Imperative Programming Styles

You will have some leeway in your deep learning programming to use either a symbolic or an imperative style of programming. In a *symbolic* style of programming, you declare all the relationships up front without having them computed at the time of declaration. In contrast, in an *imperative* style of programming, the computation takes place when it is coded, line by line, without waiting to account for what will come later. There are pluses and minuses to each style, and there is no ironclad division between them. Here are some points to keep in mind when you are thinking about what you prefer, assuming you have the choice:

- Symbolic programming tends to be more efficient because you leave room for the framework to optimize computations rather than perform them when and how they are ordered.

- Imperative programming tends to be easier to debug and understand.

- TensorFlow and MXnet tend toward a symbolic programming style, while Torch has a more imperative flavor.

Data, Symbols, Operations, Layers, and Graphs

Deep learning frameworks often focus on some notion of a graph and building that graph. The idea is that any architecture should be describable in terms of its individual components and their relationship to one another. Additionally, the idea of variables as divorced from actual values is also quite important. So you may have a

---

symbol, A, and a symbol, B, and then a third symbol, C, which is the result of matrix multiplying A and B:

```
# pseudo code
symbol A;
symbol B;
symbol C = matmul(A, B);
```

The distinction between symbols and data is essential to every framework because of the relationship between the symbol and the data. The symbol is used to learn a more general relationship; the data may have noise. There is only one symbol A, even as we may have millions or even billions of values to feed in for A, each paired with a respective B and C.

Popular Frameworks

As with many other tech domains, the landscape of offerings in the deep learning world is constantly changing. And just as with the business world, there are often indiscriminate mergers and acquisitions, as with the merging of TensorFlow and Theano, and of MXNet and Caffe. No framework has particularly established itself as a leader for time series deep learning, although MXNet may be the best bet for the near future given Amazon's own interest in forecasting and its newly unveiled deep learning for time series forecasting services. As someone who often needs to write custom data iterators for time series analysis projects in deep learning, I have found this most easily done in MXNet.

For someone new to the field, at the time of writing the most prominent and widely used libraries, and their respective tech behemoth parents, are:

- TensorFlow (Google)
- MXNet (Amazon)
- Torch (Facebook)

However, there are often academic and purely open source upstarts offering new innovations that can take over a domain, or at least make a substantial contribution to an existing framework. Dominance comes and goes quickly, although as the tech giants begin to roll out hardware to match their software, the market leaders may gain more endurance than has been seen in past years.

As we take a step back and think about what we do with data, these are the operations. We might add or multiply symbols together, and we can think of these as operations. We can also perform univariate operations, such as changing the shape of a symbol (perhaps converting symbol A from a 2×4 matrix to an 8×1 matrix) or passing values through an activation function, such as computing $\tanh(A)$.

Taking a step even further back, we can think of layers as those traditional units of processing we associate with common architectures, such as a fully connected layer, which we explored in the previous section. Taking into account the activation function and bias as well as the core matrix operation, we can say:

layer L = tanh(A × B + bias)

In many frameworks this layer may be expressible as one or two layers rather than several operations, depending on whether a combination of operations is popular enough to warrant its own designation as a unit of a layer.

Finally, we can relate multiple layers to one another, with one passing to the next:

layer L1 = tanh(A × B + bias1)

layer L2 = tanh(L1 × D + bias2)

The entire conglomerate of these symbols, operations, and layers results in a graph. A graph need not be fully connected—that is, it will not necessarily be the case that all symbols have dependencies on one another. A graph is used precisely to sort out what symbols depend on what other symbols and how. This is essential so that when we are computing gradients for the purposes of gradient descent, we can tweak our weights with each iteration for a better solution. The nice thing is that most modern deep learning packages do all this for us. We don't need to specify what depends on what and how the gradient changes with each layer added—it's all baked in.

What's more, as I mentioned before, we can use high-level APIs such that we do not need to spell out matrix multiplications as I did in the preceding example. Rather, if we want to use a fully connected layer, which is the equivalent of a matrix multiplication, followed by a matrix addition, and then an element-wise activation function of some kind, we don't need to write out all the math. For example, in mxnet, we can accomplish this with the following simple code:

```python
## python
>>> import mxnet as mx
>>> fc1 = mx.gluon.nn.Dense(120, activation='relu')
```

This will give us a fully connected layer that converts inputs to an output dimension of 120. Furthermore, this one line represents all that I just mentioned: a matrix multiplication, followed by a matrix addition, followed by an element-wise activation function.

You can verify this in the documentation (*https://perma.cc/8PQW-4NKY*), or by trying sample outputs and inputs when you know the weights. It's really quite impressive how much the API doesn't require you to do. You are not required to specify the input shape (although this must be a constant once you have built your graph so that it can be deduced). You are not required to specify a data type—this will default to the very sensible and most commonly used value of float32 (float64 is overkill for deep learning). If your data happens to come in a non-1D shape per example, this layer will also automatically "flatten" your data so that it is in the appropriate shape to be input into a fully connected/dense layer. This is helpful for a beginner, but once you develop a minimal level of competence it's good to revisit the documentation to understand just how much is being decided and taken care of for you even in a simple deep learning model.

This simple code, of course, does not require everything you need even for a short deep learning task. You will need to describe your inputs, your targets, and how you want to measure loss. You also need to situate your layer inside a model. We do this in the following code:

```python
## python
>>> ## create a net rather than a freestanding layer
>>> from mx.gluon import nn
>>> net = nn.Sequential()
>>> net.add(nn.Dense(120, activation='relu'),
>>>         nn.Dense(1))
>>> net.initialize(init=init.Xavier())
>>>
>>> ## define the loss we want
>>> L2Loss = gluon.loss.L2Loss()
>>>
>>> trainer = gluon.Train(net.collect_params(), 'sgd',
>>>                                 {'learning_rate': 0.01})
```

Finally, assuming we have also set up our data as needed, we can run through an epoch of training (that is, one pass through all the data) as follows:

```python
## python
>>> for data,target in train_data:
>>>     ## calculate the gradient
>>>     with autograd.record():
>>>         out = net(data)
>>>         loss = L2Loss(output, data)
>>>     loss.backward()
>>>     ## apply the gradient to update the parameters
>>>     trainer.step(batch_size)
```

It is also easy in most packages, as it is with mxnet, to save your model and accompanying parameters for later production use or additional training:

```python
## python
>>> net.save_parameters('model.params')
```

In the examples that follow, we will use mxnet's Module API to give you another flavor of creating graphs and training models.

As a deep learning practitioner, you will eventually want to become familiar with all the major packages and their various APIs (usually packages have, at the least, a very high-level API and a low-level API) so that you are easily able to read sample code. Working knowledge of all major deep learning modules is necessary to keep up with the latest industry practices and academic research, because this information is most easily learned through open source code.

Automatic Differentiation

The magic of a deep learning model—apart from its sometimes impressive performance—is that it can even learn at all. Standard models can have thousands or even millions of parameters, and these parameters must be tuned to the problem at hand. Generally this involves taking advantage of sophisticated symbolic differentiation techniques made accessible by deep learning frameworks. In past decades, most mathematicians, physicists, and others needing to optimize functions relied on numerical optimization and differentiation, which means that a gradient was computed empirically (say, via $f(x + \delta) - f(x)/\delta$). The upside of this was there was no need to teach computers how to perform differentiation explicitly, but the downside is that it was very slow and there was no taking advantage of the fact that the mathematical relationships are known or even designed.

Automatic differentiation uses the fact that, in computing the relationships between variables in different layers, we actually know exactly how the variables are mathematically related, and we can differentiate to get a direct expression for the value of a gradient rather than performing intensive numerical computations. More concretely, automatic differentiation makes use of the fact that, particularly in a computer more so than for math more generally, any sequence of mathematical operations is limited to a few simple operations for which the rules of differentiation are known and the chain rule can be applied. Automatic differentiation does this to come up with computationally tractable expressions for gradients. Once these gradients are available, it is possible to tune the weights over time and thereby improve and train the model.

Building a Training Pipeline

Throughout this section we will model the same data set: a measurement of hourly electric use in a variety of locations for several years. We'll preprocess this data so that we look at the change in electric use from hour to hour, a more difficult task than predicting the overall values because we are looking at the most unpredictable part of the time series.

Inspecting Our Data Set

We use an open data repository of hourly electric measurements provided in a code base demonstrating a new neural network architecture (which we will discuss later in this post).[1] To get a sense of what this data looks like, we read it in R and make some fast plots:

```R
## R
> elec = fread("electricity.txt")
> elec
     V1 V2  V3  V4  V5   V6 V7   V8  V9 V10 V11 V12 V13 V14 V15  V16 V17 V18
 1: 14 69 234 415 215 1056 29  840 226 265 179 148 112 171 229 1001  49 162
 2: 18 92 312 556 292 1363 29 1102 271 340 235 192 143 213 301 1223  64 216
 3: 21 96 312 560 272 1240 29 1025 270 300 221 171 132 185 261 1172  61 197
 4: 20 92 312 443 213  845 24  833 179 211 170 149 116 151 209  813  40 173
 5: 22 91 312 346 190  647 16  733 186 179 142 170  99 136 148  688  29 144
```

Already from a quick inspection, we can see how many rows and columns of data we have, as well as that there are no timestamps. We have been independently told these timestamps are hourly, but we don't know when exactly these measurements took place:

```R
## R
> ncol(elec)
[1] 321
> nrow(elec)
[1] 26304
```

We can also plot some random samples of the data to get a sense of what it looks like (see Figure 10-2). Since this is hourly data, we know if we plot 24 data points we will have a full day's worth of data:

```R
## R
> elec[125:148, plot(V4,   type = 'l', col = 1, ylim = c(0, 1000))]
> elec[125:148, lines(V14,  type = 'l', col = 2)]
> elec[125:148, lines(V114, type = 'l', col = 3)]
```

We also make a weekly plot (see Figure 10-3):

```R
## R
> elec[1:168, plot(V4,   type = 'l', col = 1, ylim = c(0, 1000))]
> elec[1:168, lines(V14,  type = 'l', col = 2)]
> elec[1:168, lines(V114, type = 'l', col = 3)]
```

---

1 The file download (*https://github.com/laiguokun/multivariate-time-series-data/raw/master/electricity/electricity.txt.gz*) comes from data providers at GitHub.

Remember that we don't know what the local hour is for any particular index, although we know the relation between them. However, consistent with our discussions back in Chapter 2, we could probably guess at what parts of the day represent standard human schedules based on the patterns of electric use. We could likely also identify the weekend. We don't do that here, but it could be a good idea to include models relating to time of day and day of the week for a more in-depth analysis and modeling of this data set.

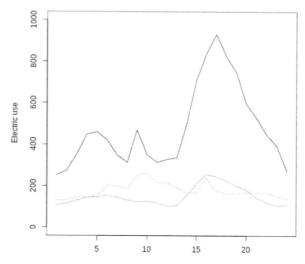

Figure 10-2. 24 hours sampled from the data set for 3 different locations out of the 321 locations available in the data set. While we do not have a sense of which local hours these indices correspond to, we do see a coherent daily pattern.

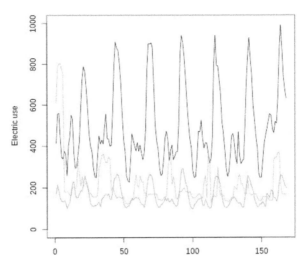

Figure 10-3. A full seven-day cycle of the sampled data for the same three locations as pictured in the daily plot. Our sense of a daily pattern is confirmed with this broader look at the data, indicating that one large peak per day along with some smaller features seems consistent in the behavior.

While we could predict the absolute values of the data as our forecasting technique, this has already been done in both academic papers and blog posts. Instead, we are going to predict the difference of the data. Predicting differences rather than total values of a time series tends to be more challenging because the data is noisier, and we can see this even in analogous plots of what we have done earlier (see Figure 10-4):[2]

```
## R
> elec.diff[1:168, plot( V4, type = 'l', col = 1, ylim = c(-350, 350))]
> elec.diff[1:168, lines(V14, type = 'l', col = 2)]
> elec.diff[1:168, lines(V114, type = 'l', col = 3)]
```

---

2 I do not include the differencing code, as we have covered this in Chapter 11 and it is a matter of a few lines.

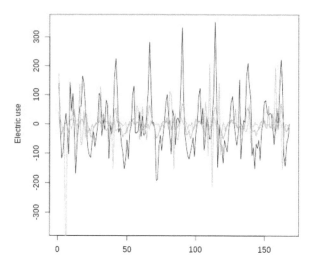

Figure 10-4. A week-long sample of the same three sites' differenced power time series, representing the change from hour to hour of electric use. While this series still exhibits a pattern, as did the original series, the unpredictable components of the series are made more apparent as they represent a larger portion of the value of the differenced series than they do of the original series.

If we were going to run a traditional statistical model, a state space model, or even a machine learning model, we would need to do a fair amount of analysis at this point to look at correlations between different electricity-consuming sites in the data set. We would want to see whether there was drift over time in the data and assess stationarity.

You should do these things too for deep learning so you can assess appropriate models for a data set and also establish expectations for how well you think your model can perform. However, the beauty of deep learning is that we can move forward even with somewhat messier data and even without passing any specific statistical tests for the quality of our data. In a production environment, we would spend more time on data exploration, but for the purposes of this chapter we will move on to our modeling options.

Steps of a Training Pipeline

In general when we are modeling with a neural network, there are a few common steps our scripts will always have. These scripts can often be more challenging to write than when we are fitting statistical or traditional machine learning models because our data sets tend to be larger. Also we fit these deep models in batches so that we are using iterators rather than arrays of the full data set.

A data pipeline will include the following steps:

- Make our code easily configurable by importing a preset list of parameters for default training values, which is particularly convenient since there are otherwise so many values to set.
- Pull the data into memory and preprocess it.
- Shape the data into the appropriately expected format.
- Build iterators appropriate to the deep learning module you are using.
- Build a graph that uses these iterators to know what shape of data to expect; this includes building out the entire model.
- Set training parameters, such as your optimizer, your learning rate, and how many epochs you will train for.
- Establish some recordkeeping system both for your weights and for your results from epoch to epoch.

Making our code easily configurable

The following code shows how we have accomplished these tasks. We start with a list of standard imports:

```python
## python
>>> from math import floor
>>>
>>> ## for archiving
>>> import os
>>> import argparse
>>>
>>> ## deep learning module
>>> import mxnet as mx
>>>
>>> ## data processing
>>> import numpy as np
>>> import pandas as pd
>>>
>>> ## custom reporting
>>> import perf
```

Then, we use a mix of hardcoded variables and adjustable parameters. To some extent these are a matter of experience and your priorities when training. Don't expect the variables to make much sense yet, as many of these parameters apply to neural network components we will discuss later in the chapter. The main thing is to notice the adjustability of the parameters:

```python
## python
>>> ## some hyperparameters we won't tune via command line inputs
>>> DATA_SEGMENTS    = { 'tr': 0.6, 'va': 0.2, 'tst': 0.2}
```

```
>>> THRESHOLD_EPOCHS = 5
>>> COR_THRESHOLD    = 0.0005
>>>
>>> ## set up parser
>>> parser = argparse.ArgumentParser()
>>>
>>> ## DATA SHAPING
>>> parser.add_argument('--win',       type=int,   default=24*7)
>>> parser.add_argument('--h',         type=int,   default=3)
>>>
>>> ## MODEL SPECIFICATIONS
>>> parser.add_argument('--model',     type=str,   default='rnn_model')
>>> ## cnn components
>>> parser.add_argument('--sz-filt',   type=str,   default=8)
>>> parser.add_argument('--n-filt',    type=int,   default=10)
>>> ## rnn components
>>> parser.add_argument('--rnn-units', type=int,   default=10)
>>>
>>> ## TRAINING DETAILS
>>> parser.add_argument('--batch-n',   type=int,   default=1024)
>>> parser.add_argument('--lr',        type=float, default=0.0001)
>>> parser.add_argument('--drop',      type=float, default=0.2)
>>> parser.add_argument('--n-epochs',  type=int,   default=30)
>>>
>>> ## ARCHIVE WORK
>>> parser.add_argument('--data-dir',  type=str,   default='../data')
>>> parser.add_argument('--save-dir',  type=str,   default=None)
```

It's crucial to have many adjustable parameters because training a deep learning model always involves hyperparameter searches to improve your model from your baseline. Commonly tuned hyperparameters affect all aspects of training, from data preparation (how far back in time to look) to model specification (how complex and what kind of a model to build) and training (how long to train, and how large a learning rate to use).

There are distinct categories of parameters in the preceding code. First, the data shaping relates to how we want to take our raw input, in this case a CSV full of parallel differenced time series of electricity use at 321 stations. To shape our data, we need two parameters. The window variable is how far back in time we will allow our models to look when attempting to make predictions. The horizon variable is how far forward we want to predict. Notice that these aren't specific to units, such as "5 minutes," but rather to time steps, consistent with our practices in earlier chapters. Like other statistical and machine learning models, neural networks care about our computational representation and don't have any sense of 5 minutes versus 5 eons when looking at data.

The penultimate section, the training details, will often be the most important for hyperparameter optimization and the most commonly adjusted. It's crucial at the outset to experiment with learning rate and make sure you have not picked something wildly off. A good rule of thumb is to start with 0.001 and then adjust up and down by orders of magnitude. It's not so important to have just the right learning rate, but it is important to have the right order of magnitude.

The model specifications allow us to specify a variety of models (such as an RNN versus a CNN) and architectural details about those models. In general, we will want to tune hyperparameters.

For the current examples, we use the following hyperparameters fed to our script at the command line:

```
--drop=0.2 --win=96 --batch-n=128 --lr=0.001 --n-epochs=25
--data-dir=/data/elec --save-dir=/archive/results
--model=model_will_vary
```

Prepping our input data

Once we have configurable parameters, there is a way to provide information to our script such as where to find a file, how far ahead we want to predict, and how far back in time we want to include in our inputs for a given horizon. These configurable parameters are important even at the preliminary step of reading in data and shaping it properly. We also need to arrange infrastructure to work through the data, because neural networks are trained via variations of stochastic gradient descent, which means that training takes place on small batches of data at a time, with one epoch meaning that all data has been used for training (although not all at the same time).

Next we discuss both the high-level process of providing data for training via iterators and the low-level details of shaping the data that goes into the iterators.

Shaping the input data.   In the preceding section, we saw how to form iterators from NumPy arrays, and we took the provision of those NumPy arrays for granted. Now we will discuss how the data is shaped, first conceptually, and then with a code example. We will discuss two data formats, NC and NTC.

We begin our discussion of different input data formats with a worked example that has nothing to do with coding. Let's imagine we have multivariate time series data, with columns A, B, and C.

Time	A	B	C
$t-3$	0	−1	−2
$t-2$	3	−2	−3
$t-1$	4	−2	−4
t	8	−3	−9

We want to build a model that predicts one step ahead, and we want to use the data from the previous two points in time to predict. We want to use the data from A, B, and C to predict A, B, and C. We will call our inputs X and our outputs Y.

We seek to predict Y at time t. At time t, these are the actual values of Y = [A, B, C]:

t 8 –3 –9

We specified that we would have the previous two time points available for all variables to make the prediction. That amounts to the following:

A, $t-1$	A, $t-2$	B, $t-1$	B, $t-2$	C, $t-1$	C, $t-2$
4	3	–2	–2	–4	–3

Likewise, to predict Y at time $t-1$, we have the following data as our target:

$t-1$ 4 –2 –4

and we expect to have the following values available to make the prediction:

A, $t-2$	A, $t-3$	B, $t-2$	B, $t-3$	C, $t-2$	C, $t-3$
3	0	–2	–1	–3	–2

If we wanted to store the inputs for both these data points in one data format, it would look like this:

Time	A, time – 1	A, time – 2	B, time – 1	B, time – 2	C, time – 1	C, time – 2
$t-1$	3	0	–2	–1	–3	–2
t	4	3	–2	–2	–4	–3

This is called the NC data format, where N indicates individual samples and C indicates channels, which is another way of describing multivariate information. We will use this data format for training a fully connected neural network, and it is the first option in the method we will eventually discuss that takes input data in CSV format and converts it to NumPy arrays of the proper shape and dimensionality.

On the other hand, we can shape the data differently, in a way that creates a specific axis for time. This is commonly done by putting the data in NTC format, which designates the number of samples × time × channels. In this case, the sample is each row of the original data—that is, each slice in time for which we want to make a prediction (and for which we have available data to do so). The time dimension is how far back in time we will look, which in this example is two time steps (and which is specified by --win in our sample script for this chapter).

In the NTC format, the input data we formatted before would look something like this for predicting the $t - 1$ horizon:

Time	A	B	C
$t-1$	0, 3	−1, −2	−2, −3

Or if we wanted to represent the input data for both of the samples we produced earlier, we could do so compactly as follows:

Time	A	B	C
$t-1$	0, 3	−1, −2	−2, −3
t	3, 4	−2, −2	−3, −4

The data we prepared in NTC format above has the corresponding labels indicated below. Also remember that the time labels on the side represent which sample, of the N samples in a batch, is in that row, so the row label "$t - 1$" is indicating that this is the sample prepared for a y measured at $t - 1$:

Time	A	B	C
$t-1$	4	−2	−4
t	8	−3	−9

Neither of these representations is more accurate than the other, but one convenient aspect of the NTC representation is that time has meaning with an explicit temporal axis.

The reason we form two shapes of inputs is that some models prefer one format and some the other. We will use the NTC format for input into convolutional and recurrent neural networks, which we discuss later in this chapter.

NTC Versus TNC

There are a few things to note about the NTC data storage format just shown. One is that it is highly repetitive. You can avoid this if the order of your data is not important, namely by using the TNC format rather than the NTC format. Within TNC, you can use your "batch" axis to run through different parts of the data in parallel.

Consider the following sequence of numbers ordered from left to right and top to bottom. (The bold is to help the reader find some of the starting values in the series delineated below in the TNC and NTC formats and does not otherwise have significance.)

56 **29** 56 94 10 92 52 32 19 59 88 94 6 57 73 59 95 79 97 38 65 51 27
18 **77** 39 4 19 60 38 73 51 65 4 96 96 6 12 62 59 21 49 65 37 64 69

---

36 **32** 48 97 33 44 63 99 10 56 75 20 61 53 71 48 41 2 58 18 4 10 17
66 **64** 53 24 36 23 33 38 1 17 59 11 36 43 61 96 55 21 45 44 53 26 55
99 22 10 26 25 82 54 82

If you prepared these in NTC format, with just one channel your data would look like this:

Time	Values
t	56, 29, 56, 94
$t+1$	29, 56, 94, 10
$t+2$	56, 94, 10, 92

There is a lot of repetition, but the good news is that the algorithm will train in the actual chronological order of the data, as all the batches are closely related in time. In contrast, if we elect the TNC format (N being the number of samples per batch), and if we supposed a batch size of 4 (unrealistically small), we could imagine our data looking like this:

Time	Batch member 1	Batch member 2	Batch member 3	Batch member 4
t	29	77	32	64
$t+1$	56	39	48	53
$t+2$	94	4	97	24

In this case we can see there is no data repetition in preparing our inputs. Whether we want to train on data sequenced in chronological order or not will depend on the data set. Those who have worked with natural language processing (NLP) in deep learning will recognize that format, which is quite successful for various NLP tasks but not always so helpful in numerical time series data.

Building iterators. Broadly, to provide data to a training routine, we need to provide iterators. Iterators are not unique to deep learning or to Python but rather reflect the general idea of an object that works its way through some kind of collection, keeping track of where it is and indicating when it has worked through an entire collection. Forming an iterator is simple in the case of training data coming from a NumPy array. If X and Y are NumPy arrays, we see that forming iterators is simple:

```python
## python
>>> ##############################
>>> ## DATA PREPARATION ##
>>> ##############################
>>>
>>> def prepare_iters(data_dir, win, h, model, batch_n):
>>>     X, Y = prepared_data(data_dir, win, h, model)
```

```
>>>
>>>     n_tr = int(Y.shape[0] * DATA_SEGMENTS['tr'])
>>>     n_va = int(Y.shape[0] * DATA_SEGMENTS['va'])
>>>
>>>     X_tr, X_valid, X_test = X[              : n_tr],
>>>                             X[n_tr          : n_tr + n_va],
>>>                             X[n_tr + n_va   : ]
>>>     Y_tr, Y_valid, Y_test = Y[              : n_tr],
>>>                             Y[n_tr          : n_tr + n_va],
>>>                             Y[n_tr + n_va   : ]
>>>
>>>     iter_tr = mx.io.NDArrayIter(data       = X_tr,
>>>                                 label      = Y_tr,
>>>                                 batch_size = batch_n)
>>>     iter_val = mx.io.NDArrayIter(data      = X_valid,
>>>                                  label     = Y_valid,
>>>                                  batch_size = batch_n)
>>>     iter_test = mx.io.NDArrayIter(data     = X_test,
>>>                                   label    = Y_test,
>>>                                   batch_size = batch_n)
>>>
>>>     return (iter_tr, iter_val, iter_test)
```

Here we have a method to prepare the iterators for the data set, and these iterators wrap numpy arrays that are received by a method called prepared_data() (more on this in a moment). Once the arrays are available, they are broken down into training, validation, and testing data sources, with the training data being the earliest, the validation being used as a way to tune hyperparameters with out-of-sample feedback, and testing data held until the end for true testing.[3]

Notice that the initializer for an iterator takes the input (data), the target value (label), and batch_size parameter, which reflects how many examples will be used per iteration to compute gradients and update the model weights.

Shaping the data in code

Now that we know the two shapes of data we want to create, we can look at the code that shapes it:

```python
## python
>>> def prepared_data(data_dir, win, h, model_name):
>>>     df = pd.read_csv(os.path.join(data_dir, 'electricity.diff.txt'),
>>>                      sep=',', header=0)
```

---

3 As we discussed earlier in the book, the gold standard is to train and roll models forward across many segments of time, but we avoid this complication for this code base. In a production code base you would want to have rolling validation and testing to optimize your model and get a better sense of real performance; the technique here introduces an additional lag between testing and training data and also means the testing data used to judge the model ultimately reflects only one time period of the whole history.

```
>>>    x  = df.as_matrix()
>>>    ## normalize data. notice this creates a lookahead since
>>>    ## we normalize based on values measured across the data set
>>>    ## so in a less basic pipeline we would compute these as
>>>    ## rolling statistics to avoid the lookahead problem
>>>    x = (x - np.mean(x, axis = 0)) / (np.std(x, axis = 0))
>>>
>>>    if model_name == 'fc_model': ## NC data format
>>>        ## provide first and second step lookbacks in one flat input
>>>        X = np.hstack([x[1:-1], x[:-h]])
>>>        Y = x[h:]
>>>        return (X, Y)
>>>    else:                          ## TNC data format
>>>        # preallocate X and Y
>>>        # X shape = num examples * time win * num channels (NTC)
>>>        X = np.zeros((x.shape[0] - win - h, win, x.shape[1]))
>>>        Y = np.zeros((x.shape[0] - win - h, x.shape[1]))
>>>
>>>        for i in range(win, x.shape[0] - h):
>>>            ## the target/label value is h steps ahead
>>>            Y[i-win] = x[i + h - 1     , :]
>>>            ## the input data are the previous win steps
>>>            X[i-win] = x[(i - win) : i , :]
>>>
>>>        return (X, Y)
```

After reading our data in from a text file, we standardize each column. Notice that each column is standardized on its own rather than across the whole data set homogeneously. This is because we saw even in our brief data exploration that different electric sites had very different values (see the plots in Figures 10-2 and 10-4):

```
## python
>>> x = (x - np.mean(x, axis = 0)) / (np.std(x, axis = 0))
```

NC data format. Producing the NC data format is fairly straightforward:

```
## python
>>> if model_name == 'fc_model': ## NC data format
>>>     ## provide first and second step lookbacks in one flat input
>>>     X = np.hstack([x[1:-h], x[0:-(h+1)]])
>>>     Y = x[(h+1):]
```

To generate the X inputs representing time $t - h$, to make a prediction at t, we take x and remove the last h rows (since that input data would require label values later in time than the latest data we have). Then we shift this data back along the time axis to produce further lagged values, and we need to make sure the NumPy arrays representing different lags have the same shape so they can be stacked together. This is what leads to the preceding formulation. It's worth working through this on your own computer and proving that it works out. You might also think about how to generalize the expression for an arbitrarily long lookback.

We check our work by setting a `Pdb` breakpoint and verifying that the values in X and Y match their expected counterparts in x:

```python
## python
(Pdb) X[0, 1:10] == x[1, 1:10]
array([ True,  True,  True,  True,  True,  True,  True,  True,  True])
(Pdb) X[0, 322:331] == x[0, 1:10]
array([ True,  True,  True,  True,  True,  True,  True,  True,  True])
(Pdb) Y[0, 1:10] == x[4, 1:10]
array([ True,  True,  True,  True,  True,  True,  True,  True,  True])
```

The first half of the columns in X represent the last points in time we make available for prediction, and the label/target for the prediction is three steps ahead of this. That's why `X[0, 1:10]` should match `x[1, 1:10]`, and `Y[0, 1:10]` should match `x[4, 1:10]`, because it should be three time steps ahead (our input set the horizon to 3).

It can be confusing that time and samples (data point indices) often have the same label, but they are separate things. There is the time ahead we are forecasting, there is the time at which we are taking the snapshot of inputs to make a forecast, and there is the time back we look at to collect data to make the forecast. These values are necessarily interrelated, but it's a good idea to keep the concepts separate.

NTC data format. Producing the NTC format is also not too bad:

```python
## python
>>> # preallocate X and Y
>>> # X shape = num examples * time win * num channels (NTC)
>>> X = np.zeros((x.shape[0] - win - h, win, x.shape[1]))
>>> Y = np.zeros((x.shape[0] - win - h, x.shape[1]))
>>>
>>> for i in range(win, x.shape[0] - h):
>>>         ## the target/label value is h steps ahead
>>>         Y[i-win] = x[i + h - 1     , :]
>>>         ## the input data are the previous win steps
>>>         X[i-win] = x[(i - win) : i , :]
```

For any given example (the N dimension, i.e., the first dimension), we took the last win rows of the input data across all columns. That is how we create three dimensions. The first dimension is essentially the data point index, and the values provided in that data point amount to 2D data, namely time × channels (here, electricity sites).

As before, we set a `Pdb` breakpoint to test this code. Note also that we confirm our own understanding of what we have done to test the code. Often the code to test a data format is more illuminating than the actual code because we use concrete numbers to do spot checks:

```python
## python
(Pdb) Y[0, 1:10] == x[98, 1:10]
array([ True,  True,  True,  True,  True,  True,  True,  True,  True])
```

```
(Pdb) X.shape
(26204, 96, 321)
(Pdb) X[0, :, 1] == x[0:96, 1]
array([ True,   True,   True,   True,   True,   True,   True,   True,   True,
        True,   True,   True,   True,   True,   True,   True,   True,   True,
        True,   True,   True,   True,   True,   True,   True,   True,   True,
        True,   True,   True,   True,   True,   True,   True,   True,   True,
        True,   True,   True,   True,   True,   True,   True,   True,   True,
        True,   True,   True,   True,   True,   True,   True,   True,   True,
        True,   True,   True,   True,   True,   True,   True,   True,   True,
        True,   True,   True,   True,   True,   True,   True,   True,   True,
        True,   True,   True,   True,   True,   True,   True,   True,   True,
        True,   True,   True,   True,   True,   True,   True,   True,   True,
        True,   True,   True,   True,   True,   True])
```

We see that the first data point we have prepared in X and Y (that is, the first row) corresponds to rows 0:96 (because we set our configurable window lookback to be 96 time steps in our parser inputs), and the forward horizon 3 time steps ahead corresponds to row 98 (because x ends at 95; remember, indexing on a slice excludes the last number in the slice, so x represents all the rows from 0 to 95 inclusive or 0 to 96 exclusive).

Data processing code is bug-prone, messy, and slow. However, you will find that the more times you write it and work through it, the more sense it makes. Nonetheless, it's good to test your data processing code thoroughly and then keep it somewhere safe so you can avoid having to work through the sample problems every time you need to shape your data. It's also wise to keep this code in a versioning system and have some way to track which version of your code was used in training a particular model.

Setting training parameters and establishing a recordkeeping system

We will discuss the details of various models in the coming sections, so for now we bypass the portion of the code that relates to graph building and go directly to training and recordkeeping logistics.

This is how we implement training in the simple examples we will focus on:

```python
## python
>>> def train(symbol, iter_train, valid_iter, iter_test,
>>>           data_names, label_names,
>>>           save_dir):
>>>     ## save training information/results
>>>     if not os.path.exists(args.save_dir):
>>>         os.makedirs(args.save_dir)
>>>     printFile = open(os.path.join(args.save_dir, 'log.txt'), 'w')
>>>     def print_to_file(msg):
>>>         print(msg)
>>>         print(msg, file = printFile, flush = True)
>>>     ## archiving results header
```

```
>>>     print_to_file('Epoch     Training Cor     Validation Cor')
>>>
>>>     ## storing prior epoch's values to set an improvement threshold
>>>     ## terminates early if progress slow
>>>     buf     = RingBuffer(THRESHOLD_EPOCHS)
>>>     old_val = None
>>>
>>>     ## mxnet boilerplate
>>>     ## defaults to 1 gpu, of which index is 0
>>>     devs = [mx.gpu(0)]
>>>     module = mx.mod.Module(symbol,
>>>                            data_names=data_names,
>>>                            label_names=label_names,
>>>                            context=devs)
>>>     module.bind(data_shapes=iter_train.provide_data,
>>>                 label_shapes=iter_train.provide_label)
>>>     module.init_params(mx.initializer.Uniform(0.1))
>>>     module.init_optimizer(optimizer='adam',
>>>                           optimizer_params={'learning_rate':
>>>                                             args.lr})
>>>
>>>     ## training
>>>     for epoch in range( args.n_epochs):
>>>         iter_train.reset()
>>>         iter_val.reset()
>>>         for batch in iter_train:
>>>             # compute predictions
>>>             module.forward(batch, is_train=True)
>>>             # compute gradients
>>>             module.backward()
>>>             # update parameters
>>>             module.update()
>>>
>>>         ## training results
>>>         train_pred  = module.predict(iter_train).asnumpy()
>>>         train_label = iter_train.label[0][1].asnumpy()
>>>         train_perf  = perf.write_eval(train_pred, train_label,
>>>                                       save_dir, 'train', epoch)
>>>
>>>         ## validation results
>>>         val_pred  = module.predict(iter_val).asnumpy()
>>>         val_label = iter_val.label[0][1].asnumpy()
>>>         val_perf = perf.write_eval(val_pred, val_label,
>>>                                    save_dir, 'valid', epoch)
>>>
>>>         print_to_file('%d        %f       %f ' %
>>>                     (epoch, train_perf['COR'], val_perf['COR']))
>>>
>>>         # if we don't yet have measures of improvement, skip
>>>         if epoch > 0:
>>>             buf.append(val_perf['COR'] - old_val)
>>>         # if we do have measures of improvement, check them
```

```
>>>        if epoch > 2:
>>>            vals = buf.get()
>>>            vals = [v for v in vals if v != 0]
>>>            if sum([v < COR_THRESHOLD for v in vals]) == len(vals):
>>>                print_to_file('EARLY EXIT')
>>>                break
>>>        old_val = val_perf['COR']
>>>
>>>    ## testing
>>>    test_pred  = module.predict(iter_test).asnumpy()
>>>    test_label = iter_test.label[0][1].asnumpy()
>>>    test_perf = perf.write_eval(test_pred, test_label,
>>>                                 save_dir, 'tst', epoch)
>>>    print_to_file('TESTING PERFORMANCE')
>>>    print_to_file(test_perf)
```

Why Use a Ring Buffer?

A *ring buffer* is a place to put values such that the first in is the first out when you retrieve a value. It also has the property of growing only up to a preset value. While performance and memory size are not an issue here, using this class helps us avoid boilerplate.

Python doesn't have a built-in ring buffer, but you can find simple examples online. They are handy whenever you want to keep track of a small number of values in the order they were received.

The preceding code accomplishes a diverse set of tasks, all routine. First, we set up values to track the history of the validation accuracy scores as a way to make sure training is seeing improvements. If a model is not training sufficiently quickly, we don't want to keep spinning our GPUs, wasting both time and electricity.

The MXNet boilerplate uses the Module API (as opposed to the Gluon API we saw earlier in this chapter):

```
## python
>>>    ## mxnet boilerplate
>>>    ## defaults to 1 gpu of which index is 0
>>>    devs = [mx.gpu(0)]
>>>    module = mx.mod.Module(symbol,
>>>                            data_names=data_names,
>>>                            label_names=label_names,
>>>                            context=devs)
>>>    module.bind(data_shapes=iter_train.provide_data,
>>>                label_shapes=iter_train.provide_label)
>>>    module.init_params(mx.initializer.Uniform(0.1))
>>>    module.init_optimizer(optimizer='adam',
>>>                            optimizer_params={'learning_rate':
>>>                                                args.lr})
```

These four lines of code accomplish the following:

1. Set up the raw component of a neural network as a computational graph.

2. Set up the data shapes so the network knows what to expect and can optimize.

3. Initialize all weights in the graph to random values (this is an art, not purely a random set of numbers drawn from infinite possibilities).

4. Initialize an optimizer, which can come in a variety of flavors and for which we explicitly set the initial learning rate depending on our input parameters.

Next we use our training data iterator to increment our way through the data as we train:

```python
## python
>>> for epoch in range( args.n_epochs):
>>>     iter_train.reset()
>>>     iter_val.reset()
>>>     for batch in iter_train:
>>>         module.forward(batch, is_train=True) # compute predictions
>>>         module.backward()                    # compute gradients
>>>         module.update()                      # update parameters
```

We then measure the predicted results both for the training and validation sets (in the same outer for loop as before):

```python
## python
>>> ## training results
>>> train_pred  = module.predict(iter_train).asnumpy()
>>> train_label = iter_train.label[0][1].asnumpy()
>>> train_perf  = evaluate_and_write(train_pred, train_label,
>>>                                  save_dir, 'train', epoch)
>>>
>>> ## validation results
>>> val_pred  = module.predict(iter_val).asnumpy()
>>> val_label = iter_val.label[0][1].asnumpy()
>>> val_perf  = evaluate_and_write(val_pred, val_label,
>>>                                save_dir, 'valid', epoch)
```

The loop concludes with some logic for early stopping:

```python
## python
>>> if epoch > 0:
>>>     buf.append(val_perf['COR'] - old_val)
>>> if epoch > 2:
>>>     vals = buf.get()
>>>     vals = [v for v in vals if v != 0]
>>>     if sum([v < COR_THRESHOLD for v in vals]) == len(vals):
>>>         print_to_file('EARLY EXIT')
>>>         break
>>> old_val = val_perf['COR']
```

This clunky code does some recordkeeping and simple logic, recording each successive correlation value between predicted and actual values. If the correlation from epoch to epoch has failed to improve sufficiently for enough epochs (or has even gotten worse), training will stop.

Evaluation metrics

Our function, evaluate_and_write, both records the correlations per epoch and the raw value for both the target value and estimated value. We do the same for the testing at the end of all training:

```python
## python
>>> def evaluate_and_write(pred, label, save_dir, mode, epoch):
>>>     if not os.path.exists(save_dir):
>>>         os.makedirs(save_dir)
>>>
>>>     pred_df  = pd.DataFrame(pred)
>>>     label_df = pd.DataFrame(label)
>>>     pred_df.to_csv( os.path.join(save_dir, '%s_pred%d.csv'
>>>                                  % (mode, epoch)))
>>>     label_df.to_csv(os.path.join(save_dir, '%s_label%d.csv'
>>>                                  % (mode, epoch)))
>>>
>>>     return { 'COR': COR(label,pred) }
```

This in turn makes use of a correlation function we define as follows:

```python
## python
>>> def COR(label, pred):
>>>     label_demeaned = label - label.mean(0)
>>>     label_sumsquares = np.sum(np.square(label_demeaned), 0)
>>>
>>>     pred_demeaned = pred - pred.mean(0)
>>>     pred_sumsquares = np.sum(np.square(pred_demeaned), 0)
>>>
>>>     cor_coef =  np.diagonal(np.dot(label_demeaned.T, pred_demeaned)) /
>>>                 np.sqrt(label_sumsquares * pred_sumsquares)
>>>
>>>     return np.nanmean(cor_coef)
```

Occasionally in this data set there are cases of zero variance, which can create a NAN in a column, so we opt to use np.nanmean() rather than np.mean().

Notice that one basic functionality we don't include here is to save the weights of the model, checkpointing throughout the training process. If we were training for production and needed to be able to reload the model and deploy it, we would want to use Module.save_checkpoint (to save the weights) and Module.load (to load a model back into memory, from which point you can continue training or deploy your model to production). There is a lot to learn to get started with a proper deep learning pipeline, but here we are keeping it basic.

Putting it together

We put our pipeline components together in the body of our __main__ scope:

```python
>>> if __name__ == '__main__':
>>>     # parse command line args
>>>     args = parser.parse_args()
>>>
>>>     # create data iterators
>>>     iter_train, iter_val, iter_test = prepare_iters(
>>>         args.data_dir, args.win, args.h,
>>>         args.model, args.batch_n)
>>>
>>>     ## prepare symbols
>>>     input_feature_shape = iter_train.provide_data[0][1]
>>>
>>>     X = mx.sym.Variable(iter_train.provide_data[0].name )
>>>     Y = mx.sym.Variable(iter_train.provide_label[0].name)
>>>
>>>     # set up model
>>>     model_dict = {
>>>         'fc_model'            : fc_model,
>>>         'rnn_model'           : rnn_model,
>>>         'cnn_model'           : cnn_model,
>>>         'simple_lstnet_model' : simple_lstnet_model
>>>         }
>>>     model = model_dict[args.model]
>>>
>>>     symbol, data_names, label_names = model(iter_train,
>>>                                         input_feature_shape,
>>>                                         X, Y,
>>>                                         args.win, args.sz_filt,
>>>                                         args.n_filt, args.drop)
>>>
>>>     ## train
>>>     train(symbol, iter_train, iter_val, iter_test,
>>>         data_names, label_names, args.save_dir)
```

Here we make use of the infrastructure we have just arranged. First, we parse our command-line arguments. Then we create iterators with the configurable inputs, including our prediction horizon, our lookback window, our batch size, and the name of the model we want to build. We create MXNet symbols and also record the input shape, and these are passed along as we create the model. Finally, we pass information about the model, along with our iterators and our save directory, to the training function, which does the interesting part: train a model and output its performance metrics.

So in this case, we see a minimal but fully functional training pipeline, featuring data ingestion and reshaping, model construction, model training, and logging of important values for model assessments.

By the way, our `print_to_file()` method is just a handy wrapper for `print()`:

```python
## python
def print_to_file(msg):
    print(msg, file = printFile, flush = True)esfa
print_to_file(args)
```

You will want to create a record of your model as it trains. It's possible that your pre-ferred model weights will not be the ones where you completed your training run but sometime earlier. Having a record of how training progressed will help you tune hyperparameters related to both model structure and training, ranging from how many parameters to have (to avoid underfitting or overfitting) to what learning rate to train at (to avoid making overly large or small adjustments during training).

We now have a full minimal viable pipeline, except that we are missing our models. Now we will look at some basic kinds of models that can be applied to time series data, and we will train each to see its relative performance.

Feed Forward Networks

This book is quite unusual in presenting feed forward networks in the context of time series analysis. Most modern time series analysis problems are undertaken with recurrent network structures, or, less commonly, convolutional network structures. We, however, begin with the feed forward neural network, as it's the simplest and most historic network structure. There are several reasons this is a good place to begin:

- Feed forward networks are highly parallelizable, which means they are quite per-formant. If you can find a reasonably good feed forward model for your pur-poses, you can compute it very quickly.

- Feed forward networks are good tests for whether there really are complex time-axis dynamics in your sequence. Not all time series really are time series in the sense of having time-axis dynamics, where earlier values have a specific relation-ship to later values. It can be good to fit a feed forward neural network as one baseline, apart from a simpler linear model.

- Feed forward network components are often integrated into larger and more complex time series deep learning architectures. Therefore, you need to know how these work even if they will not form an entire model in your work.

A Simple Example

A feed forward network is the simplest kind of neural network, so we begin by apply-ing a feed forward architecture to a time series example. There is nothing in the structure of a standard feed forward neural network that indicates a temporal relationship, but the idea is that the algorithm may nonetheless learn something of how past inputs predict future inputs. We saw an example of a feed forward network in Figure 10-1. A feed forward network is a series of fully connected layers, meaning the input to each layer connects to every node in the graph.

We start with this simple example, which uses the first kind of data formatting we saw earlier. That is, our input, X, will simply be two-dimensional data of N × C (i.e., samples × channels), where the time components has been flattened into channels. That corresponds to this data formatting:

A, $t-2$	A, $t-3$	B, $t-2$	B, $t-3$	C, $t-2$	C, $t-3$
3	0	−2	−1	−3	−2

As a reminder, in code that represents this branch of the data:

```python
## python
>>> if model_name == 'fc_model':
>>>     ## provide first and second step lookbacks in one flat input
>>>     X = np.hstack([x[1:-1], x[:-2]])
>>>     Y = x[2:]
>>>     return (X, Y)
```

We then build a fully connected model with MXNet's Module API:

```python
## python
>>> def fc_model(iter_train, window, filter_size, num_filter, dropout):
>>>     X = mx.sym.Variable(iter_train.provide_data[0].name)
>>>     Y = mx.sym.Variable(iter_train.provide_label[0].name)
>>>
>>>     output = mx.sym.FullyConnected(data=X, num_hidden = 20)
>>>     output = mx.sym.Activation(output, act_type = 'relu')
>>>     output = mx.sym.FullyConnected(data=output, num_hidden = 10)
>>>     output = mx.sym.Activation(output, act_type = 'relu')
>>>     output = mx.sym.FullyConnected(data = output, num_hidden = 321)
>>>
>>>     loss_grad = mx.sym.LinearRegressionOutput(data  = output,
>>>                                               label = Y)
>>>     return loss_grad, [v.name for v in iter_train.provide_data],
>>>                       [v.name for v in iter_train.provide_label]
```

Here we build a three-layer, fully connected network. The first layer has 20 hidden units, and the second layer has 10 hidden units. After the first two layers, there is an "activation" layer. This is what makes the model nonlinear, and without this it would simply be a series of matrix multiplications that would come down to a linear model. A fully connected layer effectively is:

$$Y = X W^T + b$$

W is a set of weights corresponding to those of the correction dimensions for a matrix multiplication, with the output resulting in a vector of dimensions equal to the hidden units. This is then added with a bias. The weights in W and b are both trainable. However, training one such set of weights, or even a series of such weights, would not be very interesting—it would be a roundabout way of running a linear regression. However, it is the activation applied after this set of matrix operations that creates the interest. The activation layer applies various nonlinear functions, such as the tanh and ReLU, which are pictured in Figures 10-5 and 10-6, respectively.

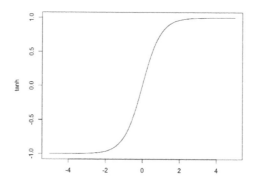

Figure 10-5. The tanh function exhibits nonlinear behavior at small values before becoming functionally constant, with a zero derivative, at higher values.

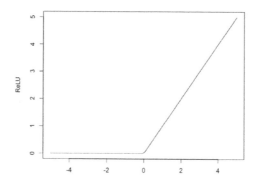

Figure 10-6. The ReLU function is easy to compute while also introducing a nonlinearity.

In the preceding code, you will notice we use the ReLU function as our activation function. It often makes sense to test a variety of activation functions (ReLU, tanh, sigmoid) in your early stages of model construction and hyperparameter tuning.

Now let's train with this model and see how it performs:

```
Epoch     Training Cor     Validation Cor
0           0.578666        0.557897
1           0.628315        0.604025
2           0.645306        0.620324
3           0.654522        0.629658
4           0.663114        0.636299
5           0.670672        0.640713
6           0.677172        0.644602
7           0.682624        0.648478
8           0.688570        0.653288
9           0.694247        0.657925
10          0.699431        0.663191
11          0.703147        0.666099
12          0.706557        0.668977
13          0.708794        0.670228
14          0.711115        0.672429
15          0.712701        0.673287
16          0.714385        0.674821
17          0.715376        0.674976
18          0.716477        0.675744
19          0.717273        0.676195
20          0.717911        0.676139
21          0.718690        0.676634
22          0.719405        0.677273
23          0.719947        0.677286
24          0.720647        0.677451

TESTING PERFORMANCE
{'COR': 0.66301745}
```

Is this good performance? It's hard to know in a void. When you progress to a deep learning model, you should put it into production only if it can outperform the substantially simpler models we have looked at in previous chapters. A deep learning model will take longer to compute a prediction and generally has higher overhead, so it should be used only when it can justify the additional costs. For this reason, when you are approaching a time series problem with deep learning tools, you usually want to have previous models you have fit to serve as an objective standard to beat.

In any case, we see that even a model without any temporal awareness can train and produce predictions that are reasonably correlated with measured values.

Using an Attention Mechanism to Make Feed Forward Networks More Time-Aware

Although feed forward networks have not made great strides in time series problems, there is still research being done to imagine architectural variations that could enhance their performance on sequential data. One idea is *attention*, which describes mechanisms added to a neural network that allow it to learn what part of a sequence to focus on and what part of an incoming sequence might relate to desirable outputs.

The idea of attention is that a neural network architecture should provide a mechanism for a model to learn which information is important when. This is done through *attention weights*, which are fit for each time step so that the model learns how to combine information from different time steps. These attention weights multiply what would otherwise be either the output or hidden state of a model, thereby converting that hidden state to a *context vector*, so-called because the hidden state is now better contextualized to account for and relate to all the information contained in the time series over time, hopefully including the temporal patterns.

The Softmax Function

You might not be familiar with the softmax function; here's the functional form: $softmax(y_i) = exp(y_i) / \sum_j exp(y_j)$, which serves two purposes:

- Outputs will now sum to 1, as a complete set of probabilities would. This is often the interpretation of outputs of a softmax.

- Keeping in mind the general shape of the exponential function, larger inputs (y_i) tend to get even larger, while smaller inputs tend to get relatively smaller. This accentuates strong activations/weights and de-emphasize weaker signals.

The softmax function is also known as the *normalized exponential* function, with the name clearly related to the functional form we see here.

Attention first came into use within recurrent neural networks (more on these later in this chapter), but its use and implementation is actually easier to understand in a variant proposed for a feed forward architecture in a research paper (*https://arxiv.org/pdf/1512.08756.pdf*). The paper proposes a way to apply feed forward neural networks to sequential data such that the network's reaction to each step of the sequence can be used as an input to the ultimate output of that network. This approach also allows variable-length sequences, just as variable-length time series are common in real-world problems.

Figure 10-7 shows an example architecture of how a feed forward neural network can be used for a task that involves processing sequential data.

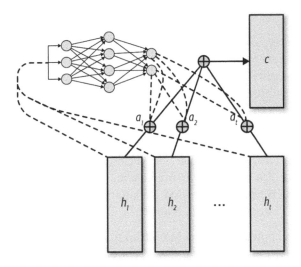

Figure 10-7. A "feed forward attention mechanism."

A feed forward network is applied to the inputs at each time step individually, generating a "hidden state" for each time step $h_1...h_T$. This effectively creates a series of hidden states. A second neural network, developed to learn $a(h_t)$, is depicted in the corner, and this is the attention mechanism. The attention mechanism allows the network to learn how much to weight each input, where each input represents a state coming from a different point in time. This allows the mechanism to determine which time steps to weight more or less heavily in the final summation of inputs.

Then the hidden states of different times are combined with their attention coefficients before final processing to produce whatever target/label is sought. The researchers who designed this network found that it performed respectably on a variety of tasks usually thought to necessitate a recurrent neural network to meet the requirements of remembering earlier inputs and combining them with later ones.

This is a great example of how there are no simple architectures in deep learning models because there are so many possibilities for adapting even a basic model to a complex time series question.

As noted, feed forward neural networks are not the leading networks for time series problems. Nonetheless, they are a useful starting point to establish the performance of a relatively simple model. Interestingly, if we do not include activation functions, we can use these to code up AR and VAR models using the MXNet framework. This can sometimes come in handy in its own right. More importantly, there are architectural variations of fully connected models that can be quite accurate for certain time series data sets.

CNNs

If you already have exposure to deep learning, you are likely quite familiar with convolutional neural networks (CNNs). Most of the jaw-dropping, record-breaking, human-like functionality attributable to computers in the last years has been thanks to extremely sophisticated and complex convolutional architectures. All that notwithstanding, the idea of a convolution is fairly intuitive and long predates deep learning. Convolutions have long been used in more transparent human-driven forms of image processing and image recognition studies, starting with something as simple as a Gaussian blur. If you are not familiar with the concepts of image processing kernels and don't have an idea of how you might code one up, Stack Overflow (*https://perma.cc/8U8Y-RBYW*) provides a good explanation of how this can be done both with a high-level API or manually via NumPy.

Convolution means applying a kernel (a matrix) to a larger matrix by sliding it across the larger matrix, forming a new one. Each element of the new matrix is the sum of element-wise multiplication of the kernel and a subsection of the larger matrix. This kernel is applied repeatedly as it slides across a matrix/image. This is done with a pre-specified number of kernels so that different features can emerge. A schematic of how this works on many layers is shown in Figure 10-8.

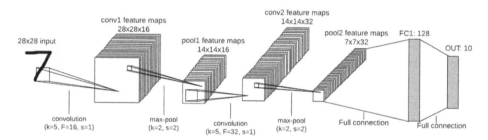

Figure 10-8. A convolutional network. Many two-dimensional windows of specified kernel size slide across the original image, producing many feature maps out of the trainable weights applied to the image. Often these are pooled and otherwise post-processed with an activation function. The process is repeated several times over several layers to collapse many features down into a smaller range of values, ultimately leading to, for example, classification ratings.

For many reasons, traditional convolution is a poor match to time series. A main feature of convolutions is that all spaces are treated equally. This makes sense for images but not for time series, where some points in time are necessarily closer than others. Convolutional networks are usually structured to be scale invariant so that, say, a horse can be identified in an image whether it's a larger or smaller portion of the image. However, in time series, again, we likely want to preserve scale and scaled features. A yearly seasonal oscillation should not be interpreted in the same way or "triggered" by the same feature selector as a daily oscillation, although in some contexts this might be helpful.

It is this double-edge nature of convolutions—that their strengths are their weaknesses from a time series perspective—that leads to them being used most often as a component of a network for time series analysis rather than the entire network. Also, it has led to their being studied more often for purposes of classification than forecasting, though, both uses are found in the field.

Some uses of convolutional networks for time series applications include:

- Establishing a "fingerprint" for an internet user's browsing history, which helps detect anomalous browsing activity
- Identifying anomalous heartbeat patterns from EKG data
- Generating traffic predictions based on past recordings from multiple locations in a large city

Convolutions are not all that interesting per se to apply to a univariate time series. Multichannel time series can be more interesting because then we can develop a 2D (or even 3D) image where time is only one axis.

There are other architectural innovations worth mentioning, as we will discuss two examples of convolutions for time series in the following section.

A Simple Convolutional Model

We can fit a convolutional model to our data rather than a fully connected one by swapping in a different model. In this case (and for the rest of the examples), we take the data in NTC format, which, as a reminder, looks like this:

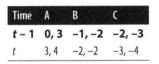

Time	A	B	C
$t-1$	0, 3	−1, −2	−2, −3
t	3, 4	−2, −2	−3, −4

This is N × T × C. However, the data expected by a convolutional layer is batch_size, channel, height × width:

```python
## python
>>> def cnn_model(iter_train, input_feature_shape, X, Y,
>>>               win, sz_filt, n_filter, drop):
>>>     conv_input = mx.sym.reshape(data=X, shape=(0, 1, win, -1))
>>>     ## Convolution expects 4d input (N x channel x height x width)
>>>     ## in our case channel = 1 (similar to a black and white image
>>>     ## height = time and width = channels slash electric locations
>>>
>>>     cnn_output = mx.sym.Convolution(data=conv_input,
```

```
>>>                                 kernel=(sz_filt,
>>>                                             input_feature_shape[2]),
>>>                                     num_filter=n_filter)
>>>         cnn_output = mx.sym.Activation(data=cnn_output, act_type='relu')
>>>         cnn_output = mx.sym.reshape(mx.sym.transpose(data=cnn_output,
>>>                                             axes=(0, 2, 1, 3)),
>>>                             shape=(0, 0, 0))
>>>         cnn_output = mx.sym.Dropout(cnn_output, p=drop)
>>>
>>>         output = mx.sym.FullyConnected(data=cnn_output,
>>>                             num_hidden=input_feature_shape[2])
>>>         loss_grad = mx.sym.LinearRegressionOutput(data=output, label=Y)
>>>         return (loss_grad,
>>>                 [v.name for v in iter_train.provide_data],
>>>                 [v.name for v in iter_train.provide_label])
```

Notice that, again, this does not include any explicit temporal awareness. What is different is that now time is laid out along a single axis, giving it some ordering.

Does this temporal awareness improve performance? Possibly not:

```
0           0.330701        0.292515
1           0.389125        0.349906
2           0.443271        0.388266
3           0.491140        0.442201
4           0.478684        0.410715
5           0.612608        0.564204
6           0.581578        0.543928
7           0.633367        0.596467
8           0.662014        0.586691
9           0.699139        0.600454
10          0.692562        0.623640
11          0.717497        0.650300
12          0.710350        0.644042
13          0.715771        0.651708
14          0.717952        0.651409
15          0.712251        0.655117
16          0.708909        0.645550
17          0.696493        0.650402
18          0.695321        0.634691
19          0.672669        0.620604
20          0.662301        0.597580
21          0.680593        0.631812
22          0.670143        0.623459
23          0.684297        0.633189
24          0.660073        0.604098

TESTING PERFORMANCE
{'COR': 0.5561901}
```

It can be difficult to determine why one model doesn't improve on another. In fact, often even a model's creator can be wrong about why the model works so well. There are few analytic proofs, and given how much can depend on the contours of a data set, this can also add confusion. Does the CNN not doing any better reflect the fact that most of the important information is in the nearest points in time? Or does it reflect a difference in the number of parameters? Or perhaps it reflects a failure to choose good hyperparameters. It would be important in a real use case to understand whether the performance seemed reasonable given the structure of the model, the structure of the data, and the overall number of parameters available.

We also see in the code a fault with our early stopping logic. It appears that it was too lenient. In this case, I revisited the issue and noticed that changes in correlation could look like the following over a series of epochs:

```
[-0.023024142, 0.03423196, -0.008353353, 0.009730637, -0.029091835]
```

This means that the change in correlation could be terrible—even negative—many times, so long as there was improvement, even a small one, every once in a while. This leniency turns out to be a bad decision, so it would be a good idea to backtrack to our pipeline and have a more stringent early stopping condition. This is an example of the give-and-take you will find as you fit deep learning models. It helps to do many trial runs as you build your pipeline to get an idea of what parameters work for your data set and models of interest—bearing in mind that these will vary substantially from one data set to the next.

Alternative Convolutional Models

The simple convolutional model we just saw did surprisingly well even though it included no modifications to be time aware. Now we discuss two approaches from research and industry to use convolutional architectures in time series problems.

Why is this attractive? There are a number of reasons. First, convolutional architectures are tried-and-true methods, with a number of best practices that are well known by practitioners. This makes convolutional models appealing because they are a known quantity. Additionally, convolutional models have few parameters since the same filters are repeated over and over, meaning there are not too many weights to train. And, finally, large portions of convolutional models can be computed in parallel, meaning these can be very fast for purposes of inference.

Causal convolutions

Causal convolutions are best understood with an image, as this can intuitively express how convolutions are modified to produce causality and a sense of time. Figure 10-9 shows an example of a *dilated causal convolution*. The causality part refers to the fact that only points earlier in time go into any given convolutional filter. That is why the image is not symmetric: earlier points flow into the convolutions used at later times, but not vice versa.

The dilation part refers to the fact that points are skipped in the arrangement of the convolutional filters, such that any given point goes only into one convolutional filter in each level of layering. This promotes model sparsity and reduces redundant or overlapping convolutions, allowing the model to look further back in time while keeping overall computations reasonably contained.

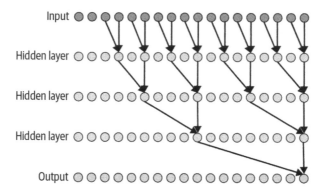

Figure 10-9. This graph depicts an important component of WaveNet (https:// perma.cc/Z4KZ-ZXBQ)'s architecture. Here we see how convolutional neural networks have architectural modifications appropriate to time series.

This example of dilated causal convolution introduces the notion of temporal causality by permitting only data from prior time points. That is, the convolutions in this image are not equal opportunity; they allow data to flow only from the past to the future and not the other way around. Every data point in the original input does have an impact on the final point. Note what dilation means here, which is that increasingly "deep" layers in the convolution skip between increasing numbers of data points in previous layers. The way the dilation here was set up, every data point in the original input is included in the final input, but only once. This is not required for dilation but just what was used in this case. Dilation could also be used to skip time points altogether.

While causal convolutions sound complicated and theoretical, they are surprisingly easy to perform. Simply add padding onto the left side of the matrix—that is, add earlier time step stand-ins that are zero, so that they won't contribute to the value—and then set padding to "valid" so that the convolution will only run against the actual boundaries of the matrix without including imaginary empty cells. Given that convolutions work as the sum of element-wise products, adding the zeros to the left side of Figure 10-9 would mean that we could run a standard convolution and that the extra zero cells would not change the final outcome.

Causal convolutions have had great success in the wild, most particularly as part of the model used for Google's text-to-speech and speech-recognition technologies.

Converting a time series into pictures

Convolutional models are known to work very well on image analysis, so a good idea in trying to make them relevant to time series is to find a way to turn a time series into a picture. There are many ways to do this, one of which is interesting because it can turn even a univariate time series into a picture: constructing a recurrence plot (see Figure 10-10).

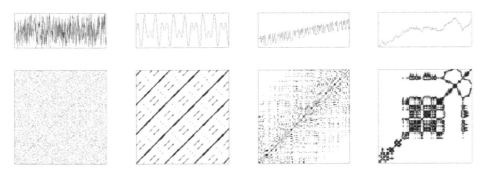

Figure 10-10. A beautiful and insightful visualization of four kinds of time series, from left to right: (1) white noise, (2) harmonic/seasonal series with two frequencies, (3) chaotic data with a trend, and (4) an autoregressive process. Source: Wikipedia (https://perma.cc/4BV2-57T4), provided by Norbert Marwan in 2006.

A recurrence plot is a way of describing, in phase-state space, when a time series revisits approximately the same phase and state it was in at an earlier moment in time. This is defined via the binary recurrence function. Recurrence is defined as $R(i, j) = 1$ if $f(i) - f(j)$ is sufficiently small; 0 otherwise. This results in a binary black-and-white image, such as the ones we see depicted in Figure 10-10. Notice that i and j refer to the time values, and the time axis is not bounded or in any way restricted.

While it is relatively easy to code up your own recurrence plot, recurrence plot functionality is also available in packages such as `pyts` (*https://perma.cc/4K5X-VYQR*), and the source code (*https://perma.cc/VS2Z-EJ8J*) for the plotting is easy to find and understand.

We can imagine extending this idea beyond a univariate time series by treating different variables of a multivariate time series as different "channels" of an image. This is just one example of how deep learning architectures and techniques are quite versatile and open to variation.

RNNs

Recurrent neural networks (RNNs) are a broad class of networks wherein the same parameters are applied again and again even as the inputs change as time passes. This sounds much like the feed forward neural network we covered previously, and yet RNNs make up—in whole or in part—many of the most successful models in academia and industry for sequence-based tasks, language, forecasting, and time series classification. The important differences between an RNN and a feed forward network are:

- An RNN sees time steps one at a time, in order.
- An RNN has state that it preserves from one time step to another, and it is this state as well as its static parameters that determine its response updates to each new piece of information at each step.
- An RNN thus has parameters that help it "update" its state, including a hidden state, from time step to time step.

Often when RNNs are introduced, they are presented via the paradigm of "unrolling" because a recurrent architecture is cell-based. This describes the way the same parameters are used again and again, so that the number of parameters is fairly small even for very long time sequences (see Figure 10-11).

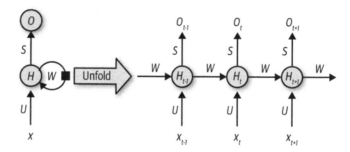

Figure 10-11. How a recurrent neural network architecture is unrolled once for each time step when it is applied to data.

The easiest way to understand how an RNN works, however, may just be to see an example. My particular favorite RNN cell is the Gated Recurrent Unit (GRU). Sometimes it can be more intimidating to see the expressions in mathematical form than in code, so here I provide an implementation of a GRU in Python with the help of NumPy. As you can see, there are two activation functions used: sigmoid and tanh. Other than this, all we do is perform matrix multiplication and addition as well as element-wise matrix multiplication (the *Hadamard product*):

```python
## python
>>> ## this code is designed to work with weights exported
>>> ## from TensorFlow's
>>> ## https://www.tensorflow.org/api_docs/python/tf/contrib/cudnn_rnn/CudnnGRU
>>> ## but can easily be repurposed to accommodate other weights
>>> def calc_gru(X, weights, num_inputs, num_features):
>>>     Us = weights[:(3*num_features*num_inputs)]
>>>     Us = np.reshape(Us, [3, num_features, num_inputs])
>>>
>>>     Ws = weights[(3*num_features*num_inputs):(3*num_features*num_features +
>>>     3*num_features*num_inputs)]
>>>     Ws = np.reshape(Ws, [3, num_features, num_features])
>>>
>>>     Bs = weights[(-6 * num_features) :]
>>>     Bs = np.reshape(Bs, [6, num_features])
>>>     s = np.zeros([129, num_features])
>>>     h = np.zeros([129, num_features])
>>>
>>>     for t in range(X.shape[0]):
>>>         z = sigmoid(np.matmul(Us[0, :, :], X[t, :]) +
>>>     np.matmul(Ws[0, :, :], s[t, :]) + Bs[0, :] + Bs[3, :])
>>>         r = sigmoid(np.matmul(Us[1, :, :], X[t, :]) +
>>>     np.matmul(Ws[1, :, :], s[t, :]) + Bs[1, :] + Bs[4, :])
>>>         h[t+1, :] = np.tanh(np.matmul(Us[2, :, :], X[t, :]) +
>>>     Bs[2, :] +
>>>     r*(np.matmul(Ws[2, :, :], s[t, :]) + Bs[5, :]))
>>>         s[t+1, :] = (1 - z)*h[t + 1, :] + z*s[t, :]
>>>
>>>     return h, s
```

A GRU is currently one of the most widely used RNN cells. It is a simpler version of the Long Short-Term Memory (LSTM) cell, which operates in a similar way. The differences between a GRU and an LSTM are as follows:

- A GRU has two "gates," while an LSTM has three. These gates are used to determine how much new information is allowed in, how much old information is preserved, and so on. Because an LSTM has more gates, it has more parameters.

- An LSTM tends have better performance, but a GRU is faster to train (due to the number of parameters). However, there are published results where a GRU outperforms an LSTM. A GRU is particularly likely to outperform an LSTM for nonlanguage tasks.

As you can see, the difference is more a matter of the degree of complexity appropriate for your training resources and of what you are trying to understand and predict.

It is important to be familiar with the matrix implementation of the GRU and LSTM so that you get a sense of how they work. Once you do, you may also develop an intuition when they don't train on a particular data set as to why they don't. There may be something about the dynamics that is not easily recognized by the format shown here.

Note that both GRUs and LSTMs help solve the problem that was first encountered when RNNs were used, namely exploding and vanishing gradients. Because of the recurrent application of the same parameters, it was often the case that gradients would quickly go to zero (not helpful) or to infinity (also not helpful), meaning that backpropagation was difficult or even impossible as the recurrent network was unrolled. This problem was addressed with GRU and LSTM because these tend to keep inputs and outputs from the cell in tractable value ranges. This is due both to the form of the activation function they use and to the way that the update gate can learn to pass information through or not, leading to reasonable gradient values being much more likely than in a vanilla RNN cell, which has no notion of a gate.

While both GRUs and LSTMs are reasonably easy to DIY, as just demonstrated, in reality you would not want to do this. The most important reason is that many of the matrix multiplication operations can be fused. The most efficient and accessible implementation for taking care of this and exploiting the hardware is NVIDIA's cuDNN, which fuses the matrix multiplication operations needed for both GRU and LSTM cells. Using the cuDNN interface rather than another implementation is substantially faster, and has even been cited by some Kaggle contest winners as the difference between winning and not even coming close because it helps so much in speeding up training. All the main deep learning frameworks offer access to this implementation, although in some cases (such as TensorFlow's `tf.con trib.cudnn_rnn`), you need to use the specially designated interface. In other cases, such as MXNet, you will default to using cuDNN so long as you don't do anything fancy with custom unrolled cells.

Continuing Our Electric Example

We can work through the electric forecasting example with an RNN as well. Again, we begin with the TNC data format as an input. This is the expected format for an RNN, so we don't even need to change it:

```python
## python
>>> def rnn_model(iter_train, window, filter_size, num_filter,
>>>               dropout):
>>>     input_feature_shape = iter_train.provide_data[0][1]
>>>     X = mx.sym.Variable(iter_train.provide_data[0].name)
>>>     Y = mx.sym.Variable(iter_train.provide_label[0].name)
>>>
>>>     rnn_cells = mx.rnn.SequentialRNNCell()
>>>     rnn_cells.add(mx.rnn.GRUCell(num_hidden=args.rnn_units))
>>>     rnn_cells.add(mx.rnn.DropoutCell(dropout))
>>>     outputs, _ = rnn_cells.unroll(length=window, inputs=X,
>>>                                   merge_outputs=False)
>>>
```

```
>>>     output = mx.sym.FullyConnected(data=outputs[-1],
>>>                                     num_hidden =
                                        input_feature_shape[2])
>>>     loss_grad = mx.sym.LinearRegressionOutput(data  = output,
>>>                                               label = Y)
>>>
>>>     return loss_grad, [v.name for v in iter_train.provide_data],
>>>                       [v.name for v in iter_train.provide_label]
```

The performance of this model is rather disappointing given that it is designed to handle temporal data:

```
Epoch    Training Cor    Validation Cor
0        0.072042        0.069731
1        0.182215        0.172532
2        0.297282        0.286091
3        0.371913        0.362091
4        0.409293        0.400009
5        0.433166        0.422921
6        0.449039        0.438942
7        0.453482        0.443348
8        0.451456        0.444014
9        0.454096        0.448437
10       0.457957        0.452124
11       0.457557        0.452186
12       0.463094        0.455822
13       0.469880        0.461116
14       0.474144        0.464173
15       0.474631        0.464381
16       0.475872        0.466868
17       0.476915        0.468521
18       0.484525        0.477189
19       0.487937        0.483717
20       0.487227        0.485799
21       0.479950        0.478439
22       0.460862        0.455787
23       0.430904        0.427170
24       0.385353        0.387026

    TESTING PERFORMANCE
    {'COR': 0.36212805}
```

It would be much more investigation to see why the performance is underwhelming. Did we not provide enough parameters to the RNN? Would it make sense to add some related model architectures that are commonly used, such as attention? (This applies to RNNs as well as feed forward networks, as was discussed earlier.)

The model reached its peak performance quite a bit earlier in the training process than did the feed forward or convolutional models. This is a lack of sufficient parameters to describe the data set, or that an RNN is specially tailored to this data or something else. You would need to do additional experiments and tuning to get a sense of the reason.

The Autoencoder Innovation

Occasionally you may happen upon a data set where a very simple model already does spectacular work for the mission you hope to accomplish.[4] But sometimes thinking outside the box can prove fruitful as well. For example, early on it was discovered that a simple innovation with an RNN could also substantially improve performance in sequence-like modeling. While this model was originally developed for language learning and machine translation, it has often proven quite successful at more numeric tasks, such as predicting electric load forecasting or stock prices. Known as the *autoencoder*, or alternately as the *seq2seq* model, this is a very commonly used model and one you should make a go-to part of your time series deep learning toolkit (see Figure 10-12). We will deploy it in several of the example chapters later in the book, where we analyze real-world time series data.

The autoencoder is two recurrent layers, but not in the traditional sense where each layer processes each input successively. Rather, the first layer runs to completion. Its hidden state is then passed onto the second layer, which takes this hidden state and its own outputs as new inputs for the next step. This model was particularly formulated with the idea of machine translation, whereby outputting a prediction at each time step was not meaningful given that in different languages, word and concept ordering can be drastically different and say the same thing. The idea was that once the first layer had fully processed the time series, its hidden state could come to have a kind of summarizing functionality. This summary would then be injected into the new model, which would gradually unroll the meaning into the new language by combining this summary with its own output at each time step so that it would know what it had said.

As mentioned, despite the natural language processing origins of this model, it is also useful for more traditional time series tasks, such as predicting univariate or multivariate time series. For example, in a Kaggle competition to predict web traffic in Wikipedia posts, the first-place winner, after trying many hyperparameters and architectural components, eventually settled on an autoencoder model. It seems that the winner's advantage was likely in smart training and hyperparameter search as well as a good study of useful feature generation and selection.

---

4 Rumor has it that at some point Google Translate was powered by a fairly vanilla seven-layer LSTM. However, the largeness of their data set undoubtedly helped, as would clever and careful training. Not all vanilla models are equal!

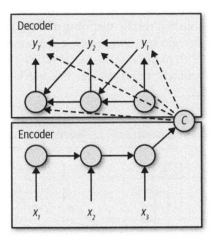

Figure 10-12. An autoencoder, also often referred to as a seq2seq model, is very popular for language processing and modeling but has also shown significant success in time series analysis.

Combination Architectures

Very often, successful industry and contest applications of deep learning for time series forecasting will use some kind of novel architecture, be it a new spin on applying traditional LSTM cells or combining different components. One example of this is a 2018 neural network architecture proposed by researchers at Carnegie Mellon University. The researchers looked to exploit the strengths of both convolutional and recurrent architectures, but they added other innovations as well.

They developed a "skip recurrent" layer whereby the recurrent model could be tuned to pay attention to periodicities (such as yearly, weekly, daily, depending on the nature of the data set) present in the data (and this periodicity could itself be explored as a hyperparameter).

They recognized that many time series have trends that are not well modeled by non-linear deep learning models. This meant that a deep learning model alone would not show the substantial scale changes over time that some time series data sets demonstrate. The researchers adapted by using a traditional linear model, namely the autoregressive model we discussed in Chapter 6.

The final model, the *modified LSTNet*, was a sum of the outputs from the AR model and a model built with a traditional recurrent layer and a skip recurrent layer in parallel. The inputs fed into each recurrent layer were the outputs of convolutional layers that convolved both along the time and channel axis (see Figure 10-13).

The researchers found that in three out of the four data sets they tried, they outperformed state-of-the-art published results spanning a broad range of topic areas. They failed only in the case of foreign exchange rates, a domain of finance that is notoriously difficult to predict, with high noise-to-signal ratios and a highly efficient market whereby any signal quickly dissipates as the market becomes efficient and investors try to find an edge.

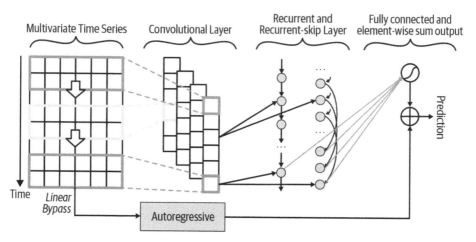

Figure 10-13. In the modified LSTNet architecture, we can see there is an autoregressive component (bottom of image) in parallel to a neural network architecture. The neural network architecture puts a convolutional element and a recurrent element in consecutive order, operating on the same inputs one after the other.

The inspiration for the next elaboration comes from the researchers' paper, and we will be working with code (*https://perma.cc/3W4Y-E8E2*)[5] modified from the MXNet package catalog of examples, described in great detail by the example's author, Oliver Pringle, in a blog post (*https://perma.cc/9KM2-RNPK*).

Here, as noted, we apply code based on modifications to the MXNet repository, simplifying it by removing the seasonal/skip connections and also by using only one convolutional filter size. We apply a convolutional layer, just as we did in the cnn_model example earlier:

```python
## python
>>> ## must be 4d or 5d to use padding functionality
>>> conv_input = mx.sym.reshape(data=X, shape=(0, 1, win, -1))
>>>
>>> ## convolutional element
>>> ## we add padding at the end of the time win
```

---

5 Code is also available in Oliver Pringle's personal GitHub repository (*https://oreil.ly/L-7ri*).

```python
>>> cnn_output = mx.sym.pad(data=conv_input,
>>>                         mode="constant",
>>>                         constant_value=0,
>>>                         pad_width=(0, 0,
>>>                                    0, 0,
>>>                                    0, sz_filt - 1,
>>>                                    0, 0))
>>> cnn_output = mx.sym.Convolution(data=cnn_output,
>>>                                 kernel=(sz_filt,
>>>                                         input_feature_shape[2]),
>>>                                 num_filter=n_filter)
>>> cnn_output = mx.sym.Activation(data=cnn_output, act_type='relu')
>>> cnn_output = mx.sym.reshape(mx.sym.transpose(data=cnn_output,
>>>                                               axes=(0, 2, 1, 3)),
>>>                             shape=(0, 0, 0))
>>> cnn_output = mx.sym.Dropout(cnn_output, p=drop)
```

We then apply a RNN not to the original inputs but to the convolutional component, as follows:

```python
## python
>>> ## recurrent element
>>> stacked_rnn_cells = mx.rnn.SequentialRNNCell()
>>> stacked_rnn_cells.add(mx.rnn.GRUCell(num_hidden=args.rnn_units))
>>> outputs, _ = stacked_rnn_cells.unroll(length=win,
>>>                                        inputs=cnn_output,
>>>                                        merge_outputs=False)
>>>
>>> rnn_output    = outputs[-1]
>>> n_outputs     = input_feature_shape[2]
>>> cnn_rnn_model = mx.sym.FullyConnected(data=rnn_output,
>>>                                       num_hidden=n_outputs)
```

Finally, in parallel to this CNN/RNN combination, we train an AR model, with one AR model per electric station location (so 321 distinct AR models, one per column/ variable/electric site), as shown here. This uses every time point for every station, with the model being specified one station at a time:

```python
## python
>>> ## ar element
>>> ar_outputs = []
>>> for i in list(range(input_feature_shape[2])):
>>>     ar_series = mx.sym.slice_axis(data=X,
>>>                                   axis=2,
>>>                                   begin=i,
>>>                                   end=i+1)
>>>     fc_ar = mx.sym.FullyConnected(data=ar_series, num_hidden=1)
>>>     ar_outputs.append(fc_ar)
>>> ar_model = mx.sym.concat(*ar_outputs, dim=1)
```

The full code is as follows:

```python
## python
>>> def simple_lstnet_model(iter_train,  input_feature_shape, X, Y,
>>>                     win, sz_filt, n_filter, drop):
>>>     ## must be 4d or 5d to use padding functionality
>>>     conv_input = mx.sym.reshape(data=X, shape=(0, 1, win, -1))
>>>
>>>     ## convolutional element
>>>     ## we add padding at the end of the time win
>>>     cnn_output = mx.sym.pad(data=conv_input,
>>>                         mode="constant",
>>>                         constant_value=0,
>>>                         pad_width=(0, 0,
>>>                                 0, 0,
>>>                                 0, sz_filt - 1,
>>>                                 0, 0))
>>>     cnn_output = mx.sym.Convolution(data = cnn_output,
>>>                             kernel = (sz_filt,
>>>                                     input_feature_shape[2]),
>>>                             num_filter = n_filter)
>>>     cnn_output = mx.sym.Activation(data     = cnn_output,
>>>                             act_type = 'relu')
>>>     cnn_output = mx.sym.reshape(mx.sym.transpose(data = cnn_output,
>>>                                             axes = (0, 2, 1, 3)),
>>>                             shape=(0, 0, 0))
>>>     cnn_output = mx.sym.Dropout(cnn_output, p = drop)
>>>
>>>     ## recurrent element
>>>     stacked_rnn_cells = mx.rnn.SequentialRNNCell()
>>>     stacked_rnn_cells.add(mx.rnn.GRUCell(num_hidden = args.rnn_units))
>>>     outputs, _ = stacked_rnn_cells.unroll(length = win,
>>>                                     inputs = cnn_output,
>>>                                     merge_outputs = False)
>>>     rnn_output    = outputs[-1]
>>>     n_outputs     = input_feature_shape[2]
>>>     cnn_rnn_model = mx.sym.FullyConnected(data=rnn_output,
>>>                                     num_hidden=n_outputs)
>>>     ## ar element
>>>     ar_outputs = []
>>>     for i in list(range(input_feature_shape[2])):
>>>         ar_series = mx.sym.slice_axis(data=X,
>>>                                 axis=2,
>>>                                 begin=i,
>>>                                 end=i+1)
>>>         fc_ar = mx.sym.FullyConnected(data     = ar_series,
>>>                                 num_hidden = 1)
>>>         ar_outputs.append(fc_ar)
>>>     ar_model = mx.sym.concat(*ar_outputs, dim=1)
>>>
>>>     output = cnn_rnn_model + ar_model
>>>     loss_grad = mx.sym.LinearRegressionOutput(data=output, label=Y)
>>>     return (loss_grad,
```

```
>>>            [v.name for v in iter_train.provide_data],
>>>            [v.name for v in iter_train.provide_label])
```

Notice that the performance is strikingly better in this model than any other:

```
Epoch      Training Cor     Validation Cor
0          0.256770         0.234937
1          0.434099         0.407904
2          0.533922         0.506611
3          0.591801         0.564167
4          0.630204         0.602560
5          0.657628         0.629978
6          0.678421         0.650730
7          0.694862         0.667147
8          0.708346         0.680659
9          0.719600         0.691968
10         0.729215         0.701734
11         0.737400         0.709933
12         0.744532         0.717168
13         0.750767         0.723566
14         0.756166         0.729052
15         0.760954         0.733959
16         0.765159         0.738307
17         0.768900         0.742223
18         0.772208         0.745687
19         0.775171         0.748792
20         0.777806         0.751554
21         0.780167         0.754034
22         0.782299         0.756265
23         0.784197         0.758194
24         0.785910         0.760000

TESTING PERFORMANCE
{'COR': 0.7622162}
```

This model obviously does quite a bit better than the other models, so there is something special and helpful about this architecture that uses convolutional images as the sequential data on which the recurrent layer trains. The traditional statistical tool of an AR model also adds quite a bit of functionality.[6] This model is far and away the best one, and this is a good lesson for the small amount of data exploration and training we did. It's worth trying a variety of models, and you don't even need to spend a lot of time training all models to be able to discover a clear leader.

---

6 Try training without the AR component if you don't believe me—it's not difficult to modify the code to remove this component.

Summing Up

One interesting aspect of our training in this chapter is that the models' performances did not turn out the way we might have expected. The simple feed forward network substantially outperformed some models that seem more conceptually complicated. However, this doesn't necessarily settle the question of which models are better or worse for our data set for a number of reasons:

- We did not check the number of parameters used in each model. It's possible that different genres of model would have drastically different performance with different numbers of parameters. We could play with model complexity, such as number of convolutional/recurrent layers or number of filters/hidden units.

- We did not tune hyperparameters. Sometimes getting the right hyperparameters can make an enormous difference in the performance of a model.

- We did not explore our data enough to have a prior idea of what model we would expect to do better or worse given the correlations over time and between different columns/electric sites.

How to Develop Your Skills

If you begin to browse academic journals, conferences, and arXiv postings looking specifically for deep learning resources for time series, you will often be frustrated by a few conventions in the literature:

- Much time series research sees itself as domain-specific first and time series second. This means you should really be looking in related domains, such as economics, healthcare, and climate studies.

- Deep learning researchers often think of "deep learning" or "neural networks" as convolutional networks. So if you are really trying to dig in specifically to "recurrent neural networks," make sure to look for these words in the title. Most papers lacking these words (or similar) in their title won't have RNNs in the body of the research.

- Unfortunately, no one has yet written a deep learning model to help you find deep learning papers, but that would be a worthy project.

More Resources

- Historical documents:

Sepp Hochreiter and Jürgen Schmidhuber, "Long Short-Term Memory," Neural Computation 9, no. 8 (1997):1735–80, https://perma.cc/AHR3-FU5H.
    This 1997 seminal work introduced the Long Short-Term Memory cell (LSTM) and also proposed several experimental benchmarks that remain in use today for studying neural network performance on sequence analysis.

Peter G. Zhang, Eddy Patuwo, and Michael Hu, "Forecasting with Artificial Neural Networks: The State of the Art," International Journal of Forecasting 14, no. 1 (1998): 35–62, https://perma.cc/Z32G-4ZQ3.
    This article gives an overview of the state of the art for time series and deep learning in 1998.

- On RNNs:

Aurélien Geron, "Processing Sequences Using RNNs and CNNs," in Hands-On Machine Learning with Scikit-Learn, Keras, and TensorFlow, 2nd Edition (Sebastopol: O'Reilly Media, Inc., 2019).
    Aurélien Geron offers an exceptionally extensive set of examples on how to apply deep learning to time series data in this popular book. If you are comfortable with these tools, this Jupyter notebook (*https://perma.cc/D3UG-59SX*) offers an excellent example of many different kinds of models applied to sequence data, including some exercises with solutions.

Valentin Flunkert, David Salinas and Jan Gasthaus, "DeepAR: Probabilistic Forecasting with Autoregressive Recurrent Networks," unpublished paper, 2017, https://perma.cc/MT7N-A2L6.
    This groundbreaking paper illustrated Amazon's model that was developed to fit time series for its retail data that occurs at a large variety of scales, and with trends. One particular innovation of the work was the ability to make probabilistic forecasts rather than the usual point estimates that tend to be the result of deep learning analysis.

Lingxue Zhu and Nikolay Laptev, "Deep and Confident Prediction for Time Series at Uber," paper presented at the 2017 IEEE International Conference on Data Mining Workshops (ICDMW), New Orleans, LA, https://perma.cc/PV8R-PHV4.
    This paper illustrates another example of probability and statistically inspired modifications to typical RNNs. In this case, Uber proposed a novel Bayesian deep model that provided both a point estimate and an uncertainty estimation that could be deployed in production with reasonably fast performance.

Zhengping Che et al., "Recurrent Neural Networks for Multivariate Time Series with Missing Values," Scientific Reports 8, no. 6085 (2018), https://perma.cc/4YM4-SFNX.

> This paper provides a strong example of state-of-the-art work on medical time series. It demonstrates the use of a GRU, combined with novel architectures to account for missing data and make missingness into an informative attribute. The authors showcase neural networks that beat all existing clinical metrics currently deployed to make predictions about patient health and hospital stay statistics. This is a great example of how an intuitive and easy-to-understand modification to a simple and widely used RNN structure (the GRU) can lead to groundbreaking results on a good data set.

- On CNNs:

Aäron van den Oord and Sander Dieleman, "WaveNet: A Generative Model for Raw Audio DeepMind," DeepMind blog, September 8, 2016, https://perma.cc/G37Y-WFCM.

> This blog provides an extremely well done and accessible description of a groundbreaking CNN architecture that was used to enhance both text-to-speech and speech-to-text technologies in a variety of languages and with different speakers. The new architecture led to a significant boost in performance and was subsequently deployed on other sequence-related AI tasks, particularly with respect to time series forecasting.

- On the application of deep learning:

Vera Rimmer et al., "Automated Website Fingerprinting Through Deep Learning," paper presented at NDSS 2018, San Diego, CA, https://perma.cc/YR2G-UJUW.

> This paper illustrates a way in which deep learning can be used to uncover private information about a user's internet browsing content throught Website Fingerprinting. In particular the authors highlighted a way in which various neural network architectures could be used to formulate Website Fingerprinting attacks to pierce user privacy protections.

CPMP, "Second Place Solution to the Kaggle Web Traffic Forecasting Competition," Kaggle blog, 2017, https://perma.cc/UUR4-VNEU.

> This blog, written before the conclusion of the competition, describes the second-place winner's thinking in designing a mixed machine learning/deep learning solution to forecasting. There is also some retrospective commentary available in a related blog post (*https://perma.cc/73M3-D7DW*). This is a great example of a mix of modern packages and a relevant can-do coding style. The GitHub repository for the first-place solution is also available (*https://perma.cc/K6RW-KA9E*) alongside a discussion of its neural network based architecture (*https://perma.cc/G9DW-T8LE*).

## CHAPTER 11

Measuring Error

In previous chapters we used a variety of measures to compare models or judge how well a model performed its task. In this chapter we examine best practices for judging forecast accuracy, with emphasis placed on concerns specific to time series data.

For those new to time series forecasting, it's most important to understand that standard cross-validation usually isn't advisable. You cannot select training, validation, and testing data sets by selecting randomly selected samples of the data for each of these categories in a time-agnostic way.

But it's even trickier than that. You need to think about how different data samples relate to one another in time even if they appear independent. For example, suppose you are working on a time series classification task, so that you have many samples of separate time series, where each is its own data point. It could be tempting to think that in this case you can randomly pick time series for each of training, validation, and testing, but you would be wrong to follow through on such an idea. The problem with this approach is that it won't mirror how you would use your model, namely that your model will be trained on earlier data and tested on later data.

You don't want future information to leak into your model because that's not how it will play out in a real-world modeling situation. This in turn means that the forecast error you measure in your model will be lower during testing than in production because in testing you will have cross-validated your model in a way that yielded future information (i.e., feedback on which model to use).

Here's a concrete scenario demonstrating how this could happen. Imagine you are training an air quality monitor for major cities in the western United States. In your training set you include all 2017 and 2018 data for San Francisco, Salt Lake City, Denver, and San Diego. In your testing set, you include the same range of dates for Las Vegas, Los Angeles, Oakland, and Phoenix. You find that your air quality model does particularly well for Las Vegas and Los Angeles, but also extremely well in 2018 overall. Great!

You then try to replicate your model training process on data from earlier decades and find that it does not test as well as it trains in other training/testing runs. You then remember the record-breaking wildfires in Southern California of 2018 and realize that they were "baked in" to the original testing/training run because your training set gave you a window into the future. This is an important reason to avoid standard cross-validation.

There can be times where it is not a problem for information from the future to propagate backward into the choice of a model. For example, if you are merely looking to understand the dynamics of a time series by testing how well you can forecast it, you are not looking to forecast, per se, so much as to test the best possible fit of a given model to data. In this case, including data from the future is helpful to understanding the dynamics, although you would want to be wary of overfitting. So arguably, even in this case, maintaining a valid test set—which requires not letting information leak backward from the future—would still warrant concerns about time series and cross-validation.

With this general commentary complete, we turn to an example of the concrete mechanics of divvying up data for training, validating, and testing a model. We then talk more generally about how to determine when a forecast is good enough, or as good as it reasonably can be. We also look at how to estimate uncertainty of our forecast for when we use techniques that do not directly produce an uncertainty or error measure as part of the output. We conclude with a list of gotchas that can be helpful to review from time to time when you are constructing a time series model or preparing to put one into production. It may help you avoid some embarrassment!

The Basics: How to Test Forecasts

The most important element of generating a forecast is to make sure that you are building it solely with data you could access sufficiently in advance for that data to be used in generating the forecast. For this reason, you need to think not only about when events happen but also when the data would be available to you.[1]

While this seems simple enough, remember that common preprocessing, such as exponential smoothing, may inadvertently lead to leakage from the training period to the testing period. You can test this in an example yourself by fitting a linear regression first to an autoregressive time series and then to an exponentially smoothed

---

[1] Incidentally, modeling with or without data available at certain times to show the importance of timely data delivery can motivate your data engineers and managers to prioritize specific data inputs if you can demonstrate that it would make a difference to your model's accuracy and/or to your organization's bottom line.

autoregressive time series. You will see that the more you smooth the time series, and the longer your smoothing half-life, the "better" your predictions become. This is because in fact you are having to make less and less of a prediction as more and more of your value is composed of an exponential average of past values. This is quite a dangerous lookahead and is so insidious it still turns up in academic papers!

Because of these dangers and other hard-to-spot ways of feeding the future into the past, or vice versa, the gold standard for any model should be backtesting with training, validation, and testing periods that roll forward.

In backtesting, a model is developed for one set of dates, or a range of dates, but then it is tested extensively on historical data, preferably representing the full range of possible conditions and variability. Also important, practitioners need to have principled reasons for backtesting a particular model and should try to avoid trying out too many models. As most data analysts are aware, the more models you test, the more likely you are to overfit to your data—that is, choose a model that has picked out overly specific details about the current data set rather than generalizing in a robust way. Unfortunately for time series practitioners, this is tricky balancing act and one that can lead to embarrassing results when you put models into production.

Concretely, how do we implement some form of backtesting? We do this in a way that preserves a cross-validation-like structure, albeit a temporally aware one. The common paradigm, assuming you have data that represents time passing sequentially "in alphabetical order," is as follows:

```
Train with [A]          test with [B]
Train with [A B]        test with [C]
Train with [A B C]      test with [D]
Train with [A B C D]    test with [E]
Train with [A B C D E]  test with [F]
```

Figure 11-1 illustrates this testing structure.

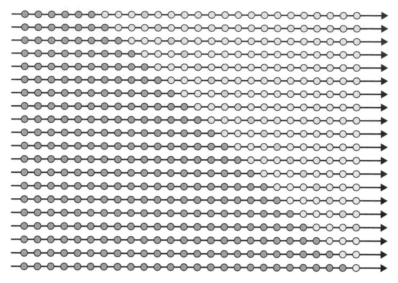

Figure 11-1. The gold standard for evaluating the performance of a time series model, namely rolling forward your training, validation, and testing windows.

It is also possible to move the training data window rather than expand it. In this case, your training could look something like this:

```
Train with [A B]   test with [C]
Train with [B C]   test with [D]
Train with [C D]   test with [E]
Train with [D E]   test with [F]
```

Which method you choose depends in part on whether you think the behavior of your series is evolving over time. If you think it does, you are better served by a moving window so that all testing periods are tested with a model trained to the most relevant data. Another consideration could be that you want to avoid overfitting, where using an expanding window will better discipline your model than a fixed-length window.

Because this kind of rolling split is a common training need, both R and Python offer easy ways to generate it:

- In Python, one easy way to generate data splits is with `sklearn.model_selection.TimeSeriesSplit`.

- In R, `tsCV` from the `forecast` package will roll a model forward in time using this backtesting schema and report the errors.

There are other packages in both R and Python that will do the same. You can also write your own functions to split your data if you have specific ideas about how you would implement this model testing for a particular project. You might want to skip certain time periods because they exhibited anomalous dynamics, or you might want to weight performance of certain time periods more.

For example, suppose you work with financial data. Depending on your goals, it could be worthwhile to exclude data from extraordinary times, such as the financial crisis of 2008. Or if you work with retail data, you might want to weight model performance most heavily for Christmas shopping season even if you sacrifice some accuracy in the predictions of lower-volume seasons.

Model-Specific Considerations for Backtesting

Consider the dynamics of the model you are training when structuring your backtesting, particularly with respect to what it means to train a model with a particular temporal range of data.

With traditional statistical models, such as ARIMA, all data points are factored in equally when selecting model parameters, so more data may make the model less accurate if you think the model parameters should be varied over time. This also is true for machine learning models for which all training data is factored in equally.[2]

Batched stochastic methods, on the other hand, can result in weights and estimates that evolve over time. So, neural network models trained with typical stochastic gradient descent methods will, to some extent, take the temporal nature of the data into account if you train on the data in chronological order. The most recent gradient adjustments to the weight will reflect the most recent data. In most cases, time series neural network models are trained on the data in chronological order, as this tends to produce better results than models trained on the data in random order.

Don't make holes in your data. The very difficulty and beauty of time series data is that the data points are autocorrelated. For this reason, we cannot randomly select out points within our time series for validation or testing because this will destroy some of the autocorrelation in our data. As a result, our model's recognition of an autoregressive component of our data would be undermined, a most undesirable result.

---

2 Note that you can write your own custom weighting function for your loss function as a way to weight more recent data more heavily, but this requires above-average programming skill and knowledge about numerical optimizations.

State space models also offer opportunities for the fitting to adapt over time with the mode. This cuts in favor of a longer training window because a long time window will not keep the posterior estimate from evolving over time.

Snapshot Your Model Over Time

Because you will test your models by rolling them forward in time after fitting them on past data, you need an easy way to save a model with a timestamp so that you know the earliest point in time that you could appropriately have used that model. This will help you avoid inadvertently testing a model on its own training data. It will also give you an opportunity to apply several different models from different time periods to testing data. This can be a way of seeing whether it matters how recently a model was trained with respect to that model's accuracy on test data, which can ultimately help you choose a rhythm for how often your time series models need to be refit in production code.

When Is Your Forecast Good Enough?

When your forecast is good enough depends on your overall goals, the minimum quality you can "get away with" for what you need to do, and the limits and nature of your data. If your data itself has a very high noise-to-signal ratio, you should have limited expectations for your model.

Remember, a time series model isn't perfect. But you should hope to do as well as, or slightly better than, alternative methods, such as solving a system of differential equations about climate change, asking a well-informed stock trader for a tip, or consulting a medical textbook that shows how to classify EEG traces. As you assess performance, keep in mind the known domain expert limits to forecasting as such measures indicate—for now, the upper limit to performance in many forecasting problems.

There are times when you know that your model is not yet good enough and that you can do better. Here are some actions you can take to identify these opportunities:

Plot the outputs of your model for the test data set
> The distribution produced by the model should match the distribution of the values you are seeking to forecast, assuming there is no expected regime change or underlying trend. As an example, if you are trying to predict stock prices, and you know that stock prices go up and down about equally often, you have an inadequate model if the model always predicts a price increase. Sometimes the distributions will obviously be off, whereas at other times you can apply a test statistic to compare your model's output with your actual targets.

Plot the residuals of your model over time

If the residuals are not homogenous over time, your model is underspecified. The temporal behavior of the residuals can point you toward additional parameters needed in your model to describe the temporal behavior.

Test your model against a simple temporally aware null model

A common null model is that every forecast for time t should be the value at time $t - 1$. If your model is not beating such a simple model, there isn't any justification for it. If a simple, all-purpose naive model can beat the model you have crafted, it's a fundamental problem with your model, loss function, or data preprocessing rather than a matter of a hyperparameter grid search. Alternately, it can be a sign of data that has a lot of noise relative to signal, which also suggests that your model is useless for its intended purposes.[3]

Study how your model handles outliers

In many industries, outliers simply are what they are called: outliers. There is likely no way these events could have been predicted,[4] which means the best your model can do is ignore these outliers rather than fit them. In fact, your model predicting outliers very well may even be a sign of overfitting or poor loss function selection. This depends on the model you chose and the loss functions you have employed, but for most applications you will want a model whose predictions are not as extreme as the extreme values in your data set. Of course this advice does not hold when the cost of outlier events is high and where the forecasting task is mainly to warn of outlier events when possible.

Conduct temporal sensitivity analysis

Are qualitatively similar behaviors in related time series producing related outcomes in your model? Using your knowledge of the underlying dynamics of your system, ensure that this is true and that your model recognizes and treats similar temporal patterns in the same way. For example, if one time series shows a trend upward with a drift of 3 units per day and another shows a trend upward with a drift of 2.9 units per day, you'd likely want to be sure that the forecasts ultimately made for these series were similar. Also you'd want to make sure that the ranking of the forecasts as compared to the input data was sensible (greater drift should result in a larger forecast value). If this is not the case, your model may be overfit.

This is not a full list of the many ways you can test your time series model, but it can be a starting point that you build on as you gain experience in a particular domain.

---

3 A simple null model can be surprisingly difficult to beat.

4 In some cases, there may not be anything to learn for future decisions.

Estimating Uncertainty in Your Model with a Simulation

One advantage of traditional statistical time series analysis is that such analyses come with well-defined analytical formulas for the uncertainty in an estimate. However, even in such cases—and certainly in the case of nonstatistical methods—it can be helpful to understand the uncertainty associated with a forecasting model via computational methods. One very intuitive and accessible way to do this is a simple simulation.

Suppose we have conducted an analysis of what we believe to be an AR(1) process. As a reminder, an AR(1) process can be expressed as:

$$y_t = \phi \times y_{t-1} + e_t$$

Following a model fit, we want to study how variable our estimate of the ϕ coefficient could be. In such a case, one way to study this is to run a number of Monte Carlo simulations. This can be accomplished easily in R so long as we remember what we learned about AR processes in Chapter 6:

```R
## R
> require(forecast)
>
> phi          <- 0.7
> time_steps   <- 24
> N            <- 1000
> sigma_error  <- 1
>
> sd_series    <- sigma_error^2 / (1 - phi^2)
> starts       <- rnorm(N, sd = sqrt(sd_series))
> estimates    <- numeric(N)
> res          <- numeric(time_steps)
>
> for (i in 1:N) {
>   errs = rnorm(time_steps, sd = sigma_error)
>   res[1]  <- starts[i] + errs[1]
>
>   for (t in 2:time_steps) {
>     res[t] <- phi * tail(res, 1) + errs[t]
>   }
>   estimates <- c(estimates, arima(res, c(1, 0, 0))$coef[1])
> }
>
> hist(estimates,
>      main = "Estimated Phi for AR(1) when ts is AR(1)",
>      breaks = 50)
```

This leads to the histogram shown in Figure 11-2 for the estimated ϕ.

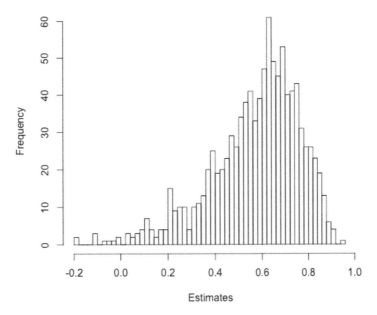

Figure 11-2. The distribution of estimates for φ.

We can also get a sense of the range of the estimates and the quantiles with the `summary()` function applied to `estimates`:

```
## R
> summary(estimates1)
    Min. 1st Qu.  Median    Mean 3rd Qu.    Max.
 -0.3436  0.4909  0.6224  0.5919  0.7204  0.9331
```

We can use bootstrapping to ask more complicated questions, too. Suppose we want to know what numerical costs we may pay for oversimplifying our model compared to ground truth. Imagine the process studied is an AR(2) even though we diagnosed it as an AR(1) process. To find out how this affects our estimate, we can modify the preceding R code as follows:

```
## R
> ## now let's assume we have a true AR(2) process
> ## because this is more complicated, we switch over to arima.sim
> phi_1 <- 0.7
> phi_2 <- -0.2
>
> estimates <- numeric(N)
> for (i in 1:N) {
>     res <- arima.sim(list(order = c(2,0,0),
>                        ar = c(phi_1, phi_2)),
```

```
>                       n = time_steps)
>    estimates[i] <- arima(res, c(1, 0, 0))$coef[1]
> }
>
> hist(estimates,
>        main = "Estimated Phi for AR(1) when ts is AR(2)",
>        breaks = 50)
```

We see the resulting distribution in Figure 11-3. As we can see, the distribution is not as smooth and well defined for this misspecified model as it was for the appropriately specified model from Figure 11-2.

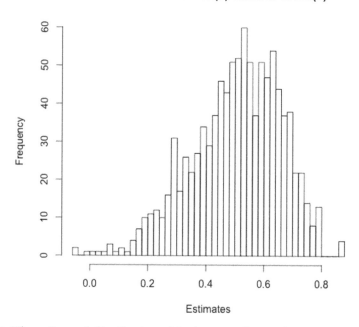

Figure 11-3. The estimated distribution of the lag 1 coefficient for an AR(1) model fitted to a process for which the true description is an AR(2) process.

It may strike you that these distributions are not all that different, and you are right. We confirm this with the summary statistics:[5]

```
## R
> summary(estimates)
    Min. 1st Qu.  Median    Mean 3rd Qu.    Max.
 -0.5252  0.4766  0.6143  0.5846  0.7215  0.9468
```

---

5 Whether the misfit is a problem will depend on the specific coefficients. Here the impact of our model misspecification is not too bad, but in other cases it could be catastrophic.

We can see that the range of estimates is wider when the model is misspecified and that the estimate for the first order term is slightly worse than it was when the model was properly specified, but the deviation is not too large.

This can address concerns that underestimating the order of the model will affect our estimate of ϕ. We can run a variety of simulation scenarios to address potential issues and understand the range of likely misestimation given some imagined possibilities.

Predicting Multiple Steps Ahead

While one-step-ahead forecasting is what we've covered in most of the past chapters, you might also want to predict multiple time steps ahead. This happens, among other reasons, when the time series data you have is at a greater temporal resolution than the time series values you would like to predict. For example, you might have daily stock prices available, but you would like to predict monthly stock prices so that you can execute a long-term strategy for your retirement savings. Alternately, you might have brain electrical activity readings that are taken every minute but you would like to predict a seizure at least five minutes in advance to give your users/patients as much warning as possible. In such cases, you have numerous options for generating multi-step-ahead forecasts.

Fit Directly to the Horizon of Interest

This is as simple as setting your y (target) value to reflect the forecast horizon of interest. So if your data consisted of minute-by-minute indicators but you wanted a five-minute forward horizon for your forecast, you would cut off your model inputs at time t and train them to a label generated with data up to time $t + 5$. Then you would fit to this data as what you were trying to predict, be it via a simple linear regression or a machine learning model or even a deep learning network. This effectively functions as:

$$model(X) = Y$$

where you can choose Y to have any time horizon you want. So each of these would be a legitimate scenario depending on your forward horizon of interest (be it 10 steps or 3 steps as shown here):

- $model_1(X_t)$ is fit to Y_{t+10}
- $model_2(X_t)$ is fit to Y_{t+3}

Recursive Approach to Distant Temporal Horizons

Using a recursive approach to fitting a variety of horizons, you build one model but prepare to feed its own output back into it as an input to predict more distant horizons. This idea should feel familiar because we demonstrated how to make multi-step-ahead forecasts with ARIMA modeling using this very strategy. Suppose we developed a model to fit one step ahead, by training on $model(X_t) = Y_{t+1}$. If we wanted to fit to horizons three steps ahead, we would do the following:

- $model(X_t) \rightarrow estimate\,Y_{t+1}$
- $model(X_t$ with estimate of $Y_{t+1}) \rightarrow$ estimate of Y_{t+2}
- $model(X_t$ with estimate of Y_{t+1} and estimate of $Y_{t+2}) \rightarrow$ estimate of Y_{t+3}

The expected error for our estimate of Y_{t+3} would necessarily be greater than that of our estimate for Y_{t+1}. How much greater? This could be complicated to think about. A great option to get a sense would be to run a simulation, as discussed earlier in this chapter.

Multitask Learning Applied to Time Series

Multitask learning is a general concept from deep learning that can be applied with a particular meaning for time series analysis. More generally, multitask learning describes the idea that a model can be built to serve multiple purposes at once or to learn to generalize by trying to predict several different but related targets at the same time. Some think of this as a means of regularization, encouraging the model to be more general by teaching it related tasks.

In the time series context, you can apply multitask learning by setting up targets for different time horizons in a forecasting context. In this case, fitting your model could look like this:

- $model(X_t) = (Y_{t+1}, Y_{t+10}, Y_{t+100})$
- $model(X_t) = (Y_{t+1}, Y_{t+2}, Y_{t+3})$

In training such a model, you could also think about how to see the loss function: would you want to weight all forecasts equally or would you want to privilege certain forecast horizons over others?

If you were seeking to make a very distant forecast, you could use multitask horizons to teach your model by including shorter-term horizons that might point to salient features useful for the longer-time horizon but difficult to suss out directly from the low signal-to-noise data of far-future forecasts. Another scenario for multitask modeling would be to fit to several time windows in the future, all in the same season but perhaps different points in time (so springtime over several years or Mondays over several weeks). This would be one way to fit both for seasonality and a trend at the same time.

Model Validation Gotchas

These are the most important issues to consider when thinking about whether you have appropriately tested your model as compared to expected production implementations:

Lookahead

Lookahead is emphasized throughout this book because it is so difficult to avoid and so potentially catastrophic and embarrassing when a model is put into production. Whenever possible you want to save yourself the danger of putting out models that suddenly stop validating in production. This is the hallmark of not being aware of lookahead in your system. Don't let this happen to you!

Structural Changes

Identifying structural changes is partly a matter of judgment, partly a matter of subject matter, and partly a matter of the quality of your data. The dynamics underlying a time series can change over time, and they can change enough that a model well suited to one portion of a time series may not be well suited to another. This is one of the reasons exploratory analysis is important—to make sure you are not training the same model across structural changes where doing so is neither justifiable nor sensible.

More Resources

Christoph Bergmeir, Rob J. Hyndman, and Bonsoo Koo, "A Note on the Validity of Cross-Validation for Evaluating Autoregressive Time Series Prediction," Computational Statistics & Data Analysis 120 (2018): 70–83, https://perma.cc/YS3J-6DMD.

The authors of this article (including the estimable and highly prolific Rob Hyndman) discuss cases where cross-validation will produce good results even for time series analysis. This is a helpful article in particular because it offers insights into why errors emerge from cross-validation and also what kinds of models you can use with a standard cross-validation breakdown rather than backfitting.

Angelo Canty, "tsboot: Bootstrapping of Time Series," n.d., https://perma.cc/MQ77-U5HL.

This description of the `tsboot()` function included in the `boot` package makes for educational reading, as it implements multiple widely accepted variations on block sampling as a form of bootstrapping. Exploring the options available in this function is a good way to get familiar with various ways to bootstrap time series, and familiarity with this function in turn will enable you to make convenient, reliable estimates of test statistics and assess the uncertainty of your estimation even in the case of small data sets.

Hans R. Kunsch, "The Jackknife and the Bootstrap for General Stationary Observations," Annals of Statistics 17, no. 3 (1989): 1217–41, https://perma.cc/5B2T-XPBC.
This classic and widely cited article demonstrates a statistical approach to extending jackknife (less commonly used) and bootstrapping methods to time series data, by eliminating the assumption of such methods that data points are independently distributed. This article introduces the idea of thinking about and sampling from time series with units of blocks of points rather than individual points.

Christian Kleiber, "Structural Changes in (Economic) Time Series," in Complexity and Synergetics, ed. Stefan Müller et al. (Cham, Switzerland: Springer International, 2018), https://perma.cc/7U8N-T4RC.
This chapter summarizes a variety of canonical methods for identifying structural changes in time series, and it is written from a business and economic time series perspective, although the lessons apply more broadly. It is particularly useful because it provides a listing of the R packages that can be used to apply these widely accepted structural change identification methods and because it offers helpful visualizations of how methods behave somewhat differently when analyzing the same time series.

Robert Stambaugh, "Predictive Regressions," Journal of Financial Economics 54 (1999): 375–421, https://perma.cc/YD7U-RXBM.
This widely cited article demonstrates how to perform a Bayesian posterior estimation of regression coefficients and how estimations of quantities from finite samples can differ substantially given different assumptions about the underlying dynamics of financial time series models. This is an excellent, albeit domain-specific, example of how much starting assumptions can affect estimates of uncertainty in your estimates, parameters.

## CHAPTER 12

Performance Considerations in Fitting and Serving Time Series Models

In the literature on machine learning and statistical analyses, the overwhelming focus tends to be on performance of models in terms of accuracy. While accuracy should usually be the primary concern when evaluating a model, sometimes computational performance considerations matter tremendously in the face of large data sets or widely deployed models to serve large populations of client applications.

Time series data sets get so large that analyses can't be done at all—or can't be done properly—because they are too intensive in their demands on available computing resources. In such cases, many organizations treat their options as follows:

- Upsize on computing resources (expensive and often wasteful both economically and environmentally).
- Do the project badly (not enough hyperparameter tuning, not enough data, etc.).
- Don't do the project.[1]

None of these options are satisfying, particularly when you are just starting out with a new data set or a new analytical technique. It can be frustrating not to know whether your failures are the result of poor data, an overly difficult problem, or a lack of resources. Hopefully, we will find some workarounds to expand your options in the case of very demanding analyses or very large data sets.

This chapter is designed to guide you through some considerations of how to lessen the computing resources you need to train or infer using a particular model. For the

---

1 Yes, this happens all the time in the real world.

most part, such questions are specific to a given data set, the resources you have available, and both your accuracy and speed targets. You will see this reality echoed in the concerns detailed in this chapter, but the hope is that they will partly cover the problems you run into and can provide inspiration for further brainstorming. These are considerations to come when you have completed your first rounds of analysis and modeling and should not be a priority when you are first attaching a problem. However, when it's time to put something into production or extend a small research project into a larger one, you should revisit these concerns often.

Working with Tools Built for More General Use Cases

One challenge of time series data is that most tools, particularly those for machine learning, are built for a more general use case, with most illustrative examples showcasing uses of cross-sectional data. There are a few ways these machine learning methods then fail to be as efficient as possible with time series data. The solutions to your individual problems will vary, but the general ideas are the same. In this section I discuss common problems and potential solutions.

Models Built for Cross-Sectional Data Don't "Share" Data Across Samples

In many cases where you are feeding discrete samples of time series data into an algorithm, most often with machine learning models, you will notice that large chunks of the data you are feeding between samples overlap. For example, suppose you have the following data on monthly sales:

Month	Widgets sold
Jan 2014	11,221
Feb 2014	9,880
Mar 2014	14,423
Apr 2014	16,720
May 2014	17,347
Jun 2014	22,020
Jul 2014	21,340
Aug 2014	25,973
Sep 2014	11,210
Oct 2014	11,583
Nov 2014	12,014
Dec 2014	11,400
Jan 2015	11,539
Feb 2015	10,240

You are seeking to make predictions by mapping each "shape" to a nearest neighbor curve. You prepare many shapes from this data. Just a smattering of those data points are listed here, as you might want to use six-month-long curves as the "shapes" of interest (note we are not doing any data preprocessing to normalize or create additional features of interest, such as moving average or smoothed curves):[2]

```
11221, 9880, 14423, 16720, 17347, 22020
9880, 14423, 16720, 17347, 22020, 21340
14423, 16720, 17347, 22020, 21340, 25973
```

Interestingly, all we have managed to do with this preparation of inputs is to make our data set six times larger, without including any additional information. This is a catastrophe from a performance standpoint, but it's often necessary for inputs into a variety of machine learning modules.

If this is a problem you run into, there are a few solutions you should consider.

Don't use overlapping data

Consider only producing a "data point" so that each individual month makes its way into only one curve. If you do so, the preceding data could look like the following table:

```
11221, 9880, 14423, 16720, 17347, 22020
21340, 25973, 11210, 11583, 12014, 11400
```

Note that this would be particularly simple because it amounts to a simple array reshape rather than a custom repetition of data.

Employ a generator-like paradigm to iterate through the data set

Employing a generator-like paradigm to iterate through the data set, resampling from the same data structure as appropriate, is particularly easy to code up in Python but can also be done with R and other languages. If we imagine that the original data is stored in a 1D NumPy array, this could look something like the following code (note that this would have to be paired with a machine learning data structure or algorithm that accepts generators):

```python
## python
>>> def array_to_ts(arr):
>>>     idx = 0
```

---

2 We extracted many time series samples from one large time series in the case of both the machine learning and the deep learning models discussed in earlier chapters.

```
>>>    while idx + 6 <= arr.shape[0]:
>>>        yield arr[idx:(idx+6)]
```

Note that designing data modeling code that does not unnecessarily blow up a data set is desirable from both a training and production standpoint. In training this will allow you to fit more training examples in memory, and in production you can make multiple predictions with fewer training resources in the case of predictions (or classifications) on overlapping data. If you are making frequent predictions for the same use case, you will likely be working with overlapping data, so this problem and its solutions will be quite relevant.

Models That Don't Precompute Create Unnecessary Lag Between Measuring Data and Making a Forecast

Usually machine learning models do not prepare for or take into account the possibility of precomputing part of a result in advance of having all the data. Yet for time series this is a very common scenario.

If you are serving your model in a time-sensitive application, such as for medical predictions, vehicle location estimations, or stock price forecasting, you may find that the lag of computing a forecast only after all the data becomes available is too great. In such a case, you should consider whether the model you have chosen can actually be partly precomputed in advance. A few examples of where this is possible are as follows:

- If you are using a recurrent neural network that takes several channels of information over 100 different time steps, you can precompute/unroll the neural network for the first 99 time steps. Then when the last data point finally comes in, you only need to do 1 final set of matrix multiplications (and other activation function computations) rather than 100. In theory this could speed up your response time 100-fold.

- If you are using an AR(5) model, you can precompute all but the most recent term in the sum that constitutes the mode. As a reminder, an AR(5) process looks like the equation that follows. If you are about to output a prediction, it means that you already know the values of y_{t-4}, y_{t-3}, y_{t-2}, and y_{t-1}, which means that you can have everything other than $ph\, i_0 \times y_t$ ready to go in advance of knowing y_t:

$$y_{t+1} = ph\, i_4 \times y_{t-4} + ph\, i_3 \times y_{t-3} + ph\, i_2 \times y_{t-2} + ph\, i_1 \times y_{t-1} + ph\, i_0 \times y_t$$

- If you are using a clustering model to find nearest neighbors by summarizing features of a time series (mean, standard deviation, max, min, etc.), you can compute these features with a time series with one less data point and run your model with that time series to identify several nearest neighbors. You can then update these features once the final value rolls in and rerun the analysis with only the nearest neighbors found in the first round of analysis. This will actually require more computation resources overall but will result in a lower lag time between taking the final measurement and delivering the forecast.

In many cases your model may not be nearly as slow as the network lag or other factors, so precomputation is a worthwhile technique only where feedback time is extremely important and where you are confident that model computation is substantially contributing to the time between an application receiving all needed information and outputting a useful prediction.

Data Storage Formats: Pluses and Minuses

One overlooked area of performance bottlenecks for both training and productionizing time series models is the method of storing data. Some common errors are:

- *Storing data in a row-based data format even though the time series is formed by traversing a column.* This results in data where time-adjacent points are not memory-adjacent.

- *Storing raw data and running analyses off this data.* Preprocessed and downsampled data is preferable to the greatest extent possible for a given model.

Next we discuss these data storage factors to keep your model training and inference as speedy as possible.

Store Your Data in a Binary Format

It is tempting to store data in a comma-separated text file, such as a CSV file. That is often how data is provided, so inertia pushes us to make this choice. Such file formats are also human readable, which makes it easy to check the data in the file against pipeline outputs. Finally, such data is usually easy to ship around different platforms.[3]

However, it's not easy for your computer to read text files (*https://perma.cc/XD3Y-NEGP*). If you are working with data sets so large that you are not able to fit all your data in memory during training, you will be dealing with I/O and related processing all tied to the file format you choose. By storing data in a binary format, you can substantially cut down on I/O-related slowdowns in a number of ways:

---

3 Although there are Unicode problems related to different platforms and devices, so you are not in the clear just because you use a text-based file format.

- Because the data is in binary format, your data processing package already "understands" it. There is no need to read a CSV and turn it into a data frame. You will already have a data frame when you input your data.

- Because the data is in binary format, it can be compressed far more than a CSV or other text-based file. This means that I/O itself will be shorter because there is less physical memory to read in a file so as to re-create its contents.

Binary storage formats are easily accessible in both R and Python. In R, use `save()` and `load()` for `data.table`. In Python, use pickling and note that both Pandas (`pd.DataFrame.load()` and `pd.DataFrame.save()`) and NumPy (`np.load()` and `np.save()`) include wrappers around pickling that you can use for their specific objects.

Preprocess Your Data in a Way That Allows You to "Slide" Over It

This recommendation is related to "Models Built for Cross-Sectional Data Don't "Share" Data Across Samples" on page 358. In this case, you should also think about how you preprocess your data and make sure the way you do so is consistent with using a moving window over that data to generate multiple test samples.

As an example, consider normalization or moving averages as preprocessing steps. If you plan to do these for each time window, this may lead to improved model accuracy (although in my experience such gains are often marginal). However, there are several downsides:

- You need more computational resources to compute these preprocessing features many, many times on overlapping data—only to end up with very similar numbers.

- You need to store overlapping data with slightly different preprocessing many, many times.

- You cannot take optimum advantage of sliding a window over your data.

Modifying Your Analysis to Suit Performance Considerations

Many of us are guilty of getting too comfortable with a particular set of analytical tools, and the accompanying suite of software and rules of thumb about how to fit a model. We also tend to assess accuracy needs once and not reassess them when we determine the computational cost of various possible model performances.

Time series data, often used to make a fast forecast, is particularly prone to needing models that can be both fit and productionized quickly. The models need to be fit

---

quickly so they can be updated with new data coming in, and they need to perform quickly so that the consumers of the models' forecasts have as much time as possible to act on those forecasts. For this reason, you may sometimes want to modify your expectations—and accompanying analysis—to make faster and more computationally streamlined processes for analysis and forecasting.

Using All Your Data Is Not Necessarily Better

One important factor in thinking about how to streamline your analysis is to understand that not all data in a time series is equally important. More distant data is less important. Data during "exceptional" times is less important to building a model for ordinary times.

There are many ways you should consider lessening the amount of data you use to train a model. While many of these options have been discussed earlier in this book, it's good to review them, particularly with respect to performance:

Downsampling
> It is often the case that you can use data that is less frequent to cover the same lookback window when making a prediction. This is a way to downsize your data by a multiplicative factor. Note that, depending on the analytical technique you use, you also have more creative options, such as downsampling at different rates depending on how far back the data goes.

Training only on recent data
> While machine learning loves data, there are many time series models where statistical or even deep learning techniques will actually do better just focusing on recent data rather than training on all data. This will help you reduce your input data by subtracting out data that is only marginally informative for your model.

Reduce the lookback window used to make your prediction
> In many time series models, the model's performance will continue to improve, if only slightly, as you look further and further back into the past. You should make some decisions about how much accuracy is really required relative to performance. It may be that you are loading far more data into memory per sample than is really necessary for acceptable performance.

Complicated Models Don't Always Do Better Enough

It can be interesting and fun to try the latest and greatest when it comes to choosing an analytical model. However, the real question is whether a fancier model "pays" for any additional computational resources required.

Almost all the computational advances in machine learning of recent years have come from throwing more computing power at a problem. With problems such as image recognition, where there is definitely a right answer and 100% accuracy, this can certainly make sense.

On the other hand, with problems such as time series predictions, where there may be physical or mathematical limits to how accurate a prediction can be, you should make sure that choosing a more complex model is not simply automatic upgrading without a cost-benefit analysis. Think about whether the accuracy gains justify the additional lag the model may produce in computing a forecast, or the additional training time that will be required, or the additional computational resources that will be spun up. It may be that a less resource-intensive method with slightly poorer accuracy is a far better "deal" than a fancy model that barely improves upon the simple version.

If you are the data analyst, this trade-off between complexity/accuracy and lag time/ computer resources is something you should analyze, thinking of it as another hyper-parameter to tune. It's your job to point out these trade-offs rather than assume that a data engineer will take care of it. People upstream or downstream in the data pipeline cannot substitute your judgment for model selection, so you have to acknowledge the engineering side of data science while weighing pros and cons.

A Brief Mention of Alternative High-Performance Tools

If you have fully explored the previous options, you can also consider changing your underlying code base, more specifically by moving away from slower scripting languages, such as Python and R. There are a number of ways to do this:

- Go all in with C++ and Java. Even if you haven't looked at these languages, just learning the basics can sometimes speed up the slow parts of your pipeline enough to transform impossible tasks into manageable ones. C++ in particular has evolved tremendously in terms of usability and standard libraries applicable to data processing. The STL and C++ 17 syntax now offers many options quite comparable to Python for operating on sets of data in a variety of data structures. Even if you hated C++ and Java years ago, you should revisit them.[4]

- In Python, you can use several different modules wherein you can write Python code that gets compiled into C or C++ code, speeding up execution time. This can be especially useful for very repetitive code with many for loops, which are slow in Python and can become much more performant in C or C++ without the need for clever design—simply implementing the same code in the faster

---

4 There is a steep learning curve, but once you have the basic compilation infrastructure figured out, this can be a huge bonus to your organization.

language can do the trick. `Numba` and `Cython` are both accessible drop-in Python modules that can help you speed up slow chunks of Python code this way.

- Likewise, in R, you can use `Rcpp` for similar functionality.

More Resources

- On models performing equally:

Anthony Bagnall et al., "The Great Time Series Classification Bake Off: An Experimental Evaluation of Recently Proposed Algorithms," Data Mining and Knowledge Discovery 31, no. 3 (2017): 606–60, https://perma.cc/T76B-M635.
  This article runs an extensive array of experiments to assess performance of modern time series classification methodologies, comparing their performance on a wide variety of publicly available data sets. The computational complexity of the data sets ultimately ends up varying quite a bit more than does the actual performance of the methods tried. As the authors highlight, it remains an art and an area of research to determine, without trying them all, which method will work best on a given data set. From a computing resources perspective, the lesson to learn here is that computational complexity should factor in significantly in methodological decisions. Unless you have a very compelling use case for a complicated and resource-intensive algorithm, choose something simpler.

- On building simpler models:

Yoon Kim and Alexander M. Rush, "Sequence Level Knowledge Distillation," in Proceedings of the 2016 Conference on Empirical Methods in Natural Language Processing, ed. Jian Su, Kevin Duh, and Xavier Carreras (Austin, TX: Association for Computational Linguistics, 2016), 1317–27, https://perma.cc/V4U6-EJNU.
  This article applies the general concept of "distillation" to sequence learning, as applied to a machine translation task. The concept of distillation is a broadly useful one. The idea is that first a complex model is designed and trained on the original data, and then a simpler model is trained on the outputs of the complex model. The outputs of the complex model, rather than the data itself, reduces noise and simplifies the learning problem, making it easier for a simpler model to learn approximately the same relationship by cutting out the noise. While such a technique will not lessen training time, it should produce a model that executes faster and demands fewer resources when put into production.

## CHAPTER 13

Healthcare Applications

In this chapter we will look at time series analysis in the healthcare context with two case studies: Flu forecasting and nowcasting and Blood glucose forecasting. >These are both important healthcare applications for common health problems. Also, in both cases these are not solved problems but rather topics of ongoing research in academia and in the healthcare industry.

Predicting the Flu

Predicting the flu rate from week to week in a given geographic area is a longstanding and ongoing problem. Infectious disease specialists and global security professionals alike agree that infectious diseases pose a significant risk to human welfare (*https://perma.cc/ZDA8-AKX6*). This is particularly the case for the flu, which strikes the vulnerable worldwide, inflicting hundreds of fatalities every year, mostly among the very young and very old. It's crucial from both a health and national security standpoint to develop accurate models of how the flu will run its course in a given season. Flu prediction models are useful both to predict the virus specifically and also to help researchers explore general theories for how infectious diseases travel geographically.

A Case Study of Flu in One Metropolitan Area

We'll look at a data set of weekly flu reports for a variety of administrative regions in France for the years of 2004 through 2013. We will predict the flu rate for Île de France, the Paris metropolitan region. The data can be downloaded from Kaggle (*https://perma.cc/W9VQ-UUJC*)[1] and is also available in the code repository for this book.

---

1 While this is not a public data set, it is possible to access the data by registering for the competition.

Data exploration and some cleanup

We begin by familiarizing ourselves with the raw data, first examining it in tabular form:

```
## R
> flu = fread("train.csv")
> flu[, flu.rate := as.numeric(TauxGrippe)]
> head(flu)
     Id   week region_code    region_name TauxGrippe flu.rate
1: 3235 201352          42          ALSACE          7        7
2: 3236 201352          72       AQUITAINE          0        0
3: 3237 201352          83        AUVERGNE         88       88
4: 3238 201352          25 BASSE-NORMANDIE         15       15
5: 3239 201352          26       BOURGOGNE          0        0
6: 3240 201352          53        BRETAGNE         67       67
```

We also do some basic quality checks, such as looking for NA in our variables of interest. We may not know where these NA values come from, but we need to account for them:

```
## R
> nrow(flu[is.na(flu.rate)]) / nrow(flu)
[1] 0.01393243
> unique(flu[is.na(flu.rate)]$region_name)
 [1] "HAUTE-NORMANDIE"       "NORD-PAS-DE-CALAIS"   "PICARDIE"
 [4] "LIMOUSIN"              "FRANCHE-COMTE"        "CENTRE"
 [7] "AUVERGNE"              "BASSE-NORMANDIE"      "BOURGOGNE"
[10] "CHAMPAGNE-ARDENNE"     "LANGUEDOC-ROUSSILLON" "PAYS-DE-LA-LOIRE"
[13] "CORSE"
```

The overall rate of NA data points is not very high. Additionally, our region of interest, Île-de-France, is not included in the list of regions with NA values.

We perform some data cleanup, separating out the week and year portion of the timestamp column (which is currently in character format, not numerical or timestamp format):

```
## R
> flu[, year := as.numeric(substr(week, 1, 4))]
> flu[, wk   := as.numeric(substr(week, 5, 6))]
> ## style note it's not great we have 2 week columns
```

We add a `Date` class column so we can have better plotting axes for time than if we treated the data as nontimestamped:

```
## R
> flu[, date:= as.Date(paste0(as.character(flu$week), "1"), "%Y%U%u")]
```

This line of code is slightly tricky. To convert the month-week combinations into dates, we add a component indicating the day. This is the purpose of `paste0()`, which marks each date as the first day of the week, pasting a "1" onto a string that already designates year and week of the year (out of 52 weeks of the year—more on this shortly).[2] Notice the %U and %u in the format string: these have to do with marking time according to the week of the year and day of the week, a somewhat unusual timestamping format.[3]

We then subset to the data relating specifically to Paris and sort the data by date:[4]

```
## R
## let's focus on Paris
> paris.flu = flu[region_name == "ILE-DE-FRANCE"]
> paris.flu = paris.flu[order(date, decreasing = FALSE)]

> paris.flu[, .(week, date, flu.rate)]
        week        date flu.rate
  1: 200401 2004-01-05       66
  2: 200402 2004-01-12       74
  3: 200403 2004-01-19       88
  4: 200404 2004-01-26       26
  5: 200405 2004-02-02       17
 ---
518: 201350 2013-12-16       13
519: 201351 2013-12-23       49
520: 201352 2013-12-30       24
521: 200953       <NA>      145
522: 200453       <NA>       56
```

If you are paying attention, the row count should surprise you. If there are 52 weeks in a year, and we have 10 years' worth of data, why do we have 522 rows? We would have expected 52 weeks × 10 years = 520 rows. Similarly, why are there two NA dates? You will see an explanation if you go back to the original data. There seems to be a 53rd week both for 2004 and for 2009. Every few years there is a year that has 53 weeks rather than 52—it's not a mistake but rather a part of the Gregorian calendar system (https://perma.cc/4ETJ-88QR).

We then check that the data covers a full and regularly sampled date range by first making sure that each year has the same number of data points:[5]

---

2 The correct day of the week (with the options being 1 through 7) to begin with depends on what date the flu rate was calculated, but we don't know that from the data provided. We choose the first day for simplicity. The last day would be just as sensible in this situation, and it is not important to the analysis so long as we are consistent.

3 The appropriate formatting string for week of the year and day of the week can depend on your operating system. My solution worked on a macOS, whereas I used slightly different formatting on Linux.

4 If we were working in a large data set, we should do this as a first step to avoid computationally taxing operations, such as converting strings to Date objects, for irrelevant data.

5 This is not the only necessary check to make sure the data is regularly sampled and complete. I leave the rest to you.

```
## R
> paris.flu[, .N, year]
     year  N
 1: 2004 53
 2: 2005 52
...
 9: 2012 52
10: 2013 52

> paris.flu[, .N, wk]
     wk  N
 1:  1 10
 2:  2 10
 3:  3 10
...
51: 51 10
52: 52 10
53: 53  2
     wk  N
```

We can see that the data is as expected; that is, each year (other than the two just discussed) has 52 weeks, and each week-of-year label has 10 data points, one for each year (except week 53).

Now that we've considered the timestamps of the data, we inspect the actual values of the time series (so far we have considered only the time indexing). Is there a trend? Seasonality? Let's find out (see Figure 13-1):

```
## R
> paris.flu[, plot(date, flu.rate,
>                   type = "l", xlab = "Date",
>                   ylab = "Flu rate")]
```

It's clear from a simple line plot that there is substantial seasonality (something you've likely experienced in your own community). This plot suggests a strong seasonal component but does not suggest there is a temporal drift apart from the seasonality.

The seasonal behavior complicates the 53rd week. If we want to fit a seasonal model, we need to define the seasonality in terms of weeks of the year, and we cannot have variable seasonal sizes (that is what distinguishes a season from a cycle, as discussed in Chapter 3). While we could imagine some creative solutions to the problem of the 53rd week, we will take the simple option of suppressing this data:

```
## R
> paris.flu <- paris.flu[week != 53]
```

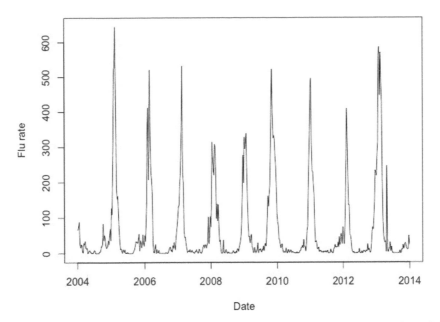

Figure 13-1. By plotting the time series of the flu rate, we can see the seasonality of the flu rate in Paris.

Whether it is a significant problem to delete a data point depends on the data set and the question we are asking. I leave it as an exercise to the reader to explore other possibilities for fitting the data while keeping the 53rd week data. There are a variety of options to do so. One is merging the 53rd week data into the 52nd week of data by averaging the two weeks. Another is to use a model that can account for cyclical behavior without having to be locked into exactly the same length cycle each year. A third option is that a machine learning model might be able to accommodate this with some creative labeling of the data to indicate seasonality to the model as an input feature.

Fitting a seasonal ARIMA model

We first consider fitting a seasonal ARIMA model to the data because of the strong seasonality. In this case the periodicity of the data is 52, since the data is sampled weekly. We want to choose a fairly parsimonious model—that is, one without too many parameters—because at 520 data points our time series is not particularly long.

This time series is a good example of how we can go wrong if we rely too strongly on autopilot. For example, we might first consider whether to difference the data, and so we might consider the plot of the autocorrelation of the flu rate and the autocorrelation of the differenced time series of the flu rate, each shown in Figure 13-2:

```
## R
> acf(paris.flu$flu.rate,          )
> acf(diff(paris.flu$flu.rate, 52))
```

Series paris.flu$flu.rate

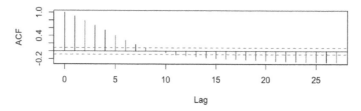

Series diff(paris.flu$flu.rate, 52)

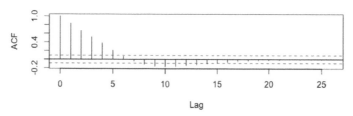

Figure 13-2. Plot of the autocorrelation function for the Paris flu rate and the differ-enced Paris flu rate. We look only at a limited range of lag values.

If we were careless, we could very well congratulate ourselves on having solved the stationarity issue for this time series by differencing once. But this doesn't make sense at all. This is weekly data, and we observed a strong seasonality. Why aren't we seeing it in our autocorrelation plot? We took the default parameters of the acf() function, and this does not take us out nearly far enough in the lag space to see the seasonal effects, which we begin at lag 52 (one year). Let's redo the acf() with an adequate window (Figure 13-3):

```
## R
> acf(paris.flu$flu.rate,            lag.max = 104)
> acf(diff(paris.flu$flu.rate, 52), lag.max = 104)
```

Series paris.flu$flu.rate

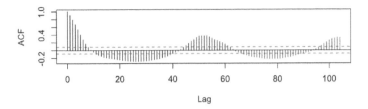

Series diff(paris.flu$flu.rate, 52)

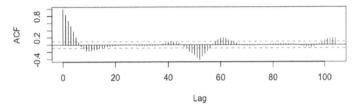

Figure 13-3. Plot of the autocorrelation function for the Paris flu rate and the differenced Paris flu rate. We now look at a more extensive range of lag values.

This gives us a more realistic picture of the autocorrelation of our series. As we can see there are substantial autocorrelations at various lags, and this makes sense (at least given my experience) living in a four-season climate. Flu rates will have a strong correlation with neighboring weeks—that is, close to their time of measurement.

Flu rates will also have a strong correlation, given the seasonality, with time periods lagged by around 52 or around 104 because this indicates the yearly seasonality. But flu rates also have a fairly strong relationship with time periods lagged by intermediary values, such as half a year, such as half a year (26 weeks) because such lags also relate to seasonal differences and predictable weather variations. For example, we know that in half a year the flu value will likely have changed quite a bit. If it was high earlier, it should be low now and vice versa, again due to the seasonality. All this is depicted in the upper plot of Figure 13-3.

We then examine the differenced series as is illustrated in the lower plot in Figure 13-3. Now we see a substantial amount of the time series autocorrelation has been diminished. However, there remains quite a bit of autocorrelation over a range of values, not just at 52 or 104 weeks (one or two years), but also at intermediate values.

While we might be tempted to continue differencing, we need to remember that real-world data will never fit a SARIMA model perfectly. We instead seek out the most reasonable way to model the data. We can consider differencing again seasonally or taking a different tactic and differencing in linear time. We plot each of these possibilities here (see Figure 13-4):

```R
## R
> plot(diff(diff(paris.flu$flu.rate, 52), 52))
> plot(diff(diff(paris.flu$flu.rate, 52), 1))
```

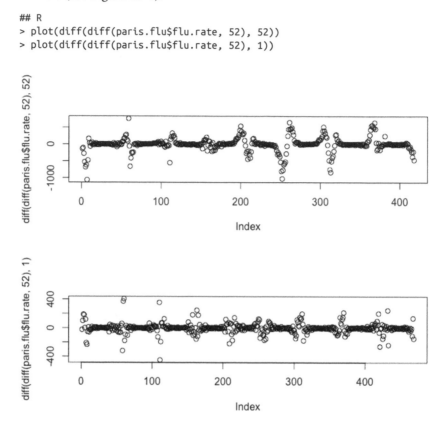

Figure 13-4. Plot of two versions of differencing our series to get a sense of seasonal behavior in the data.

While neither result is ideal, the latter, a standard first differencing of a seasonal differencing, is more satisfactory.

The decision of a fit or a parameter choice is a judgment call as much as it is a matter of applying tests. Here we choose to give weight to the seasonality we clearly observed but also to not overcomplicate the model or make it unduly opaque. So in a SARIMA (p, d, q) (P, D, Q) model, we will fit with $d = 1$ and $D = 1$. We then choose our AR and MA parameters for our standard ARIMA parameters, p and q. We do this with standard visualizations, using the following code (see Figure 13-5):

```
## R
> par(mfrow = c(2, 1))
> acf (diff(diff(paris.flu$flu.rate, 52), 1), lag.max = 104)
> pacf(diff(diff(paris.flu$flu.rate, 52), 1), lag.max = 104)
```

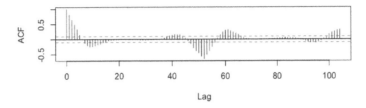

Series diff(diff(paris.flu$flu.rate, 52), 52)

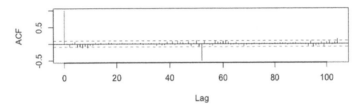

Series diff(diff(paris.flu$flu.rate, 52), 1)

Figure 13-5. Plot of the partial autocorrelation function of the differenced series we have selected.

We have a limited data set and err on the side of a simpler model. The PACF model suggests an AR(2) model could be appropriate, so we will model our data parsimoniously as a SARIMA (2, 1, 0), (0, 1, 0).

We are interested in understanding how this model would work if we fit it continuously on new data as it becomes available, the way most models built for real-world systems will function. That is, if we were modeling the flu starting several years ago, each week modeling only with the data that was available up to that point in time, how would our model do? We answer this by rolling the fitting and evaluation of the model forward, as follows:

```
## R
> ## arima fit
> ## let's estimate 2 weeks ahead
> first.fit.size <- 104
> h               <- 2
> n               <- nrow(paris.flu) - h - first.fit.size
>
> ## get standard dimensions for fits we'll produce
> ## and related info, such as coefs
> first.fit <- arima(paris.flu$flu.rate[1:first.fit.size], order = c(2, 1, 0),
>                    seasonal = list(order = c(0,1,0), period = 52))
```

```
> first.order <- arimaorder(first.fit)
>
> ## pre-allocate space to store our predictions and coefficients
> fit.preds <- array(0, dim = c(n, h))
> fit.coefs <- array(0, dim = c(n, length(first.fit$coef)))
>
> ## after initial fit, we roll fit forward
> ## one week at a time, each time refitting the model
> ## and saving both the new coefs and the new forecast
> ## caution! this loop takes a while to run
> for (i in (first.fit.size + 1):(nrow(paris.flu) - h)) {
>     ## predict for an increasingly large window
>     data.to.fit = paris.flu[1:i]
>     fit = arima(data.to.fit$flu.rate, order = first.order[1:3],
>                 seasonal = first.order[4:6])
>     fit.preds[i - first.fit.size, ] <- forecast(fit, h = 2)$mean
>     fit.coefs[i - first.fit.size, ] <- fit$coef
> }
```

We then plot these rolling results (see Figure 13-6):

```
## R
> ylim <- range(paris.flu$flu.rate[300:400],
>                    fit.preds[, h][(300-h):(400-h)])
> par(mfrow = c(1, 1))
> plot(paris.flu$date[300:400], paris.flu$flu.rate[300:400],
>      ylim = ylim, cex = 0.8,
> main = "Actual and predicted flu with SARIMA (2, 1, 0), (0, 1, 0)",
> xlab = "Date", ylab = "Flu rate")
> lines(paris.flu$date[300:400], fit.preds[, h][(300-h):(400-h)],
>       col = 2, type = "l",
>       lty = 2, lwd = 2)
```

The plot shows a helpful forecast but also highlights some limitations of this model. It is insufficiently realistic, demonstrated by the fact that it sometimes forecasts negative flu rates. The model can do so because there is nothing intrinsic to an ARIMA model to enforce constraints such as flu rates having to be nonnegative. Enforcing these kinds of physical constraints is something we must do with data transformations prior to fitting a model.

Also, the model appears more sensitive to outlier points than we would like. Early 2013 is a prime example where several times the model drastically overestimates the flu rate. When it comes to allocation of resources crucial to fighting diseases, this is not an acceptable model.

Finally, this model produces more extreme values at the peaks than are ever measured in a given year. This could lead to overallocation of resources relative to actual need, which is not a good outcome, particularly when resources are highly constrained. This is a concern about the model that is just as rooted in real-world resource constraints as it is in pure data analysis.

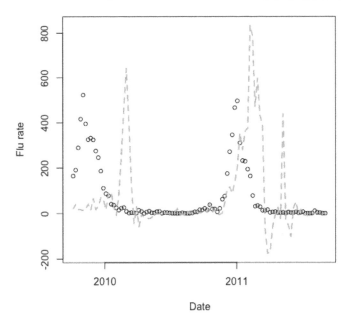

Actual and predicted flu with SARIMA (2, 1, 0), (0, 1, 0)

Figure 13-6. The flu rates (points) paired with our SARIMA predictions (dotted line). The predictions of this simple model could assist with public health planning.

Now that we have considered the fit of a basic ARIMA model to this problem, we will take a look at other modeling possibilities.

An alternative ARIMA model: Exogenous harmonic regressors instead of seasonality

Given the performance of the SARIMA model discussed in the previous section, there are a number of modifications we can make. Here we consider two modifications, each of which is independent of the other and could be applied separately.[6]

First, we would like to build in constraints to our model to prevent the prediction of negative values. One way to do this is a log transformation of the data, such that we are predicting the logarithm of the value of the time series rather than the value itself. Then, when we want to see the "true" series representing the numbers actually measured, we will use an exponential transform to back out the logarithmic transform and get the predictions in their real units.

Second, we would like to find a more transparent way of dealing with the seasonality of the data. While we described our seasonal ARIMA model as simple, in fact a

---

6 Normally, you should consider these separately, but here they are combined for succinctness.

model dealing with a seasonal recurrence of 52 is not simple at all. Seasonal ARIMA models do better on shorter seasonal cycles and less well on longer seasonal cycles (52 is a long seasonal cycle).

Here we will use *dynamic harmonic regression*. In this approach, we find a Fourier series that describes the periodicity in our data, and then use this series as an exogenous regressor that is fit alongside the ARIMA terms.[7] Because we can extrapolate the Fourier series forward in time (due to its purely periodic nature), we can also precompute the values we expect for the future when generating forecasts.

The strength of this model is that the degrees of freedom of the model can be used to explain underlying behavior in addition to the seasonal behavior rather than devoting a lot of explanatory power to the seasonal behavior.

There are a few downsides to dynamic harmonic regression. First, we assume the behavior is quite regular and recurs at exactly the same interval. Second, we assume that the seasonal behavior is not changing; that is, the period and amplitude of the seasonal behavior are not changing. These limitations are similar to the SARIMA model, although the SARIMA model shows more flexibility in how the amplitude of seasonality affects data over time.

Here, we demonstrate how to perform a dynamic harmonic regression with R code similar to what we employed before:

```
## R
> ## preallocate vectors to hold coefs and fits
> fit.preds      <- array(0, dim = c(n, h))
> fit.coefs      <- array(0, dim = c(n, 100))
>
> ## exogenous regressors
> ## that is components of Fourier series fit to data
> flu.ts         <-  ts(log(paris.flu$flu.rate + 1) + 0.0001,
>                        frequency = 52)
> ## add small offsets because small/0 vals
> ## cause numerical problems in fitting
> exog.regressors <- fourier(flu.ts, K = 2)
> exog.colnames   <- colnames(exog.regressors)
>
> ## fit model anew each week with
> ## expanding window of training data
> for (i in (first.fit.size + 1):(nrow(paris.flu) - h)) {
>   data.to.fit       <- ts(flu.ts[1:i], frequency = 52)
```

---

7 A Fourier series is the practice of describing a function as a series of sines and cosines. We have mentioned Fourier series previously in passing. If you are not familiar with Fourier series, consider taking a few minutes to read up on the background. There are well-established ways to "fit" a Fourier series to any time series, and these are widely available in R and Python packages. Ritchie Vink has one brief tutorial I like a lot (*https:// perma.cc/7HJJ-HC2T*).

```
>    exogs.for.fit    <- exog.regressors[1:i,]
>    exogs.for.predict <- exog.regressors[(i + 1):(i + h),]
>
>    fit <- auto.arima(data.to.fit,
>                      xreg = exogs.for.fit,
>                      seasonal = FALSE)
>
>    fit.preds[i - first.fit.size, ] <- forecast(fit, h = h,
>                                xreg = exogs.for.predict)$mean
>    fit.coefs[i - first.fit.size, 1:length(fit$coef)] = fit$coef
> }
```

Here we have made a few adjustments to the code from the previous section. First, we use a `ts` object.[8] With a `ts` object, we indicate explicitly the seasonality of the time series (52 weeks), when creating the `ts` object.

We log-transform the data at this point as well to ensure positive predictions of our ultimate value of interest, the flu rate:

```
## R
> flu.ts = ts(log(paris.flu$flu.rate + 1) + 0.0001, ## add epsilon
>                      frequency = 52)
```

We also add a small numerical offset (`+ 0.0001`) because numerical fitting does not play well with strict zero values or even very small values. One of our two adjustments is already accomplished with just this line of code (i.e., the log transformation to enforce a physical condition of nonnegative values).

Next we generate the exogenous harmonic regressors (that is, the Fourier approximation) we are using in place of seasonal parameters in the SARIMA. We do this via the `forecast` package's `fourier()` function:

```
## R
> exog.regressors <- fourier(flu.ts, K = 2)
> exog.colnames   <- colnames(exog.regressors)
```

We generate the accompanying harmonic series for our entire time series first, and then we subset it as appropriate to our expanding window rolling fit later in the loop.

The hyperparameter K indicates how many separate sine/cosine pairs we will include in our fit, where each one represents a new frequency used to fit the sine/cosine. In general, K will be larger for larger seasonal period lengths and smaller for smaller ones. In a more extended example, we could consider how to use an information criterion to tune K, but for this example we use K = 2 as a reasonable model.

Finally, all that is left to do is to generate new fits that take into account the exogenous Fourier components we just fit. We do the fitting as follows, where the `xreg`

---

8 These were briefly discussed in Chapter 3.

parameter takes the fit Fourier series as additional regressors, which are fit along with the standard ARIMA parameters:

```R
## R
> fit <- auto.arima(data.to.fit,
>                    xreg    = exogs.for.fit,
>                    seasonal = FALSE)
```

We set the seasonal parameter to FALSE to ensure that we will not have redundant seasonal parameters given our decision to use dynamic harmonic regression in this case.

We also need to include regressors when we generate a forecast, meaning we need to indicate what the regressors will be at the time the forecast is targeting:

```R
## R
> fit.preds[i - first.fit.size, ] <- forecast(fit, h = 2h,
>                                       xreg = exogs.for.predict)$mean
```

We plot the performance of this model, as follows (see Figure 13-7):

```R
## R
> ylim = range(paris.flu$flu.rate)
> plot(paris.flu$date[300:400], paris.flu$flu.rate[300:400],
>     ylim = ylim, cex = 0.8,
> main = "Actual and predicted flu with ARIMA +
> harmonic regressors",
> xlab = "Date", ylab = "Flu rate")
> lines(paris.flu$date[300:400], exp(fit.preds[, h][(300-h):(400-h)]),
>       col = 2, type = 'l',
> lty = 2, lwd = 2)
```

On the positive side, our predicted series no longer features negative values. However, there are many disappointing aspects of the model's performance. Most obviously, many of the predictions are quite off. The peaks are far the wrong magnitude, and they occur at the wrong time.

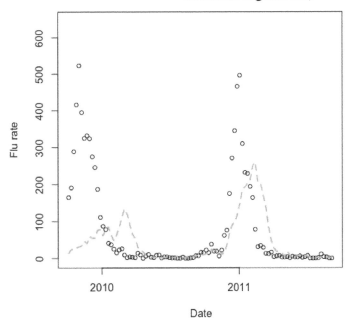

Actual and predicted flu with ARIMA + harmonic regressors,

Figure 13-7. Plot of the actual flu rates (points) compared to our ARIMA + dynamic harmonic regression model predictions (dashed line).

One explanation for the problems is that regular seasonality is not a good description of the seasonality of the flu.[9] According to the Centers for Disease Control and Prevention (CDC) (*https://perma.cc/58SP-B3YH*), in the United States flu can peak each winter season as early as December or as late as March. We can see this in the test data. Consider the following code and the plot in Figure 13-8, which identify the peaks in the testing range of the data:

```R
## R
> plot(test$flu.rate)
> which(test$flu.rate > 400)
Error in which(res$flu.rate > 400) : object 'res' not found
> which(test$flu.rate > 400)
[1]   1  59 108 109 110 111 112 113
> abline(v = 59)
```

---

9 To test whether this explanation is a good one for the situation, you could use a simulation to generate some synthetic data to test alternative data sets with this model and see whether behavior that is cyclic rather than seasonal, as the flu data seems to suggest, causes this kind of failure of the exogenous harmonic regressor model.

```
> which.max(test$flu.rate)
[1] 109
> abline(v = 109)
```

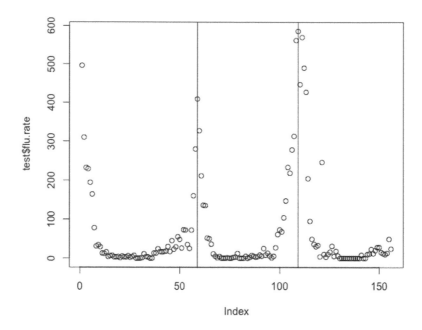

Figure 13-8. Plot of only the test values for the flu and their apparent peak locations.

The peaks occur at indices 59 and 109, that is 50 weeks (not 52 weeks) apart from one another. Also the peak that came at index 59 came at least 59 weeks after the previous peak, and possibly more since we do not have the full extent of the peak around index = 0 in our plot.

With sample peak-to-peak distances of more than 59 in one case and 50 in another, we can see quite a bit of variability from year to year. Our dynamic harmonic regression would not take this into account, and it enforces a more rigid model of seasonality than does the SARIMA model, because the latter's seasonal behavior can change over time. This mismatch between assumptions and the data could go a long way toward explaining the poor performance of this model, such that using a bad model actually drew our attention to an important feature of our data we had not noticed earlier.

Despite poor performance, this alternate model has proven useful for a few reasons. It showed us the value of building the physical limitations of the system into our data preprocessing, in this case via a log transformation. It has also shown us the value of experimenting with several classes of models with an underlying theory behind the model selection at each point. We chose this model to simplify our seasonality, but instead what we discovered is that, likely, neither the SARIMA nor the dynamic harmonic regression model of seasonality is a very good one for this system.

What Is State of the Art in Flu Forecasting?

The models we just explored were relatively simple but reasonably effective in making either forecasts or nowcasts of flu rates in the Paris metropolitan region for short forecast horizons. These models are a good start both in getting to know your data set and in recognizing the limitations of what could be predictive in a noisy but complicated system, such as flu in a given geographic region.

It's interesting to compare our simple approaches with some cutting-edge approaches currently deployed by government agencies or recently published by academics seeking to discuss the state of the art. Next I provide a brief overview of how well flu forecasts currently work and what novel approaches researchers have developed in recent years to try to improve upon traditional models.

Research in flu forecasts

The CDC actively encourages researchers to work on forecasting the flu, even sponsoring an R package (*https://perma.cc/TDX8-4Z7T*) to make its data readily available. For more than five years, the CDC has also sponsored flu forecasting competitions, although it was only for the 2017–2018 flu season that it actually incorporated a flu forecast into its official communications and bulletins. The Delphi research group from Carnegie Mellon has won the majority of competitions so far, and the group forecasts the flu with what they describe as three distinct efforts:

- An Empirical Bayes approach (*https://perma.cc/8EBA-GN2C*) that applies past flu season data with a series of manipulations to form priors for the current season's data based on the overall "shapes" of the time series from past flu seasons

- A crowdsourcing platform (*https://perma.cc/XDE9-A9Y4*) where anyone is welcome to submit a prediction of flu rates

- A nowcasting method that uses *sensor fusion* (*https://perma.cc/NGZ8-TD39*), aggregating data from many sources, such as Wikipedia access counts and relevant Twitter queries, to produce geographically localized flu predictions

This is the set of diverse approaches used in just one academic research group! If we look further afield into academia, we see even more diversity:

- Using deep convolutional networks (CNNs) to classify Instagram pictures and use the outputs of these CNNS along with text-related features from Twitter as inputs into a variety of machine learning models, including XGBoost trees. One such paper (*https://perma.cc/N39F-GSL5*) had the advantage of focusing on a small linguistic community (Finnish speakers), which enabled the use of mainstream social media platforms in a way that could still be regionally specific.

- Identifying reliable users within larger social media clusterings. One paper (*https://perma.cc/25GR-MHRK*) focuses on improving flu predictions by finding the users best placed and most trustworthy on social media platforms.

- Accessing electronic health records to have a more complete and complementary data source in addition to publicly available data. One paper (*https://perma.cc/Q8B7-5TC4*) showed that extremely large gains in forecast accuracy at a number of timescales could be obtained via the integration of electronic health records into forecasting input streams. Unfortunately, this is difficult to arrange and suggests that accurate flu forecasting abilities will go to wealthy data holders rather than to the most creative researchers (although, of course, sometimes these can be one and the same).

As we can see from the diversity of approaches to this problem, there are many routes to make a flu forecast, and none is the definitive winner just yet. This continues to be an area of active research and development, even as some of the better strategies have been applied to governmental uses and public information.

Our discussion here is only the tip of the iceberg. There is a lot of biology, sociology, medicine, and economic policy that goes into determining the course of a flu season, and there are a variety of models that are less oriented toward time series analysis and more oriented toward other aspects of the flu behavior. Time series brings one rich perspective to a very complicated topic.

Predicting Blood Glucose Levels

Another active area of machine learning research for time series data in health applications is the prediction of blood glucose levels for individual patients. Diabetes patients themselves make such predictions all the time, particularly if they have a level of disease such that they inject themselves with bolus insulin, which is specifically used at mealtimes. In such cases, diabetics need to estimate how the food they are about to eat will affect their blood sugar and adjust their dose accordingly.

Similarly, diabetes patients must time their meals and medication to optimize their blood sugar, which is best kept within a specific range, neither too high nor too low. In addition to needing to account for blood-sugar-changing activities such as eating and exercise, diabetics also need to account for specific time-of-day effects. For example, the dawn phenomenon (*https://perma.cc/GE3B-MAKY*) is a rise in blood sugar that occurs in all humans but can be problematic for those with diabetes. On the other hand, for people with type 1 diabetes, blood sugar lows during sleeping hours can be life-threatening events resulting from the failure to make an accurate prediction.

Here we look at a small data set: the self-reported continuous glucose monitor (CGM) data for one individual in several noncontiguous segments of time. This data was self-published on the internet and modified to preserve the privacy of the patient.

There are other options for obtaining diabetes data sets. In addition to large healthcare organizations and some startups with large troves of CGM data, there are numerous self-published data sets available as individuals increasingly manage their diabetes via DIY efforts, such as the Night Scout project (*https://perma.cc/N42T-A35K*). There are also several diabetes CGM data sets (*https://perma.cc/RXG2-CYEE*) open for research purposes.

In this section, we will explore the messiness of a real-world data set and attempt to forecast that data.

Data Cleaning and Exploration

The data is stored in several files available on this book's GitHub repository (*https://github.com/PracticalTimeSeriesAnalysis/BookRepo*). We first load the files and combine them into one `data.table`:

```R
## R
> files <- list.files(full.names = TRUE)
> files <- grep("entries", files, value = TRUE)
> dt     <- data.table()
> for (f in files) {
>   dt <- rbindlist(list(dt, fread(f)))
> }
>
> ## remove the na columns
> dt <- dt[!is.na(sgv)]
```

Strings of date information are available, but we don't have a proper timestamp class column, so we make one with the information we have, which includes a time zone as well as a date string:

```R
## R
> dt[, timestamp := as.POSIXct(date)]
> ## this works for me because my computer is on EST time
> ## but that might not be true for you
>
> ## proper way
> dt[, timestamp := force_tz(as.POSIXct(date), "EST")]
>
> ## order chronologically
> dt = dt[order(timestamp, decreasing = FALSE)]
```

Then we inspect the data post-processing:

```R
## R
> head(dt[, .(date, sgv)])
                date sgv
```

```
1: 2015-02-18 06:30:09 162
2: 2015-02-18 06:30:09 162
3: 2015-02-18 06:30:09 162
4: 2015-02-18 06:30:09 162
5: 2015-02-18 06:35:09 154
6: 2015-02-18 06:35:09 154
```

There are many duplicated data entries, so we need to scrub these for two reasons:

- There is no reason a priori that certain data points should be privileged or weighted more highly than others, but duplication will produce such an effect.
- If we are going to form a feature based on time series windows, these will be nonsensical if there are duplicated time points.

We first see if we can solve this problem by removing all perfectly duplicated rows:

```
## R
> dt <- dt[!duplicated(dt)]
```

However, we should not assume this resolves the problem, so we check whether nonidentical data points exist for the same timestamp:

```
## R
> nrow(dt)
[1] 24861
> length(unique(dt$timestamp))
[1] 23273
> ## we still have duplicated data as far as timestamps
> ## we will delete them
```

Since we still have duplicate time points, we do some data review to see what this means. We identify the timestamp with the most duplicate rows and examine these:

```
## R
> ## we can identify one example using
> ## dt[, .N, timestamp][order(N)]
> ## then look at most repeated data
> dt[date == "2015-03-10 06:27:19"]
   device              date                  dateString sgv direction type
1: dexcom 2015-03-10 06:27:19 Tue Mar 10 06:27:18 EDT 2015  66      Flat  sgv
2: dexcom 2015-03-10 06:27:19 Tue Mar 10 06:27:18 EDT 2015  70      Flat  sgv
3: dexcom 2015-03-10 06:27:19 Tue Mar 10 06:27:18 EDT 2015  66      Flat  sgv
4: dexcom 2015-03-10 06:27:19 Tue Mar 10 06:27:18 EDT 2015  70      Flat  sgv
5: dexcom 2015-03-10 06:27:19 Tue Mar 10 06:27:18 EDT 2015  66      Flat  sgv
6: dexcom 2015-03-10 06:27:19 Tue Mar 10 06:27:18 EDT 2015  70      Flat  sgv
7: dexcom 2015-03-10 06:27:19 Tue Mar 10 06:27:18 EDT 2015  66      Flat  sgv
8: dexcom 2015-03-10 06:27:19 Tue Mar 10 06:27:18 EDT 2015  70      Flat  sgv
## more inspection suggests this is not too important
```

Looking at these values, we can see that there are distinct blood glucose values reported, but they are not wildly different,[10] so we suppress the duplicate timestamps even though they are not exact duplicates:[11]

```R
## R
> dt <- unique(dt, by=c("timestamp"))
```

Now that we have proper timestamps and single values per timestamp, we are in a position to plot our data and get a sense of its extent and behavior (see Figure 13-9).

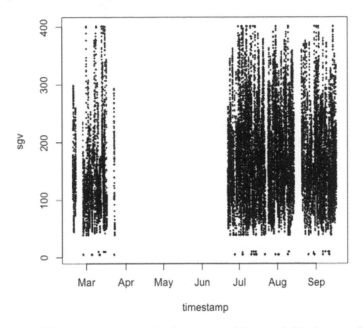

Figure 13-9. The full time domain and value range of the available data in this naive plot of the time series. Unfortunately, the data is so spread out in time and disjointed that this plot does not give us a good understanding of the system's behavior.

We zoom in on the period we see early in the time series, March 2015 (see Figure 13-10):

```R
## R
> ## let's look at some shorter time windows in the data
> start.date <- as.Date("2015-01-01")
> stop.date  <- as.Date("2015-04-01")
```

---

10 The accepted error on blood glucose meters tends to be about 15 in the US system of measurement units, which these numbers use.

11 There are other options besides suppressing. You can explore these on your own.

```
> dt[between(timestamp, start.date, stop.date),
>              plot(timestamp, sgv, cex = .5)]
```

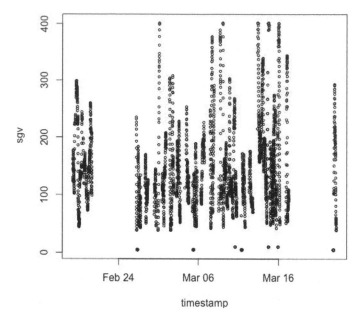

Figure 13-10. Focusing on a specific segment of the time series is more helpful but still too compressed for us to get a sense of any sort of time series dynamics. We should zoom in even further on the time axis.

Even in the plot of just the March dates, we still don't get a sense of the behavior of the time series. Is there seasonality? Drift? A daily pattern? We don't know, so we plot an even shorter time window (see Figure 13-11):

```
## R
> ## also look at a single day to see what that can look like
> ## if we had more time we should probably do 2D histograms of days
> ## and see how that works
> par(mfrow = c(2, 1))
> start.date = as.Date("2015-07-04")
> stop.date  = as.Date("2015-07-05")
> dt[between(timestamp, start.date, stop.date),
>              plot(timestamp, sgv, cex = .5)]
>
> start.date = as.Date("2015-07-05")
> stop.date  = as.Date("2015-07-06")
> dt[between(timestamp, start.date, stop.date),
>              plot(timestamp, sgv, cex = .5)]
> ## if we had more days we could probably do some "day shapes"
> ## analysis, but here we don't seem to have enough data
```

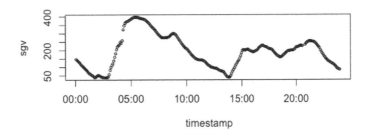

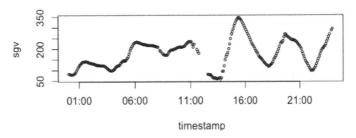

Figure 13-11. Plot of the data from two specific days in July. We can see something of a daily pattern emerge. Finally, we are looking at the data at a temporal scale where the human eye can make sense of what is going on.

By doing more of these exploratory plots (left as an exercise for the reader), you should develop an intuition of the scale of the data, the qualitative dynamics, and what might be important in describing its behavior. Some further exploratory techniques you should apply to this data set include:

- Day-length 2D histograms to look for intraday patterns, and related clustering exercises. Are there different kinds of days?

- Group-by statistics on hours of the day or seasons of the year to look for systematic difference across time.

- Smoothing of the data to look for long-term trends and particularly to compare interday relationships rather than intraday relationships. It would be valuable to develop ultra-long-range blood glucose predictions, which you can do only by looking beyond the more obvious intraday patterns.

Generating Features

Armed with some background knowledge—as well as with observations from our brief exploration of the data set—we can generate features for our data that would be helpful for predicting blood glucose.

We begin by featurizing time itself. For example, we noticed the day appears to have a structure to it, which suggests that the time of day should be relevant for making predictions. Likewise, if you plot the time series of blood glucose for different months of the data, some months show greater variance than others.[12] We generate a few time-related features here (see Figure 13-12):

```R
## R
> ## we should consider putting in a feature for month of the year.
> ## perhaps there are seasonal effects
> dt[, month := strftime(timestamp, "%m")]
>
> dt[, local.hour := as.numeric(strftime(timestamp, "%H"))]
> ## we notice some hours of the day are greatly overrepresented
> ## this person is not consistently wearing the device or perhaps
> ## the device has some time of day functionality or recordkeeping
> ## issues. since it is a weird peak at midnight, this seems to
> ## suggest the device rather than a user (who puts device both
> ## on and off at midnight?)
>
> hist(dt$local.hour)
```

We also consider qualities of the time series values and not just the timestamps of data collection. For example, we can see that the mean blood glucose values vary with the time of day (Figure 13-13):

```R
## R
> ## blood glucose values tend to depend on the time of day
> plot(dt[, mean(sgv), local.hour][order(local.hour)], type = 'l',
>         ylab = "Mean blood glucose per local hour",
>         xlab = "Local hour")
```

---

12 All these observations should be taken with a large degree of skepticism. This data set consists only of data from a single user for portions of a single year, and the data is not continuous. That means we cannot draw conclusions about seasonality even if we are tempted to speculate. Still, for purposes of fitting the limited data we have, we will consider these temporal components.

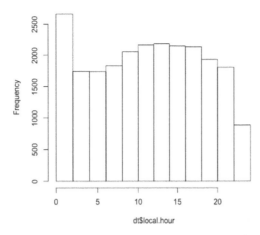

Figure 13-12. From the histogram of local hour we can see that the times the CGM is collecting and reporting data are not random. The heavy concentration around midnight additionally suggests an irregularity in reporting or device functioning that is most likely not due to user behavior (the user is not likely manipulating their device at that time of day).

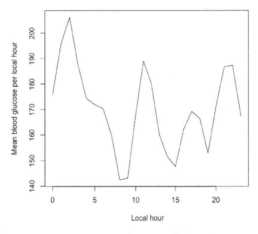

Figure 13-13. Mean blood glucose varies substantially by the time of day.

There is other interesting data about the blood glucose values. The column named direction in the data set pulls information from the CGM device, applies the manufacturer's proprietary software, and comes up with a directional trend label for the data. We can use this rather than computing our own trend statistics, and so we try to understand it a little, such as by asking whether there are different trends for different hours.

We define a function that will give us the *n*th most popular label for a given vector, and we use that first to find the most popular label for each local hour of the day:

```R
## R
> nth.pos = function(x, pos) {
>   names(sort(-table(x)))[pos]
>   ## this code snippet thanks to r user group
> }
> dt[, nth.pos(direction, 1), local.hour][order(local.hour)]
    local.hour   V1
 1:          0 Flat
 2:          1 Flat
 3:          2 Flat
 ...
21:         20 Flat
22:         21 Flat
23:         22 Flat
24:         23 Flat
    local.hour   V1
```

The most popular direction label is "flat" for all hours of the day, which is reassuring because the dynamics of the system would be questionable if the trend could be strongly predicted simply by the time of day.

However, the second most common directional label per hour of the day does vary with the time of day, as we can see here:

```R
## R
> dt[, nth.pos(direction, 2), local.hour][order(local.hour)]
    local.hour            V1
 1:          0    FortyFiveUp
 2:          1    FortyFiveUp
 3:          2    FortyFiveUp
 4:          3  FortyFiveDown
 5:          4  FortyFiveDown
 6:          5    FortyFiveUp
 7:          6  FortyFiveDown
 8:          7  FortyFiveDown
 9:          8  FortyFiveDown
10:          9    FortyFiveUp
11:         10    FortyFiveUp
12:         11  FortyFiveDown
13:         12  FortyFiveDown
14:         13  FortyFiveDown
15:         14  FortyFiveDown
16:         15  FortyFiveDown
17:         16    FortyFiveUp
18:         17    FortyFiveUp
19:         18  FortyFiveDown
```

```
20:            19      FortyFiveUp
21:            20      FortyFiveUp
22:            21      FortyFiveUp
23:            22      FortyFiveUp
24:            23    FortyFiveDown
        local.hour              V1
```

It's interesting to cross-reference these labels against the intraday data plots we included in the previous section. Do the results match? (Answering this is left as an exercise for the reader.)

Next we address a numeric label that may be predictive, namely the variance of the blood glucose measures in the most recent time window. We compute the standard deviation for a short lookback window.

To do so, we must account for the noncontiguous nature of the data. We don't want to be computing standard deviations of data points that are not adjacent in time but merely adjacent in the data.table. We will compute time deltas between target window beginnings and ends to ensure they are of the right timescale, but we need to be careful about how we do this. Here is an example of how we could easily go wrong:

```
## R
> ## FIRST don't do this the wrong way
> ## Note: beware calculating time differences naively!
> as.numeric(dt$timestamp[10] - dt$timestamp[1])
[1] 40
> as.numeric(dt$timestamp[1000] - dt$timestamp[1])
[1] 11.69274
> dt$timestamp[1000] - dt$timestamp[1]
Time difference of 11.69274 days
```

It is possible the time differences are returned in different units, which would make the results of the preceding code nonsense. The correct way to compute a time delta, as well as the way we use this time delta to determine the validity of a standard deviation computation for a given row, is as follows:

```
## R
> dt[, delta.t := as.numeric(difftime(timestamp, shift(timestamp, 6),
>                                      units = 'mins'))]
> dt[, valid.sd := !is.na(delta.t) & delta.t < 31]
> dt[, .N, valid.sd]
valid.sd      N
1:    FALSE  1838
2:     TRUE 21435
```

Once we have labeled the rows for which a standard deviation calculation is appropriate, we run the calculations. We compute across all rows, and we plan to overwrite the invalid values with a column average as a simple form of missing value imputation:

```R
## R
> dt[, sd.window := 0]
> for (i in 7:nrow(dt)) {
>   dt[i, ]$sd.window = sd(dt[(i-6):i]$sgv)
> }
> ## we will impute the missing data for the non-valid sd cases
> ## by filling in the population mean  (LOOKAHEAD alert, but we're aware)
> imputed.val = mean(dt[valid.sd == TRUE]$sd.window)
> dt[valid.sd == FALSE, sd.window := imputed.val]
```

Next, we set up a column to establish the true value we should target for our forecast. We choose to forecast blood glucose values 30 minutes in advance, which would be enough time to warn someone with diabetes if a dangerously high or low blood sugar was predicted. We put the target forecast value into a column called `target`. We also create another column, `pred.valid`, which indicates whether the data points preceding the time when we want to make a forecast are sufficiently complete (that is, regularly sampled in the previous 30 minutes at 5-minute increments):

```R
## R
> ## now we also need to fill in our y value
> ## the actual target
> ## this too will require that we check for validity as when
> ## computing sd due to nontemporally continuous data sitting
> ## in same data.table. let's try to predict 30 minutes ahead
> ## (shorter forecasts are easier)
>
> ## we shift by 6 because we are sampling every 5 minutes
> dt[, pred.delta.t := as.numeric(difftime(shift(timestamp, 6,
>                                                 type = "lead"),
>                                          timestamp,
>                                          units = 'mins'))]
> dt[, pred.valid := !is.na(pred.delta.t) & pred.delta.t < 31]
>
> dt[, target := 0]
> for (i in 1:nrow(dt)) {
>   dt[i, ]$target = dt[i + 6]$sgv
> }
```

We spot-check our work to see if it produces sensible results:

```R
## R
> ## now we should spot check our work
> i = 300
> dt[i + (-12:10), .(timestamp, sgv, target, pred.valid)]
            timestamp sgv target pred.valid
 1: 2015-02-19 16:15:05 146    158       TRUE
 2: 2015-02-19 16:20:05 150    158       TRUE
 3: 2015-02-19 16:25:05 154    151      FALSE
 4: 2015-02-19 16:30:05 157    146      FALSE
 5: 2015-02-19 16:35:05 160    144      FALSE
 6: 2015-02-19 16:40:05 161    143      FALSE
 7: 2015-02-19 16:45:05 158    144      FALSE
```

```
 8: 2015-02-19 16:50:05 158    145      FALSE
 9: 2015-02-19 17:00:05 151    149      TRUE
10: 2015-02-19 17:05:05 146    153      TRUE
11: 2015-02-19 17:10:05 144    154      TRUE
12: 2015-02-19 17:15:05 143    155      TRUE
13: 2015-02-19 17:20:05 144    157      TRUE
14: 2015-02-19 17:25:05 145    158      TRUE
15: 2015-02-19 17:30:05 149    159      TRUE
16: 2015-02-19 17:35:05 153    161      TRUE
17: 2015-02-19 17:40:05 154    164      TRUE
18: 2015-02-19 17:45:05 155    166      TRUE
19: 2015-02-19 17:50:05 157    168      TRUE
20: 2015-02-19 17:55:05 158    170      TRUE
21: 2015-02-19 18:00:04 159    172      TRUE
22: 2015-02-19 18:05:04 161    153      FALSE
23: 2015-02-19 18:10:04 164    149      FALSE
          timestamp sgv target pred.valid
```

Take a careful look at this output. Something in it should make you question whether we were too harsh in assessing "validity" for computing the standard deviation of a recent window on the blood glucose data. As an independent exercise, see if you can spot those points, and think about how you could rework the pred.valid labeling to be more correctly inclusive.

Now we have a wide array of features and a target value to use for training whatever model we will use to generate forecasts, but we are not done generating features. We should simplify some of the temporal features we have already generated to lessen the complications of our model. For example, rather than having local hour be an input of 23 binaries (one for each hour of the day minus one), we should lessen the number of "hour" categories. We do this like so:

```
## R
> ## Let's divide the day into quarters rather than
> ## into 24-hour segments. We do these based on a notion
> ## of a 'typical' day
> dt[, day.q.1 := between(local.hour,  5, 10.99)]
> dt[, day.q.2 := between(local.hour, 11, 16.99)]
> dt[, day.q.3 := between(local.hour, 17, 22.99)]
> dt[, day.q.4 := !day.q.1 & !day.q.2 & !day.q.3]
```

We also simplify the month data to a simpler set of categories:

```
## R
> ## let's have a "winter/not winter" label rather than
> ## a month label. this decision is partly based on the
> ## temporal spread of our data
> dt[, is.winter := as.numeric(month) < 4]
```

Finally, to use the direction column we need to one-hot-encode this value, just as we did the different times of the day. We also clean up some inconsistently labeled features ("NOT COMPUTABLE" versus "NOT_COMPUTABLE"):

```
## R
> ## we also need to one-hot encode a direction feature
> ## and clean that data somewhat
> dt[direction == "NOT COMPUTABLE", direction := "NOT_COMPUTABLE"]
> dir.names = character()
> for (nm in unique(dt$direction)) {
>   new.col = paste0("dir_", nm)
>   dir.names = c(dir.names, new.col)
>   dt[, eval(parse(text = paste0(new.col, " :=
                              (direction == '", nm, "')")))]
> }
```

Now we have one-hot-encoded relevant features and simplified others, so we are
ready to direct these features into a model.

Fitting a Model

At last we get to the fun part of time series analysis: making a prediction.

 How Much Time Should You Spend Modeling?

As you have seen with the two real-world models of this chapter,
real-world data is messy, and each time it is scrubbed it has to be
done so with domain knowledge and common sense. There is no
general template. However, you should always proceed carefully
and without rushing to fit a model. We only want to fit models
once we are confident we are not putting garbage into them!

Our first order of business is to create training and testing data sets:[13]

```
## R
> ## we need to set up training and testing data
> ## for testing, we don't want all the testing to come
> ## at end of test period since we hypothesized some of behavior
> ##   could be seasonal let's make the testing data
> ##   the end data of both "seasons"
> winter.data      <- dt[is.winter == TRUE]
> train.row.cutoff <- round(nrow(winter.data) * .9)
> train.winter     <- winter.data[1:train.row.cutoff]
> test.winter      <- winter.data[(train.row.cutoff + 1): nrow(winter.data)]
>
> spring.data      <- dt[is.winter == FALSE]
> train.row.cutoff <- round(nrow(spring.data) * .9)
```

---

13 In practice, and with a larger data set, you'd also want to set aside a validation test set, which would be a
special subset of your training data that was test-like without polluting your actual test data. Just as the test
data should generally be the most temporally recent data (to prevent future information leaking backward,
e.g., lookahead), the validation data set in such cases should come at the end of the training period. Consider
setting this up on your own to explore the hyperparameters of the model.

```
> train.spring      <- spring.data[1:train.row.cutoff]
> test.spring       <- spring.data[(train.row.cutoff + 1): nrow(spring.data)]
>
> train.data <- rbindlist(list(train.winter, train.spring))
> test.data  <- rbindlist(list(test.winter,  test.spring))
>
> ## now include only the columns we should be using
> ## categorical values: valid.sd, day.q.1, day.q.2, day.q.3, is.winter
> ## plus all the 'dir_' colnames
> col.names <- c(dir.names, "sgv", "sd.window", "valid.sd",
>                "day.q.1", "day.q.2", "day.q.3", "is.winter")
>
> train.X <- train.data[, col.names, with = FALSE]
> train.Y <- train.data$target
>
> test.X <- test.data[, col.names, with = FALSE]
> test.Y <- test.data$target
```

Consistent with much cutting-edge work, we choose XGBoost gradient boosted trees as our model. In recent publications, these have met or exceeded state-of-the-art metrics for glucose prediction in some use cases:[14]

```
## R
> model <- xgboost(data = as.matrix(train.X), label = train.Y,
>                     max.depth = 2,
>                     eta       = 1,
>                     nthread   = 2,
>                     nrounds   = 250,
>                     objective = "reg:linear")
> y.pred <- predict(model, as.matrix(test.X))
```

We can then inspect the results of this prediction (see Figure 13-14). We also look at our prediction for a specific day (see Figure 13-15):

```
## R
> ## now let's look at a specific day
> test.data[date < as.Date("2015-03-17")]
> par(mfrow = c(1, 1))
> i <- 1
> j <- 102
> ylim <- range(test.Y[i:j], y.pred[i:j])
> plot(test.data$timestamp[i:j], test.Y[i:j], ylim = ylim)
> points(test.data$timestamp[i:j], y.pred[i:j], cex = .5, col = 2)
```

---

14 We do not tune the hyperparameters here (which would require a validation data set), but that is something you should do with XGBoost once you have settled on a general modeling pipeline.

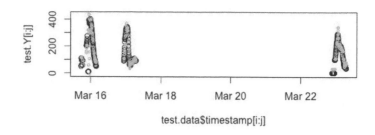

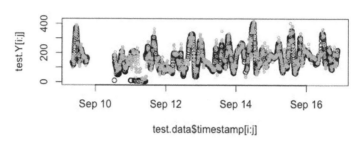

Figure 13-14. At this large scale our predictions look good, but it's hard to judge zoomed out so far. This is not how someone using the prediction would experience it.

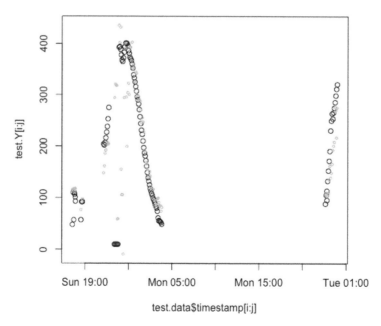

Figure 13-15. Examining single days at a time offers a better perspective on how well our prediction algorithm works.

The predictions look reasonable, but this plot suggests there may also be problems with the underlying data. The blob around midnight between Sunday and Monday seems unrealistically low, particularly as neighboring points are not low. It's more likely the device malfunctioned than that someone experienced such a drastic but short-lived low blood glucose. We might consider more data scrubbing or perhaps additional labels to indicate that particular points seem suspect.[15] It's probably a sign of overfitting that our model forecasts these likely invalid low blood glucose data points. We want our model to forecast actual blood glucose values and not device malfunctions.

At this point, we should consider the goals of our algorithm. If we merely want to show that we can do a reasonably good job of predicting blood glucose half an hour in advance—without even the benefit of knowing what someone is eating or when they are exercising—we can already say "mission accomplished." That's already quite something—and it's impressive that we can make reasonable forecasts without having relevant inputs, such as meal and exercise information.

However, we are aiming mostly for a forecast that will warn people of danger—that is, when their blood sugar will go too low or too high. Let's focus on those data points as we consider how our predictions stack up against measured values.

Importantly, we plot the high and low blood glucose value separately. If we plotted them together, we'd see a seemingly high correlation by creating an artificially dumbbell-shaped distributed of high and low points (see Figure 13-16):

```
## R
> par(mfrow = c(2, 1))
> high.idx = which(test.Y > 300)
> plot(test.Y[high.idx], y.pred[high.idx], xlim = c(200, 400),
>                                          ylim = c(200, 400))
> cor(test.Y[high.idx], y.pred[high.idx])
[1] 0.3304997
>
> low.idx = which((test.Y < 60 & test.Y > 10))
> plot(test.Y[low.idx], y.pred[low.idx], xlim = c(0, 200),
>                                        ylim = c(0, 200))
> cor(test.Y[low.idx], y.pred[low.idx])
[1] 0.08747175
```

These plots put our model's performance in a harsher light. Particularly concerning is that the model does not do very well at low blood sugar levels. This may be because dangerously low blood sugar occurrences are rare incidents, so the model does not have much opportunity to train on them. We should consider data augmentation or

---

15 One idea would be to apply outlier detection in a time series–aware manner to note that these points don't seem to match the general trend for that time window. We could also use basic physics, chemistry, and biology in some cases to say that certain drops just aren't possible.

resampling to find ways to make this phenomenon more salient to our model. We can also modify our loss function to more heavily weight the examples already in our data.

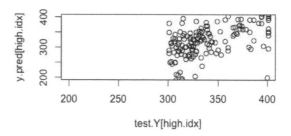

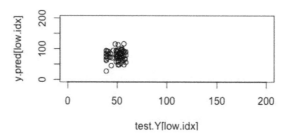

Figure 13-16. Plot of the predictions and actual values against one another individually at each end of the extreme. On top we plot the values against one another for high measured blood glucose values and on the bottom for low measured blood glucose values.

Here we have seen a case where an initial model's forecasts appear respectable but may not fulfill the most essential purpose of the model. In this case we can predict general blood glucose trends, but we should do more work to predict high and low blood sugars, with a particular emphasis on life-threateningly low levels of blood sugar.

In this case, as in the case of flu forecasting, we have seen that common sense is essential to modeling time series. We cannot blindly clean data or prepare features without thinking about what they mean in context. Similarly, the more we know about the domain of the time series, such as understanding the importance of the time of day for blood sugars in humans, the better equipped we are to clean data, prepare features, and examine our forecasts for important success or failure cases.

More Resources

Vasileios Lampos et al., "Advances in Nowcasting Influenza-Like Illness Rates Using Search Query Logs," Scientific Reports 5, no. 12760 (2015), https://perma.cc/NQ6B-RUXF.
This 2015 Nature Communications paper is highly accessible and both a good introduction to flu nowcasting as well as a historical piece that shows how nowcasting and forecasting were revolutionized by big data and real-time social media and internet data inputs. The authors work with both a traditional linear model and an innovative nonlinear model to compare different approaches and show the utility of nonlinear models in such a complex system as seasonal infectious diseases rates.

David Farrow, "Modeling the Past, Present, and Future of Influenza," doctoral thesis, Computational Biology Department, School of Computer Science, Carnegie Mellon University, 2016, https://perma.cc/96CZ-5SX2.
This thesis details some of the theory and practice of "sensor fusion" for the purposes of integrating social media sources of various geographic and temporal granularity to produce one of the Delphi group's principle influenza forecasts. It provides an extremely in-depth overview of predicting the flu, from virology to population dynamics to data science experiments on how data processing affects forecasts.

Rob J. Hyndman, "Dynamic Regression," lecture notes, n.d., https://perma.cc/5TPY-PYZS.
These lecture notes offer practical examples of how to use dynamic regression to supplement traditional statistical forecasting models with a series of alternative models for seasonality when SARIMA is not expected to be a good fit—usually because the periodicity is too complex or the periods are too long relative to the amount of data or computing resources available.

## CHAPTER 14

Financial Applications

Financial markets are the granddaddy of all time series data. If you pay for propriet-ary trading data on a high-tech exchange, you can receive terabyte-sized floods of data that can take days to process, even with high-performance computing and embarrassingly parallel processing.

High-frequency traders are among the newest and most infamous members of the finance community, and they trade on information and insights resulting from time series analysis at the microsecond level. On the other hand, traditional trading firms —looking at longer-term time series over hours, days, or even months—continue to succeed in the markets, showing that time series analysis for financial data can be conducted in a myriad of successful ways and at timescales spanning many orders of magnitude, from milliseconds to months.

Embarrassingly parallel describes data processing tasks where the results of processing one segment of data are in no way dependent on the values of another segment of data. In such cases it's embar-rassingly easy to convert data analysis tasks to run in parallel rather than in sequence to take advantage of multicore or multimachine computing options.

Consider, for example, the task of computing the daily mean of minute-by-minute returns on a given stock. Each day can be treated separately and in parallel. In contrast, computing the expo-nentially weighted moving average of daily volatility is not embar-rassingly parallel because the value for a particular day depends on the values for the days before it. Sometimes tasks that are not embarrassingly parallel can still be executed partly in parallel, but it depends on the details.

Here, we will work through a classic example of time series analysis for fun and for profit: predicting tomorrow's stock returns for the S&P 500.

Obtaining and Exploring Financial Data

It can be exceedingly difficult to obtain financial data if you have a particular product or temporal resolution you are seeking. In such cases you usually need to buy data. But historical stock prices are widely available from a variety of services, including:

- Yahoo Finance. While Yahoo has discontinued servicing its historical data API,[1] historical daily data is available for download (*https://perma.cc/RQ6D-U4JX*).

- Newer companies, such as AlphaVantage (*https://www.alphavantage.co/*) and Quandl (*https://www.quandl.com*) offer a combination of historical and real-time price information for stock data.

We limit our analysis to freely available daily stock price data from Yahoo. We download data for the S&P 500 spanning dates from 1990 through 2019. In the following code, we see what columns are available in the downloaded data set and plot the daily closing price to start exploring our data (see Figure 14-1):

```python
## python
>>> df = pd.read_csv("sp500.csv")
>>> df.head()
>>> df.tail()
>>> df.index = df.Date
>>> df.Close.plot()
```

We can see the values are notably different at the beginning and end of the date period covered by the CSV files. The change in values is even more apparent when we plot the full time series for the closing price (Figure 14-2) than it is when we look at samples from the data frame.

A look at Figure 14-2 shows that the time series is not stationary. We also see that there may be different "regimes." For reasons clearly seen in the plot, financial analysts are keen to develop models identifying regime shifts in stock prices. There appear to be different regimes even if we may not have an exact definition of where one ends and another begins.[2]

---

1 Incidentally, this created quite a bit of nonworking code in the R and Python communities.

2 Note that the S&P 500 is also tricky because it is a composite of many different stocks as inputs, with weightings periodically adjusted and with a factor that is entirely proprietary also used to divide the weighted average of the stocks. For this reason, strong domain knowledge and an understanding of how different actions taken by companies can affect their stock prices and their S&P 500 weightings would also be important to better understand the long-term behavior we see here.

```
In [9]:  df.head()
Out[9]:
```

	Date	Open	High	Low	Close	Adj Close	Volume
0	1990-02-13	330.079987	331.609985	327.920013	331.019989	331.019989	144490000
1	1990-02-14	331.019989	333.200012	330.640015	332.010010	332.010010	138530000
2	1990-02-15	332.010010	335.209991	331.609985	334.890015	334.890015	174620000
3	1990-02-16	334.890015	335.640015	332.420013	332.720001	332.720001	166840000
4	1990-02-20	332.720001	332.720001	326.260010	327.989990	327.989990	147300000

```
In [75]:  df.tail()
Out[75]:
```

	Date	Open	High	Low	Close	Adj Close	Volume
7301	2019-02-06	2735.050049	2738.080078	2724.149902	2731.610107	2731.610107	3472690000
7302	2019-02-07	2717.530029	2719.320068	2687.260010	2706.050049	2706.050049	4099490000
7303	2019-02-08	2692.360107	2708.070068	2681.830078	2707.879883	2707.879883	3622330000
7304	2019-02-11	2712.399902	2718.050049	2703.790039	2709.800049	2709.800049	3361970000
7305	2019-02-12	2722.610107	2748.189941	2722.610107	2744.729980	2744.729980	3827770000

Figure 14-1. Raw data from the beginning and end of the CSV file. Notice that the values change noticeably from 1990 to 2019—no surprise to anyone familiar with the history of the US financial markets.

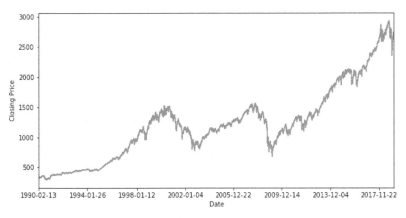

Figure 14-2. The daily closing price of the S&P 500 is not a stationary time series.

The potential for change points and different regimes suggests that it could be a good idea to break the data set into different sub-data sets to be modeled separately. However, we want to keep all the data together because daily data does not produce many data points in a few decades; we need to keep all the data we can. We consider whether we can justify keeping all this data together if we are only interested in day-ahead forecasts.

We can consider whether normalizing data can make data from different time periods comparable. Let's take a look at a week's worth of scaled closing prices for three different decades within the time series (Figure 14-3):

```python
## python
>>> ## pick three weeks (Mon - Fri) from different years
>>> ## scale closing prices for each day by the week's mean closing price
>>> ## 1990
>>> vals = df['1990-05-07':'1990-05-11'].Close.values
>>> mean_val = np.mean(vals)
>>> plt.plot([1, 2, 3, 4, 5], vals/mean_val)
>>> plt.xticks([1, 2, 3, 4, 5],
>>>     labels = ['Monday', 'Tuesday', 'Wednesday', 'Thursday', 'Friday'])
>>>
>>> ## 2000
>>> vals = df['2000-05-08':'2000-05-12'].Close.values
>>> mean_val = np.mean(vals)
>>> plt.plot([1, 2, 3, 4, 5], vals/mean_val)
>>> plt.xticks([1, 2, 3, 4, 5],
>>>     labels = ['Monday', 'Tuesday', 'Wednesday', 'Thursday', 'Friday'])
>>>
>>> ## 2018
>>> vals = df['2018-05-07':'2018-05-11'].Close.values
>>> mean_val = np.mean(vals)
>>> plt.plot([1, 2, 3, 4, 5], vals/mean_val)
>>> plt.xticks([1, 2, 3, 4, 5],
>>>     labels = ['Monday', 'Tuesday', 'Wednesday', 'Thursday', 'Friday'])
```

We plot closing prices over three weeks in three different decades, each scaled by that week's mean as per the preceding code. The relative percentage changes from day to day in the course of a week seem about the same across the decades.

These plots are promising. While the mean values and variance of the closing prices have changed substantially over time, the plots suggest similar behavior over time once we normalize the data to its mean value for a given decade.

Given this, we next consider whether we can find a way to make all data over the full time period similar enough to train meaningfully with a model. We want to know whether there is a way to transform the data that is financially meaningful but also makes the data comparable across the entire time period.

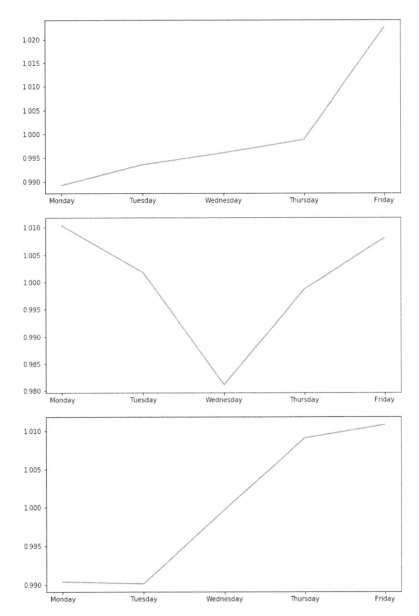

Figure 14-3. Scaled mean closing prices for a week in May 1990, 2000, and 2018.

We compute the daily return, that is, the change in price from the start to the end of each trading day (see Figure 14-4):

```python
## python
>>> df['Return'] = df.Close - df.Open
>>> df.Return.plot()
```

As we can see in Figure 14-4, this alone is not enough to make the data comparable. We will also have to find a way to normalize the data without a lookahead so that the values we are using for inputs and outputs into our model are more even throughout the time period of interest. We will see how to do this in the next section.

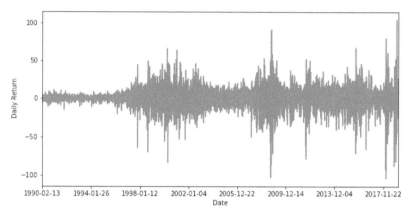

Figure 14-4. Daily returns show a near-zero mean over time, but the variance of daily returns changes markedly in different time periods. It is behavior like this that has inspired models such as GARCH, which was briefly discussed in Chapter 6.

Financial Markets as a Random Walk

Random walk is a good starting model for financial time series; indeed, stock prices are often cited as the quintessential example of a naturally occurring random walk. There are two important insights we should keep in mind from this as we analyze our data.

First, many regime shifts may just be an effect of the natural contours of a random walk. If you don't believe me, consider some of the following examples, which were quickly generated in R (see Figure 14-5):

```
## R
> ## used each of these seeds in turn to produce the plots
> ## set.seed(1)
> ## set.seed(100)
> ## set.seed(30)

> N <- 10000
> x <- cumsum(sample(c(-1, 1), N, TRUE))
> plot(x)
```

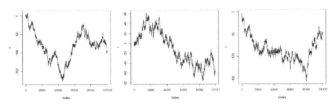

Figure 14-5. The output from a variety of random walks, all produced with different seeds from the preceding simple code.

Second, because random walks are a natural fit for stock data, our null or most basic model, against which we can begin to benchmark more complicated models, would be one in which we predict the price for tomorrow as being the price for today, consistent with our understanding of the stock market as a random walk. In such a case, we would expect a very high correlation:

```R
## R
> cor(x, shift(x), use = "complete.obs")
[1] 0.9996601
```

On the other hand, if we differenced the time series, such that the value at each time step represented the change in the value of the time series from one time step to another, the correlation would disappear because any correlation would have been due to trend rather than actual predictive capability of such a simple model:

```R
## R
> cor(diff(x), shift(diff(x)), use = "complete.obs")
[1] -0.005288012
```

So in finance, whether you produce a model with high or low correlation could have a lot to do with the kind of data you're modeling. If you model returns (what you should do) rather than stock prices (a common mistake even in industry), you'll have correlations that seem low but are much more likely to be predictive in the real world rather than models that promise the world but don't deliver in production. This is why we work to model returns rather than stock price in the next example.

Preprocessing Financial Data for Deep Learning

Our data preprocessing will be done in three steps:

1. We will form new, economically meaningful quantities of interest out of the raw inputs.

2. We will compute an exponentially weighted moving average and variance of the quantities of interest so that we can scale them without a lookahead.

3. We will package our results in a format appropriate for the recurrent deep learning model we will use to fit the data.

Financial Time Series Is Its Own Discipline

Financial time series is an entire discipline with thousands of academics diligently trying to understand how the financial markets work, both for profit and for smart regulation. There are numerous statistical models that have been developed to treat some of the tricky aspects of financial data that we have touched on here, such as the GARCH model. If you are seeking to apply statistics and machine learning to financial time series modeling, you should study the history of quantitative finance and the major classes of commonly deployed models.

Adding Quantities of Interest to Our Raw Values

We've already computed daily return in the previous section. Another quantity of interest we can form from the raw inputs is daily *volatility*, which is the difference between the highest and lowest prices recorded during the trading day. This can be easily computed given the raw data (see Figure 14-6):

```python
## python
>>> df['DailyVolatility'] = df.High - df.Low
>>> df.DailyVolatility.plot()
```

Just as the daily return prices are a nonstationary time series, so too is the daily volatility time series. This further confirms that we need to find a way to scale these appropriately. We also want to do so without introducing a lookahead.

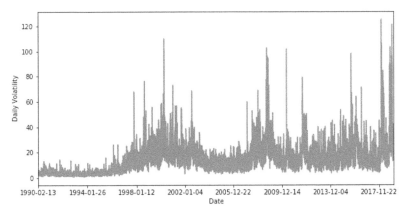

Figure 14-6. Daily volatility is a positive value time series (by definition) that shows noticeably different degrees of variance at different points in the S&P 500 time series.

Scaling Quantities of Interest Without a Lookahead

We will forecast the daily return one day ahead. Some quantities of interest that could be helpful include:

- Previous daily returns
- Previous daily volatility
- Previous daily volume

We will scale each of these quantities by subtracting out the exponentially weighted moving average and then dividing by the exponentially weighted standard deviation. Our earlier weekly data exploration showed that with appropriate preprocessing the various quantities of interest can be made into stationary time series.

First we compute the exponentially weighted moving average of every column in the data frame, and plot the daily volatility's exponentially weighted moving average (see Figure 14-8). Contrast this with the plot in Figure 14-7. This plot is much smoother due to the averaging. Note that there is a parameter here that you should effectively consider part of your model, as a hyperparameter, even though it's used in the data preprocessing step: the half-life of the exponential smoothing. Your model's behavior will certainly depend on this parameter:

```python
## python
>>> ewdf = df.ewm(halflife = 10).mean()
>>> ewdf.DailyVolatility.plot()
```

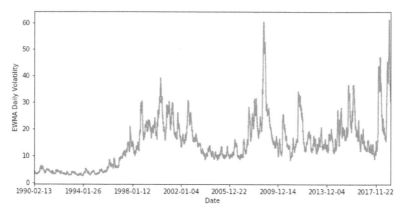

Figure 14-7. The plot of the exponentially weighted moving average of daily volatility is much smoother than the plot of the raw values but still shows a nonstationary time series.

We can then use that value, as well as the exponentially weighted moving variance calculated here, to scale values of interest in a way that results in series with more consistent behavior over time (see Figure 14-8):

```python
## python
>>> ## compute exponentially weighted moving variance
>>> vewdf = df.ewm(halflife = 10).var()
>>>
>>> ## scale by demeaning and normalizing
>>> scaled = df.DailyVolatility - ewdf.DailyVolatility
>>> scaled = scaled / vewdf.DailyVolatility**0.5
>>> scaled.plot()
```

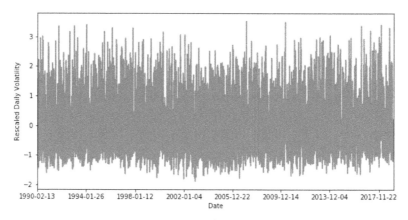

Figure 14-8. Transforming the data with exponentially weighted mean and variance results in a much more even time series with comparable values throughout the time period all the way from 1990 to 2019.

We transform all three raw inputs of interest into a scaled version as follows:

```python
## python
>>> df['ScaledVolatility'] = ((df.DailyVolatility -
>>>                            ewdf.DailyVolatility)
>>>                           / vewdf.DailyVolatility**0.5 )
>>> df['ScaledReturn']     = ((df.Return - ewdf.Return)
>>>                           / vewdf.Return**0.5 )
>>> df['ScaledVolume']     = ((df.Volume - ewdf.Volume)
>>>                           / vewdf.Volume**0.5 )
```

Finally, we drop the NA results that come from exponential smoothing:[3]

```python
## python
>>> df = df.dropna()
```

You Don't Have to Use Daily Data

Hopefully you've noticed that all the data processing has taken place at the per-day timescale. You may remember from Chapters 2 and 3 that upsampling and downsampling of data is a way to change the timescale of your analysis. Likewise, you may want to use inputs from a variety of timescales in your model. For example, I could consider computing exponentially weighted moving averages and variances at a variety of half-lives and using each of them as a different input. Likewise, I could compute quantities of interest at different timescales, such as weekly returns or monthly volatility. All of this is something to explore when you are modeling a time series, although space is limited here.

It's wise to stick to a timescale in your inputs similar to the timescale you have in mind for your targets, namely how far into the future you want to predict. The further into the future you want to predict, the further back in time you should be looking and the more you should smooth your series with a longer half-life so that you can recognize longer-term trends. This will help you make longer-term predictions.

Formatting Our Data for a Neural Network

Our data is stored in a Pandas data frame at the moment, and our planned inputs are stored alongside many raw inputs we have no intention of using. Also, for a neural network we will shape our data into the TNC format, which, you might recall, stands for time × number of samples × channels. For this reason we need to do some more preprocessing even after the rescaling work we have already discussed.

---

3 Instead of dropping these, we could also set the value of the exponentially smoothed column to the only value known at that time. In either case, this is not important.

First we portion out our data into training and testing components:[4]

```python
## python
>>> ## break our data into training and testing components
>>> train_df = df[:7000]
>>> test_df = df[7000:]
>>>
>>> ## build our pipeline variables off training data
>>> ## taking only values of interest from larger data frames
>>> horizon = 10
>>> X = train_df[:(7000 - horizon)][["ScaledVolatility", "ScaledReturn",
                                      "ScaledVolume"]].values
>>> Y = train_df[horizon:]["ScaledReturn"].values
```

Notice there is something of a problem with what we are predicting based on how we set up Y here. Think about this for a minute before you continue reading.

Diluting the Forecasting Task

What we have done with our data has diluted the forecasting task. That is, we take a future value and "dilute" it with information from the past. While this is a way of pushing the past into the future, seemingly the opposite of lookahead, it's a related problem. Information is moving along the time axis in a way we might forget about, which in turn may make us misunderstand our model's performance. We should keep track of what we've done in preprocessing to bear in mind when evaluating our model.

The problem of forecasting smoothed values is rampant in the economics and finance literature—and it's a problem that often goes unacknowledged. While it's not wrong to do this, sometimes authors overstate their results and say they have beat a benchmark when in fact they set themselves an easier task of predicting a smoothed value rather than the measured value.

The problem with this setup is that the Y value is the scaled return rather than just the return. This is better for training because the values fall within the appropriate range, but it also means that the Y we are predicting is not the actual return that interests us but rather that return adjusted by a moving average. We are forecasting how much our return differs from the exponentially weighted moving average rather than just predicting the return.

This is not wrong per se, but it does mean that we are making our task easier than the true forecasting task. We should be aware of this so that when our training looks bet-

---

4 The usual caveat that we should really have a separate validation set to avoid information leaking backward from the test set applies here, but I'm trying to keep the code simple.

ter than our model's actual performance, we'll know it's partly because in training we focus on a hybrid task, whereas ultimately making money off this model would depend only on the true forecasting.

We focus on the training data and now need to put X into the format expected by a recurrent neural network architecture, namely TNC. We do this with a series of NumPy operations.

Originally, X is two-dimensional, as it comes from a Pandas data frame. We want to add a third dimension, axis 1 (so pushing the second dimension out to axis 2, where the axes are numbered from 0 up):

```python
## python
>>> X = np.expand_dims(X, axis = 1)
```

Our temporal axis is already axis 0 because the data frame was sorted temporally. The last axis, now axis 2, is already the "channel axis" because our inputs each occupy one column of that dimension.

We will try a model that will see 10 time steps—that is 10 days of data looking backward. So we need to cut off axis 0 to be of length 10. We chop along axis 0 every 10 rows, and we reform the resulting list of submatrices such that the number of samples (i.e., length of the resulting list) becomes the dimension of the second axis:[5]

```python
## python
>>> X = np.split(X, X.shape[0]/10, axis = 0)
>>> X = np.concatenate(X, axis = 1)
>>> X.shape
(10, 699, 3)
```

Given the TNC format, we have time series of length 10, with three parallel inputs. Of these we have 699 examples. The batch size will determine how many batches make up an epoch, where an epoch is one cycle through our data.

We don't have very much data to train on, given how few examples we appear to have. How did we go from 30 years of data to not much data? The answer is that each data point has, as of now, only been included in one sample time series. However, each data point could be in 10 different time series, occupying a different position in each one.

This may not be immediately obvious, so let's look at a simple example. Assume we have the following time series:

1, 3, 5, 11, 3, 2, 22, 11, 5, 7, 9

---

5 For R users: remember Python counts from 0, so the second axis is axis 1, not axis 2.

We want to train a neural network with this time series, this time assuming a time window of length 3. If we use the data preparation we just performed, our time series examples would be:

- 1, 3, 5
- 11, 3, 2
- 22, 11, 5
- 7, 9, _

However, there is no reason to privilege the start of our data as somehow having to set the beginnings and ends of each sample time series. The windows are arbitrary. Some equally valid time series, sliced out of the whole in a window, are:

- 3, 5, 11
- 2, 22, 11
- 5, 7, 9

So if we needed more data, we would be well served to generate time series samples as we slid a window over the entire data set. That would produce more individual time series samples than we have done with our method of chopping the data into non-overlapping time series samples. Keep this in mind when preparing your own data sets. Below, you will see we use this sliding window method to preprocess our data.

Building and Training an RNN

As mentioned in the introduction to this chapter, financial time series are notoriously difficult to model and understand. Even as the finance industry continues to be a mainstay of the Western economy, experts agree that predictions are very difficult to make. For this reason, we look for a technique that is well suited for a complicated system with potentially nonlinear dynamics, namely a deep learning neural network. However, due to our lack of data, we choose a simple recurrent neural network (LSTM) architecture and training regime described by these parameters:

```python
## python
>>> ## architecture parameters
>>> NUM_HIDDEN = 4
>>> NUM_LAYERS = 2
>>>
>>> ## data formatting parameters
>>> BATCH_SIZE  = 64
>>> WINDOW_SIZE = 20
>>>
>>> ## training parameters
```

```
>>> LEARNING_RATE = 1e-2
>>> EPOCHS       = 30
```

In contrast to Chapter 10, we use the TensorFlow package rather than MXNet so you can see another example of a widely used deep learning framework. In TensorFlow, we define variables for all the quantities we will use in our network, even with changing values that represent inputs. For the inputs we use placeholders, which are a way of letting the graph know what shape to expect:

```python
## python
>>> Xinp = tf.placeholder(dtype = tf.float32,
>>>                                 shape = [WINDOW_SIZE, None, 3])
>>> Yinp = tf.placeholder(dtype = tf.float32, shape = [None])
```

Then we build our network and implement loss calculation and optimization steps:

```python
## python
>>> with tf.variable_scope("scope1", reuse=tf.AUTO_REUSE):
>>>     cells = [tf.nn.rnn_cell.LSTMCell(num_units=NUM_HIDDEN)
>>>                             for n in range(NUM_LAYERS)]
>>>     stacked_rnn_cell = tf.nn.rnn_cell.MultiRNNCell(cells)
>>>     rnn_output, states = tf.nn.dynamic_rnn(stacked_rnn_cell,
>>>                                         Xinp,
>>>                                         dtype=tf.float32)
>>>     W = tf.get_variable("W_fc", [NUM_HIDDEN, 1],
>>>                         initializer =
>>>                         tf.random_uniform_initializer(-.2, .2))
>>>
>>>     ## notice we have no bias because we expect average zero return
>>>     output = tf.squeeze(tf.matmul(rnn_output[-1, :, :], W))
>>>
>>>     loss = tf.nn.l2_loss(output - Yinp)
>>>     opt = tf.train.GradientDescentOptimizer(LEARNING_RATE)
>>>     train_step = opt.minimize(loss)
```

We have a fairly complicated way of feeding in the data due to what we discussed earlier, namely that each data point should be in multiple time series depending on which offset we use. Here we treat the same data formatting problem we discussed at greater length in Chapter 10:

```python
## python
>>> ## for each epoch
>>> y_hat_dict = {}
>>> Y_dict = {}
>>>
>>> in_sample_Y_dict = {}
>>> in_sample_y_hat_dict = {}
>>>
>>> for ep in range(EPOCHS):
>>>     epoch_training_loss = 0.0
>>>     for i in range(WINDOW_SIZE):
>>>         X = train_df[:(7000 - WINDOW_SIZE)][["ScaledVolatility",
```

```
>>>                                              "ScaledReturn",
>>>                                              "ScaledVolume"]].values
>>>         Y = train_df[WINDOW_SIZE:]["ScaledReturn"].values
>>>
>>>         ## make it divisible by window size
>>>         num_to_unpack = math.floor(X.shape[0] / WINDOW_SIZE)
>>>         start_idx = X.shape[0] - num_to_unpack * WINDOW_SIZE
>>>         X = X[start_idx:]
>>>         Y = Y[start_idx:]
>>>
>>>         X = X[i:-(WINDOW_SIZE-i)]
>>>         Y = Y[i:-(WINDOW_SIZE-i)]
>>>
>>>         X = np.expand_dims(X, axis = 1)
>>>         X = np.split(X, X.shape[0]/WINDOW_SIZE, axis = 0)
>>>         X = np.concatenate(X, axis = 1)
>>>         Y = Y[::WINDOW_SIZE]
>>>         ## TRAINING
>>>         ## now batch it and run a sess
>>>         for j in range(math.ceil(Y.shape[0] / BATCH_SIZE)):
>>>             ll = BATCH_SIZE * j
>>>             ul = BATCH_SIZE * (j + 1)
>>>
>>>             if ul > X.shape[1]:
>>>                 ul = X.shape[1] - 1
>>>                 ll = X.shape[1]- BATCH_SIZE
>>>
>>>             training_loss, _, y_hat = sess.run([loss, train_step,
>>>                                      output],
>>>                                      feed_dict = {
>>>                                          Xinp: X[:, ll:ul, :],
>>>                                          Yinp: Y[ll:ul]
>>>                                      })
>>>             epoch_training_loss += training_loss
>>>
>>>             in_sample_Y_dict[ep]     = Y[ll:ul]
>>>             ## notice this will only net us the last part of
>>>             ## data trained on
>>>             in_sample_y_hat_dict[ep] = y_hat
>>>
>>>         ## TESTING
>>>         X = test_df[:(test_df.shape[0] - WINDOW_SIZE)]
>>>                     [["ScaledVolatility", "ScaledReturn",
>>>                     "ScaledVolume"]].values
>>>         Y = test_df[WINDOW_SIZE:]["ScaledReturn"].values
>>>         num_to_unpack = math.floor(X.shape[0] / WINDOW_SIZE)
>>>         start_idx = X.shape[0] - num_to_unpack * WINDOW_SIZE
>>>         ## better to throw away beginning than end of training
>>>         ## period when must delete
>>>         X = X[start_idx:]
>>>         Y = Y[start_idx:]
>>>
```

```
>>>         X = np.expand_dims(X, axis = 1)
>>>         X = np.split(X, X.shape[0]/WINDOW_SIZE, axis = 0)
>>>         X = np.concatenate(X, axis = 1)
>>>         Y = Y[::WINDOW_SIZE]
>>>         testing_loss, y_hat = sess.run([loss, output],
>>>                             feed_dict = { Xinp: X, Yinp: Y })
>>>         ## nb this is not great. we should really have a validation
>>>         ## loss apart from testing
>>>
>>>     print("Epoch: %d   Training loss: %0.2f
>>>             Testing loss %0.2f:" %
>>>             (ep, epoch_training_loss, testing_loss))
>>>     Y_dict[ep] = Y
>>>     y_hat_dict[ep] = y_hat
```

Here we see our training and testing metrics:

```
Epoch: 0    Training loss: 2670.27   Testing loss 526.937:
Epoch: 1    Training loss: 2669.72   Testing loss 526.908:
Epoch: 2    Training loss: 2669.53   Testing loss 526.889:
Epoch: 3    Training loss: 2669.42   Testing loss 526.874:
Epoch: 4    Training loss: 2669.34   Testing loss 526.862:
Epoch: 5    Training loss: 2669.27   Testing loss 526.853:
Epoch: 6    Training loss: 2669.21   Testing loss 526.845:
Epoch: 7    Training loss: 2669.15   Testing loss 526.839:
Epoch: 8    Training loss: 2669.09   Testing loss 526.834:
Epoch: 9    Training loss: 2669.03   Testing loss 526.829:
Epoch: 10   Training loss: 2668.97   Testing loss 526.824:
Epoch: 11   Training loss: 2668.92   Testing loss 526.819:
Epoch: 12   Training loss: 2668.86   Testing loss 526.814:
Epoch: 13   Training loss: 2668.80   Testing loss 526.808:
Epoch: 14   Training loss: 2668.73   Testing loss 526.802:
Epoch: 15   Training loss: 2668.66   Testing loss 526.797:
Epoch: 16   Training loss: 2668.58   Testing loss 526.792:
Epoch: 17   Training loss: 2668.49   Testing loss 526.788:
Epoch: 18   Training loss: 2668.39   Testing loss 526.786:
Epoch: 19   Training loss: 2668.28   Testing loss 526.784:
Epoch: 20   Training loss: 2668.17   Testing loss 526.783:
Epoch: 21   Training loss: 2668.04   Testing loss 526.781:
Epoch: 22   Training loss: 2667.91   Testing loss 526.778:
Epoch: 23   Training loss: 2667.77   Testing loss 526.773:
Epoch: 24   Training loss: 2667.62   Testing loss 526.768:
Epoch: 25   Training loss: 2667.47   Testing loss 526.762:
Epoch: 26   Training loss: 2667.31   Testing loss 526.755:
Epoch: 27   Training loss: 2667.15   Testing loss 526.748:
Epoch: 28   Training loss: 2666.98   Testing loss 526.741:
Epoch: 29   Training loss: 2666.80   Testing loss 526.734:
```

The error metric we chose does not give us an idea of how well our overall data matches the results, so plotting is helpful. We plot both out-of-sample performance (important) and in-sample performance (less important), as shown in Figure 14-9:

```python
## python
>>> plt.plot(test_y_dict[MAX_EPOCH])
>>> plt.plot(test_y_hat_dict[MAX_EPOCH], 'r--')
>>> plt.show()
```

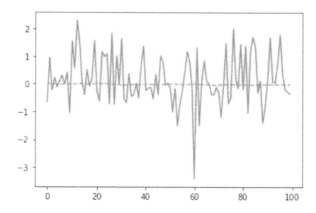

Figure 14-9. The actual return values for a subsection of the test period (solid line) are plotted against the forecasts from the neural network (dashed line). The scale of the forecast is so different from the actual data that it's difficult to assess the model.

We can see that our predicted values for the returns don't tend to be at the same value as the actual returns. We check the Pearson correlation next:

```python
## python
>>> pearsonr(test_y_dict[MAX_EPOCH], test_y_hat_dict[MAX_EPOCH])
(0.03595786881773419, 0.20105107068949668)
```

These numbers might look grim if you haven't already worked with financial time series. In this industry, our model may be useful despite the plot and despite the *p*-value. In finance, a correlation that is positive is exciting and something that can be improved upon incrementally. In fact, many research projects, when starting out, can't necessarily get to such "high" correlations.

We can get a better idea of whether the predictions at least go in the same direction by scaling the predicted returns up an order of magnitude and plotting again (Figure 14-10):

```python
## python
>>> plt.plot(test_y_dict[MAX_EPOCH][:100])
>>> plt.plot(test_y_hat_dict[MAX_EPOCH][:100] * 50, 'r--')
>>> plt.show()
```

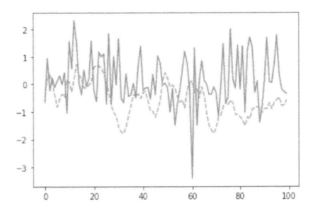

Figure 14-10. A better sense of how the model's predictions (dashed line) compare to the actual data (solid line). However, with such a low correlation we are more likely to think we see a pattern than to actually see one. For this reason, quantitative metrics will serve us better than visual assessments for noisy financial data.

If you have read blogs about using deep learning for financial time series, you may very well find the performance here disappointing. For example, you have likely seen blog posts where someone applies a simple multilayer LSTM to some daily stock data and produces predictions that look almost identical to the actual stock market data, even out of sample. There are two important reasons those results may look good but are actually not impressive:

- Preprocessing the code to scale it with an out-of-the-box scaling solution such as `sklearn.preprocessing.MinMaxScaler`. This is not ideal because it includes a lookahead by using the values across all time to scale data.

- Predicting price rather than returns. This is a much easier task—for starters, an excellent prediction of price on day $T + 1$ is price on day T. Thus, it's easy to build a model that appears to predict price reasonably well and makes impressive graphs. Unfortunately, such models can't be used to make money.

We have attempted a more realistic industry example, which means the challenge is greater and the plots won't be as satisfying.

Performance Considerations

We have used a standard LSTM `tf.nn.rnn_cell.LSTMCell` here for a number of reasons, not the least because it's a general workhorse. We also used it because this book uses code that does not require the use of a GPU so that you can run these scripts even on a standard laptop.

If you do have a GPU, you should use the API for NVIDIA's cuDNN, which is a hardware-tailored implementation available for the two main elaborations on a standard RNN: the LSTM and the GRU. Assuming your GPU is from NVIDIA, which most are, this will be substantially faster for training than the standard implementations. This is available in TensorFlow and also in the other mainstream deep learning packages, and the usage is quite similar to what we have coded here, apart from employing a different RNN cell.

This advice may change as Google's TPUs are rolled out to the mainstream and as other hardware solutions emerge for deep learning.

Needless to say, we haven't completed a full analysis of the model's performance. Doing so would give us insight into how to build our next model, what we might have overlooked, and whether a deep learning model can justify its additional complexity relative to a linear model. There are a lot of paths to take from here to improve model performance.

There are many ways we can improve this model, which you should consider as extensions of this code:

- Add more inputs from the raw data by generating additional features based on these inputs. We did not use all the raw input columns, and there are other ways to re-express these quantities that could be useful. You might consider categorical variables, such as "Did the high or the low for the day coincide with the opening or closing?" (there are several binary conditions packed into that one question).

- Integrate parallel time series for other stocks. This will add further information and data to train on.

- Use data at several different timescales. One widely cited paper that does this discusses an architecture called ClockworkRNN (*https://perma.cc/9C62-7GFK*).

- Augment your data by taking existing time series examples and adding jitter. This will help with the fact that this data set does not offer much data.

- Allow your network architecture to grow if you have scaled up the number of inputs or the amount of data. A more complicated architecture is not always a way to improve performance, but it can be appropriate if you see that your network is topping out.

- Try training the data in chronological order, rather than our approach of cycling through the data several times per epoch. Sometimes this be quite helpful (but it depends on the data set). Given that we saw the behavior of time series change over time, it might be better to end training with the last data last so that the changed behavior is reflected in the weights.

- Consider different loss functions. Here we used an L2 norm, which tends to punish larger differences much more than small differences. Given the domain, however, we might want to evaluate success differently. Perhaps we just want to get the predicted sign of the daily return right and not worry so much about its magnitude. In this case we could consider setting the targets as categorical variables: positive, negative, zero. In the case of categorical data, we usually want to use a cross-entropy measure of loss. However, given that this is not purely categorical data, since it's ranked (that is, zero is closer to negative than positive is to negative), we might want to use a custom loss function to reflect this.

- Consider building an ensemble of simple neural networks rather than a single one. Keep each individual network small. Ensembles are particularly useful for low signal-to-noise data, such as financial data.

- Determine why the scale of the forecasts is so different from the scale of the actual values. Consider starting by assessing whether the loss function we used is problematic given the strong preponderance of zero value daily returns.

As you can see, there are myriad ways to improve a network's performance or tune its functionality. How you do so will depend on the data set. It is always helpful to use visualizations of your network's performance, domain knowledge, and a firmly defined goal (which in this case is likely "make money") to drive how you refine your model. Otherwise, you can easily get lost in the overwhelming number of choices you have.

More Resources

Joumana Ghosn and Yoshua Bengio, "Multi-Task Learning for Stock Selection," Cambridge: MIT Press, 1996, https://perma.cc/GR7A-5PQ5.

This 1997 paper provides a very early example of applying a neural network to the problem of financial markets. In this case, the author used what would now be considered a quite simple network with minimal data, but they nonetheless found their network could learn to pick stocks profitably. Interestingly, this is also an early example of multitask learning.

Lawrence Takeuchi and Yu-Ying Lee, "Applying Deep Learning to Enhance Momentum Trading Strategies in Stocks," 2013, https://perma.cc/GJZ5-4V6Z.

In this paper, the authors interpret their neural network through the lens of "momentum training", a traditional way of quantitatively forecasting the financial markets in a pre-machine learning world. This paper is interesting for its discussion of how training decisions were made and model performance evaluated.

"Is anyone making money by using deep learning in trading?" Quora, https://perma.cc/Z8C9-V8FX.

In this question and answer, we see a variety of opinions regarding the extent to which deep learning has been successful in financial applications. As described in some of the answers, any profitable IP would likely be heavily protected by nondisclosure agreements and the profit incentive so that in this industry it can be very difficult to assess what state-of-the-art performance is. The answers also point to a wide rang of potential financial applications - predicting returns is just one problem among many.

## CHAPTER 15

Time Series for Government

Time series analysis is quite relevant and important for governmental applications for a number of reasons. First, governments both large and small, are the keepers of some of the most important time series data in the world, including the US jobs report, ocean temperature data (that is, global warming data), and local crime statistics. Second, governments by definition provide some of the most essential services we all rely on, and thus they need to be reasonably adept forecasters of demand if they don't want to grossly overspend on, or understaff, those services. Thus, all aspects of time series are relevant to government purposes: storage, cleaning, exploration, and forecasting.

As I mentioned back in Chapter 2 when discussing "found" time series, a very high percentage of all government data can look a lot like time series data with some restructuring. Generally, most government data sets are the result of ongoing data collection rather than a single slice of time. However, government data sets can be daunting for a number of reasons:

- Inconsistent recordkeeping (due to organizational constraints or political forces changing over time)

- Opaque or confusing data practices

- Enormous data sets with relatively low information content

Nonetheless, it can be quite interesting to look at government data sets both for intellectual interest and for many practical purposes. In this chapter we explore a governmental data set that consists of all the complaints made in New York City from 2010 to the present (*https://perma.cc/BXF6-BZ4X*) to a city-run hotline that can be reached by dialing 311. Because the data set is continuously updated, the data seen in the book will likely differ from what you see when you download it; you will have even more information than I did when preparing this chapter. Nonetheless, the results should be fairly similar. In this chapter we will go through a few topics:

- Interesting sources of government data, including the one we will analyze
- Dealing with extremely large files of plain-text data
- Online/rolling statistical analysis of large data sets and other options for analyzing data without keeping it all in memory

Obtaining Governmental Data

Governmental data sets in the "found data" category can be a nightmare from a data consistency standpoint. These data sets, although they have a timestamp, are usually released for an open data initiative rather than for a specific time series purpose. Often there is little to no information available about the timestamping conventions of the data or other recording conventions. It can be difficult to confirm that the underlying recording practices were consistent.[1]

Nonetheless, if you are adventurous or keen to be the first to identify interesting temporal features in human behaviors related to governmental activities, you are lucky to live in the age of open government and open data. Many governments at all levels have made more efforts in recent years to make their time series data transparent to the public. Here are just a few examples of where you can obtain open governmental data with a time series component:

- Monthly hospital data (*https://perma.cc/4TR3-84WA*) from the United Kingdom's National Health Services. This data set is surprisingly time series–aware: it includes having a tab named "MAR timeseries" and describes recording conventions and how they have evolved over time.

- Jamaica's open data portal also includes an appreciation and recognition of time series data, such as its timestamped data set of Chikungunya cases (*https://perma.cc/4RCP-VMY6*) for 2014 and the associated data report (*https://perma.cc/QPR6-WNMJ*), which includes an animation (i.e., a time series visualization) and an epidemic curve.

- Singapore's open data portal (*https://perma.cc/N9W4-ZDM8*) features extensive data sets and advertises the time series nature of some of that data by including two time series plots on its main page, as shown in Figure 15-1.

---

1 Note that highly sought-after time series data, such as the US jobs report, is very time series–aware and meticulously cleaned and formatted. However, such data is already quite picked over and unlikely to furnish novel time series applications to the would-be researcher or entrepreneur.

Figure 15-1. Two of the four graphs on the main page of Singapore's open data website (as accessed in spring 2019) are time series visualizations used to show important information about the country.

As you might notice, all my examples are from English-speaking areas, but of course they do not have a monopoly on the open data movement in government. For example, the city of Paris (*https://perma.cc/7V8Z-JZ4T*), the nation of Serbia (*https://perma.cc/U3SQ-WF3C*), and the African Development Bank Group (*https://perma.cc/7L6X-5B9F*) all run open data websites.[2]

In this chapter's examples we are pulling our information from the New York City Open Data portal, selected because NYC is a large and interesting place that happens to be my home. We dive into their 311 hotline data set in the next section.

---

2 Of course your access might depend on your linguistic capabilities (or those of a helpful colleague).

Exploring Big Time Series Data

When data is sufficiently large, you will be unable to fit it in memory. How large data needs to be before you reach this limit will depend on the hardware you are using.[3] You will eventually need to understand how to iterate through your data, one manageable chunk at a time. For those familiar with deep learning, you have likely already done this, particularly if you work with image processing. In deep learning frameworks there are often Python iterators that wend their way through a data set with that data set stored in specified directories, each with many files.[4]

When I downloaded the 311 data set, it was over 3 gigabytes in CSV format. There was no way I would be able to open this on my computer, so my first idea was to use standard Unix operating system options, such as head.

Unfortunately, what printed out was already so large as to be unmanageable in a Unix command-line interface, at least for someone inexpert in Unix tools:

```
## linux os command line
$ head 311.csv

Unique Key,Created Date,Closed Date,Agency,Agency Name,Complaint Type,Descripto
27863591,04/17/2014 12:00:00 AM,04/28/2014 12:00:00 AM,DOHMH,Department of Heal
27863592,04/17/2014 12:00:00 AM,04/22/2014 12:00:00 AM,DOHMH,Department of Heal
27863595,04/17/2014 10:23:00 AM,0417/2014 12:00:00 PM,DSNY,Queens East 12,Derel
27863602,04/17/2014 05:01:00 PM,04/17/2014 05:01:00 PM,DSNY,BCC - Queens East,D
27863603,04/17/2014 12:00:00 AM,04/23/2014 12:00:00 AM,HPD,Department of Housin
27863604,04/17/2014 12:00:00 AM,04/22/2014 12:00:00 AM,HPD,Department of Housin
27863605,04/17/2014 12:00:00 AM,04/21/2014 12:00:00 AM,HPD,Department of Housin
```

While the content is unwieldy, this view was sufficient to show that there were several timestamps as well as other interesting and ordered information, such as geographic coordinates. Clearly the data is very wide, so we need to be able to manipulate this information to get the columns we want.[5]

Even if you are new to Linux, you can easily learn about simple command line tools that can provide helpful information. We can get a line count of the CSV file so we

---

3 If your organization always scales up to solve the problem of not enough RAM, you're doing it wrong.

4 For inspiration, read the TensorFlow documentation on data sets and associated classes (*https://www.tensor flow.org/guide/datasets*).

5 Notice that someone expert in Unix systems could easily use awk to manipulate a CSV quite effectively on the command line or with a simple shell script. Often such tools are excellent options for big data because they are extremely efficient and well implemented, unlike, unfortunately, many commonly used data analysis tools in R and Python. If you run into problems with your favorite data processing tools, it can often be a good idea to learn some Unix command-line tools to supplement, especially in the case of big data.

---

have an idea of the scale we are looking at, namely how many data points we have. This is a one-liner:

```
## linux os command line
$ wc -l 311.csv
19811967 311.csv
```

We can see that NYC has fielded about 20 million complaints to 311 since 2010. That's more than two complaints per resident.

Armed with this knowledge, we use R's data.table knowing that its fread() function enables partial reading of files (seen in the documentation (*https://perma.cc/ ZHN9-5HD3*) when you read about the nrows and skip parameters[6]) and that data.table is extremely performant when handling large data sets. We can use this to get initial information, as in the following code:

```
## R
> df = fread("311.csv", skip = 0, nrows = 10)
> colnames(df)
 [1] "Unique Key"                        "Created Date"
 [3] "Closed Date"                       "Agency"
 [5] "Agency Name"                       "Complaint Type"
 [7] "Descriptor"                        "Location Type"
 [9] "Incident Zip"                      "Incident Address"
[11] "Street Name"                       "Cross Street 1"
[13] "Cross Street 2"                    "Intersection Street 1"
[15] "Intersection Street 2"             "Address Type"
[17] "City"                              "Landmark"
[19] "Facility Type"                     "Status"
[21] "Due Date"                          "Resolution Description"
[23] "Resolution Action Updated Date"    "Community Board"
[25] "BBL"                               "Borough"
[27] "X Coordinate (State Plane)"        "Y Coordinate (State Plane)"
[29] "Open Data Channel Type"            "Park Facility Name"
[31] "Park Borough"                      "Vehicle Type"
[33] "Taxi Company Borough"              "Taxi Pick Up Location"
[35] "Bridge Highway Name"               "Bridge Highway Direction"
[37] "Road Ramp"                         "Bridge Highway Segment"
[39] "Latitude"                          "Longitude"
[41] "Location"
```

Just from reading the first 10 lines we can already see the column names. For all the strengths I listed for NoSQL approaches to time series data, it can be nice in a large data set to know the column names from the outset. Of course there are workarounds for this with NoSQL data, but most require some effort on the part of the user rather than happening automatically.

---

6 Note that Python's Pandas offers similar functionality (*https://perma.cc/68EE-2ZZ9*).

Several columns suggest useful information:

```
"Created Date"
"Closed Date"
"Due Date"
"Resolution Action Updated Date"
```

These will likely be of character type before we convert, but once we do a conversion to a POSIXct type, we can see what the time spans between these dates look like:

```
## R
> df$CreatedDate = df[, CreatedDate := as.POSIXct(CreatedDate,
>                                     format = "%m/%d/%Y %I:%M:%S %p")
```

In the formatting string, we need to use a %I for the hour since it's expressed only in 01–12 format, and a %p because the timestamps include an AM/PM designation.

To get a sense of how these dates tend to be spaced, particularly with respect to when a complaint is created versus closed out, let's load in more rows and examine the distribution of what I'll call the *complaint lifetime* (i.e., the span between creation and closing):

```
## R
> summary(as.numeric(df$ClosedDate - df$CreatedDate,
>             units = "days"))
   Min. 1st Qu.  Median    Mean 3rd Qu.    Max.    NA's
-75.958   1.000   4.631  13.128  12.994 469.737     113
```

As we can see, it's a wide distribution. More surprising than the yearlong wait times for some complaints to be closed out is the fact that some complaints have negative, even extremely negative, times between their creation and closing date. If the negative time were around –365 days (a year), we might imagine a data entry problem, but this seems less likely with numbers such as –75 days. This is a problem we'll need to look into.

We can spot another problem by taking a range of the creation date:

```
## R
> range(df$CreatedDate)
[1] "2014-03-13 12:56:24 EDT" "2019-02-06 23:35:00 EST"
```

Given the size of this CSV file and the fact that it is supposed to be continuously updated, it is surprising that the first lines are not from the 2010 date that is supposed to mark the earliest data sets. We would have expected the CSV to be continuously appended to. More surprising still is the fact that 2019 dates are in the first rows and that 2014 and 2019 dates are both in the same first 10,000 lines. This suggests that we cannot easily determine the date ordering of the data in the file. We can visualize the date distribution from one row to the next by performing a line plot of row index versus date, as we do in Figure 15-2.

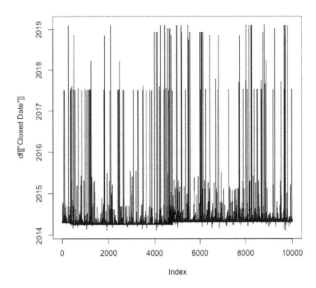

Figure 15-2. While most of the dates in the first 10,000 lines of code appear to fall in 2014, they also jump around as far forward as 2019—and often!

There is no way to avoid the problem of acting. If we want to understand how behavior changes over time, we will have to confront the unordered data. But we have a few options.

Upsample and Aggregate the Data as We Iterate Through It

One option is to upsample the data as we iterate through it to construct a condensed time series with aggregate statistics. We could choose a resolution and aggregation counts from the start of our analysis and then calculate these as we iterate through. Then we could sort our results at the end of our analysis. This would look something like having, say, a dictionary/list of all dates from 2010 to the present and then adding to the appropriate date for each row. This method would produce a list/dictionary with a relatively small number of entries, and we could then sort this at the end based on date.

The upside of this is that it would be relatively straightforward to code and would be a way to combine data cleaning, exploration, and analysis into one exploratory step. The downside is that detailed data would still be lost in the morass of the unsorted file, so that if there was a specific time period of interest we would have to search the entire unsorted file to find all relevant entries.

Since we have already done examples of upsampling in Chapter 2, I leave this as an exercise for the reader.

Sort the Data

Another option is to sort the data. This is a daunting task given the large size of the file and the relatively unordered dates from what we can observe. Even for the data within 2014 it does not seem as though the dates come in order. For this reason we have no indications that we can trust any slices of the data, so we should think of this as a pile of data in random order.

Sorting the full file would be extremely memory intensive, but there are two reasons it could be worthwhile. One is that we only need to sort once and then can save our results for whatever subsequent analysis we want to do. Second is that we can then have the full level of detail preserved so that if we spot specific time periods of interest in our analysis, we can examine the data in all its detail to understand what is happening.

Concretely, thinking about how we could accomplish this, we have a few options:

- Linux has a command-line tool (*https://perma.cc/7SNE-TQ2T*) for sorting.

- Most databases can sort data, so we could transfer the data to a database and let the database take care of it.[7]

- We could come up with our own sorting algorithm and implement it. We would need to formulate something that did not consume enormous amounts of memory. Odds are strongly against our effort matching what is available in prepackaged sorting options.

We choose to use the Linux command-line tool. While it may take some time to get this right, we will develop a new life skill as well as gaining access to a well-implemented and correct sort for this large file.

We begin by first creating a small test file to use:

```
## linux command line
$ head -n 1000 311.csv | tail -n 999  >  test.csv
```

Notice that this involves both the head (i.e., print out the start) and tail (i.e., print out the end) commands. head will include the first line of the file, which for this file provides the column names. If we include this and sort it along with the values, it will not preserve the column names as the top line in the file, so we cut it out before sorting.

If you are working on a Linux-based operating system, you can then apply the sort command as follows:

---

7 We'd have to pick one that offers this functionality—not all databases do, and especially not all time series databases, which tend to presume data comes into the database in chronological order.

```
## linux command line
$ sort --field-separator=',' --key=2,3 test.csv > testsorted.csv
```

In this case we identify the field separator and then indicate we want to sort by the second and third columns—that is, by the creation date and close date (which we only know from previously inspecting the file). We output this into a new file, as it wouldn't be very helpful to have it output to standard out.

Now we can inspect the sorted file in R, but unfortunately we will find that this also does not return a sorted file. Go back a few pages to where we ran the `head` command on the CSV file, and you will see why. We sorted according to a date column (which would be processed like a string, not in a date-aware manner). However, the current formatting of the dates begins with the month, so we will end with a column of dates by month rather than sorted by overall time, as we see when we review the "sorted" CSV resulting from our previous command:

```
  1: 02/02/2019 10:00:27 AM 02/06/2019 07:44:37 AM
  2: 03/27/2014 07:38:15 AM 04/01/2014 12:00:00 AM
  3: 03/27/2014 11:07:31 AM 03/28/2014 01:09:00 PM
  4: 03/28/2014 06:35:13 AM 03/28/2014 10:47:00 PM
  5: 03/28/2014 08:31:38 AM 03/28/2014 10:37:00 AM
  ---
995: 07/03/2017 12:40:04 PM 07/03/2017 03:40:02 PM
996: 07/04/2017 01:30:35 AM 07/04/2017 02:50:22 AM
997: 09/03/2014 03:32:57 PM 09/04/2014 04:30:30 PM
998: 09/05/2014 11:17:53 AM 09/08/2014 03:37:25 PM
999: 11/06/2018 07:15:28 PM 11/06/2018 08:17:52 PM
```

As we can see, the sorting makes sense as a string sort rather than a date sort. In fact, this points to one of the virtues of using proper ISO date formatting, which is that it will still get the sort right when sorting as a string, unlike in the preceding format. This is an example of a very common problem with "found" time series data: the timestamping format available may not be the most conducive to time series analysis.

We revisit the NYC Open Data interface to see whether there is a workaround to this formatting (see Figure 15-3).

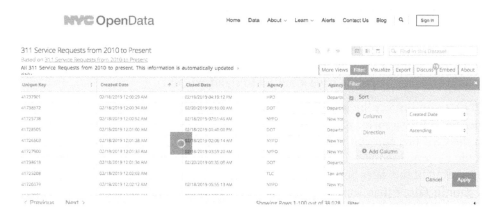

Figure 15-3. The NYC Open Data portal offers sorting via a web interface, which appears to be available to any column on this large data set. This is an impressive free resource when you think about the computational power required for that sort.

A glimpse at the web page tabular view of the data seems to concur with our CSV, in that the data does not appear to be sorted in time order. However, we see there is a sort option, so we apply this to the data set, which does indeed update the data after an understandable wait while the large data set is stored. This would be a great solution, except that, unfortunately the resulting CSV downloaded is still out of order.

There are a number of other solutions we could explore at this time. We could see whether using the Open Data's API rather than the web interface provides more digestible dates or a way to ensure sorting. We could work with more Linux command-line tools, such as awk, to extract the different parts of the timestamp into different columns or into a single rearranged column with ISO formatting.

Instead, we take a pared-back approach of seeing whether our available tools can handle this CSV if we read only certain columns. The first question that interests me about this data set is how the lag between the creation of a 311 complaint and the closing of that complaint may have varied over time. In this case, I hypothesize that I need only two columns: CreatedDate and ClosedDate. I will see whether just reading these two columns, which are a small portion of all the columns in terms of both count and character count (because some columns are quite long), is possible on my lightweight laptop. (I could also explore the lazy way of fixing the problem, which would be a hardware upgrade, either temporarily or permanently.)

So now we are able to read in all the rows of the data, and our subsequent analysis will be on the full data set rather than only the first 1,000 rows:

```
## R
> ## read in only the columns of interest
```

```
> df = fread("311.tsv", select = c("Created Date", "Closed Date"))
>
> ## use data.table's recommended 'set' paradigm to set col names
> setnames(df, gsub(" ", "", colnames(df)))
>
> ## eliminate rows with a blank date field
> df = df[nchar(CreatedDate) > 1 & nchar(ClosedDate) > 1]
>
> ## convert string date columns to POSIXct
> fmt.str = "%m/%d/%Y %I:%M:%S %p"
> df[, CreatedDate := as.POSIXct(CreatedDate, format = fmt.str)]
> df[, ClosedDate  := as.POSIXct(ClosedDate,  format = fmt.str)]
>
> ## order in chronological order of CreatedDate
> setorder(df, CreatedDate)
>
> ## calculate the number of days between creating and closing
> ## a 311 complaint
> df[, LagTime := as.numeric(difftime(ClosedDate, CreatedDate,
>                                  units = "days"))]
```

This was done on a lightweight laptop manufactured in 2015, so chances are you can do it with whatever you use for work or home, too. Those new to "big data" may be surprised that 19 million rows is not a big deal, but this is often the case. We actually need only a small slice of the data to address a relevant time series question.

Once we look at the LagTime column, we notice some surprisingly incorrect numbers —numbering in tens of thousands of days, or negative days. We weed out these numbers and put a cap on the data driven partly by a distribution of a random sample of data points:

```
## R
> summary(df$LagTime)
Min.  1st Qu.  Median    Mean  3rd Qu.    Max.     NA's
-42943.4     0.1     2.0     2.0     7.9 368961.1   609835

> nrow(df[LagTime < 0]) / nrow(df)
[1] 0.01362189

> nrow(df[LagTime > 1000]) / nrow(df)
[1] 0.0009169934

> df = df[LagTime < 1000]
> df = df[LagTime > 0]

> df.new = df[seq(1, nrow(df), 2), ]
> write.csv(df.new[order(ClosedDate)], "abridged.df.csv")
```

We discard data with negative lag times, as we lack documentation or domain knowledge to know what those are. We also reject data that we regard as having unrealistic or unhelpfully extreme values by discarding data where the lag time to close out the 311 complaint was more than 1,000 days.[8]

 Use caution when you discard data. In this exercise, we discarded around 1.3% of the data due to lag times in resolving a 311 complaint that didn't make sense, either because they were so long as to warrant an explanation we would not be able to get or because they were negative, implying a data entry error or some other issue beyond what we could likely discover from the data alone.

Given that our question relates to the overall distribution of the data, it's unlikely that such a small portion of points would affect our analysis about a distributional question. However, in the real-world you would want to investigate these data points and the possible downstream effects on your analysis task. This is not time series–specific advice but just a matter of general practice.

Now that we are able to hold all of the data of interest in memory at the same time, we could ask questions about the series holistically. However, the question of interest is whether and how the distribution of lag time may have changed over time. We could do this with a sliding or rolling window over the time series, but that is computationally taxing; we would have to do many related computations repeatedly as we slide a window over the data. Also we want to explore doing this in a way that could also work on a live stream, as we could imagine continuing this project onto current data as it comes in. It would be better not to have to store this multigigabyte data file indefinitely.

Online Statistical Analysis of Time Series Data

We are going to use a fairly straightforward online quantile estimation tool called the *P-square algorithm* (*https://perma.cc/G8LA-7738*), with one slight modification to make it time-aware. The original algorithm assumed that there was a stable distribution from which quantiles were being inferred, but we want to account for the case of distributions that change over time. As with an exponentially weighted moving average, we add this time awareness by weighting earlier observations less, and we do it in the same way by introducing a factor to scale down the weights of the previous measurements each time a new measurement is available (see Figure 15-4).

---

8 While 1,000 days may seem surprisingly high, in fact I personally have made several 311 complaints that have not been resolved in as many days. I have been waiting years for NYC to replace the tree in front of my house after the old one died—every time I call 311 I am told they are working on it.

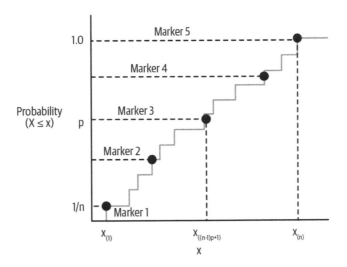

Figure 15-4. The computational structure of our online quantile estimation using the P-square algorithm. We maintain a series of markers indicating where we think the quantiles are and what the cumulative counts are for all data points less than or equal to each of those quantiles.

The algorithm requires a bit of recordkeeping, making it easier to implement in a more object-oriented programming language, so we'll switch from R to Python. Note, however, that we will use the preprocessed data from R, as the `data.table` package has substantially better performance with that kind of big data than tools available in Python.

The version of the P-square algorithm we implement forms a histogram of the values. So, as we create a `PQuantile` object, we allocate a preset number of bins for our histogram counts, bin positions, and running tallies of observations:

```python
## python
>>> ## imports and a utilit lambda
>>> import bisect
>>> import math
>>>
>>> sign = lambda x: (1, -1)[x < 0]
>>>
>>> ## beginning of class definition
>>> class PQuantile:
>>>     def __init__(self, b, discount_factor):
>>>         ## initialization
>>>         self.num_obs = 0 ## self-explanatory
>>>         ## counts per quantile
>>>         self.n = [i for i in range(self.b+1)]
>>>         self.q = [] ## the estimated quantile values
>>>
```

```
>>>          ## b is the number of quantiles,
>>>          ## including the 0th and 100th
>>>          ## (min and max values) quantile
>>>          self.b = b
>>>          ## the discount factor sets how we adjust prior counts
>>>          ## when new data is available
>>>          self.discount_factor = discount_factor
```

There are two configurable parameters: the number of evenly spaced quantiles to estimate and the discounting factor for old observations.

The other class members include a running total of the number of observations (this will be subject to time discounting by the configurable time discounting factor), the estimated quantiles, and the running count of observations less than or equal to a given quantile value.

There is only one public function, whose role is to accept the next observation. When a new observation arrives, what happens is a result of how early in the series we are. For the first self.b values, the inputs are accepted and necessarily make up the estimation of the quantile. self.q is sorted so that its values reflect the quantiles.

So, for example, imagine you input a b = 5 for the number of desired quantile values and then input the sequence 2, 8, 1, 4, 3. At the end of this sequence, self.q would be equal to [1, 2, 3, 4, 8]. self.n, the counts of values less than or equal to each of the quantiles, would be equal to [1, 2, 3, 4, 5], the value it was already initialized to in __init__:

```python
## python
>>>      def next_obs(self, x):
>>>          if self.num_obs < (self.b + 1):
>>>              self.q.append(x)
>>>              self.q.sort()
>>>              self.num_obs = self.num_obs + 1
>>>          else:
>>>              self.next_obs2(x)
>>>              self.next_obs = self.next_obs2
```

Things get interesting once you have more than self.b values. At this point, the code begins to make decisions as to how to combine values to estimate quantiles without keeping all the data points stored for repeated analysis. In this case, the P-square algorithm does this with what we call self.next_obs2:

```python
## python
>>>      def next_obs2(self, x):
>>>          ## discounting the number of observations
>>>          if self.num_obs > self.b * 10:
>>>              corrected_obs = max(self.discount_factor * self.num_obs,
>>>                                  self.b)
>>>              self.num_obs = corrected_obs + 1
>>>              self.n = [math.ceil(nn * self.discount_factor)
```

```
>>>                            for nn in self.n]
>>>
>>>            for i in range(len(self.n) - 1):
>>>                if self.n[i + 1] - self.n[i] == 0:
>>>                    self.n[i+1] = self.n[i + 1] + 1
>>>                elif self.n[i + 1] < self.n[1]:
>>>                    ## in practice this doesn't seem to happen
>>>                    self.n[i + 1] = self.n[i] - self.n[1 + 1] + 1
>>>        else:
>>>            self.num_obs = self.num_obs + 1
>>>
>>>        k = bisect.bisect_left(self.q, x)
>>>        if k is 0:
>>>            self.q[0] = x
>>>        elif k is  self.b+1 :
>>>            self.q[-1] = x
>>>            k = self.b
>>>        if k is not 0:
>>>            k = k - 1
>>>
>>>        self.n[(k+1):(self.b+1)] = [self.n[i] + 1
>>>                                   for i in range((k+1),
>>>                                               (self.b+1))]
>>>        for i in range(1, self.b):
>>>            np = (i)*(self.num_obs - 1 )/(self.b)
>>>            d = np - self.n[i]
>>>            if (d >= 1 and (self.n[i+1] - self.n[i]) > 1):
>>>                self._update_val(i, d)
>>>            elif (d <= -1 and (self.n[i-1] - self.n[i]) < -1):
>>>                self._update_val(i, d)
```

Ideally, the ith quantile value should be evenly spaced so that exactly $i/b \times$ total observations are less than it. If this is not the case, the marker is moved one position to the left or right, and its associated quantile value is modified with a formula derived from the presumption of a locally parabolic shape of the histogram. That formula dictates the criteria indicated earlier for sizing the d variable to determine whether a particular quantile value and count have to be adjusted.

If the value does have to be adjusted, there is another decision to make as to whether the parabolic or linear adjustment is appropriate. This is implemented in the following code. For more details, see the derivation in the original paper (*https://perma.cc/ G8LA-7738*). This paper is great because of the approachable mathematics used in the technique and also because it provides very clear instructions as to how to implement the method and then test your implementation:

```python
## python
>>>    ## overall update
>>>    ## as you can see both self.q and self.n are updated
>>>    ## as a quantile position is shifted
>>>    def _update_val(self, i, d):
>>>        d = sign(d)
```

```
>>>            qp = self._adjust_parabolic(i, d)
>>>            if self.q[i] < qp < self.q[i+1]:
>>>                self.q[i] = qp
>>>            else:
>>>                self.q[i] = self._adjust_linear(i, d)
>>>            self.n[i] = self.n[i] + d
>>>
>>>        ## this is the primary update method
>>>        def _adjust_parabolic(self, i, d):
>>>            new_val = self.q[i]
>>>            m1 =  d/(self.n[i+1] - self.n[i-1])
>>>            s1 = (self.n[i] - self.n[i-1] + d) *
>>>                     (self.q[i+1] - self.q[i]) /
>>>                     (self.n[i+1] - self.n[i])
>>>            s2 = (self.n[i+1] - self.n[i] - d) *
>>>
>>>        ## this is the backup linear adjustment when parabolic
>>>        ## conditions are not met
>>>        def _adjust_linear(self, i, d):
>>>            new_val = self.q[i]
>>>            new_val = new_val + d * (self.q[i + d] - self.q[i]) /
>>>                                    (self.n[i+d] - self.n[i])
>>>            return new_val
```

For perspective on the simplicity of this method, all the class code is listed here in one place:

```
## python
>>> class PQuantile:
>>>        ## INITIALIZATION
>>>        def __init__(self, b, discount_factor):
>>>            self.num_obs = 0
>>>            self.b = b
>>>            self.discount_factor = discount_factor
>>>            self.n = [i for i in range(self.b+1)]
>>>            self.q = []
>>>
>>>        ## DATA INTAKE
>>>        def next_obs(self, x):
>>>            if self.num_obs < (self.b + 1):
>>>                self.q.append(x)
>>>                self.q.sort()
>>>                self.num_obs = self.num_obs + 1
>>>            else:
>>>                self.next_obs2(x)
>>>                self.next_obs = self.next_obs2
>>>
>>>        def next_obs2(self, x):
>>>            ## discounting the number of observations
>>>            if self.num_obs > self.b * 10:
>>>                corrected_obs = max(self.discount_factor
>>>                                            * self.num_obs,
>>>                                        self.b)
```

```
>>>             self.num_obs = corrected_obs + 1
>>>             self.n = [math.ceil(nn * self.discount_factor)
>>>                                        for nn in self.n]
>>>
>>>             for i in range(len(self.n) - 1):
>>>                 if self.n[i + 1] - self.n[i] == 0:
>>>                     self.n[i+1] = self.n[i + 1] + 1
>>>                 elif self.n[i + 1] < self.n[1]:
>>>                     ## in practice this doesn't seem to happen
>>>                     self.n[i + 1] = self.n[i] - self.n[1 + 1] + 1
>>>         else:
>>>             self.num_obs = self.num_obs + 1
>>>
>>>         k = bisect.bisect_left(self.q, x)
>>>         if k is 0:
>>>             self.q[0] = x
>>>         elif k is  self.b+1 :
>>>             self.q[-1] = x
>>>             k = self.b
>>>         if k is not 0:
>>>             k = k - 1
>>>
>>>         self.n[(k+1):(self.b+1)] = [self.n[i] + 1
>>>                                 for i in range((k+1),
>>>                                                 (self.b+1))]
>>>         for i in range(1, self.b):
>>>             np = (i)*(self.num_obs - 1 )/(self.b)
>>>             d = np - self.n[i]
>>>             if (d >= 1 and (self.n[i+1] - self.n[i]) > 1):
>>>                 self._update_val(i, d)
>>>             elif (d <= -1 and (self.n[i-1] - self.n[i]) < -1):
>>>                 self._update_val(i, d)
>>>
>>>     ## HISTOGRAM ADJUSTMENTS
>>>     def _update_val(self, i, d):
>>>         d = sign(d)
>>>         qp = self._adjust_parabolic(i, d)
>>>         if self.q[i] < qp < self.q[i+1]:
>>>             self.q[i] = qp
>>>         else:
>>>             self.q[i] = self._adjust_linear(i, d)
>>>         self.n[i] = self.n[i] + d
>>>
>>>     def _adjust_parabolic(self, i, d):
>>>         new_val = self.q[i]
>>>         m1 =  d/(self.n[i+1] - self.n[i-1])
>>>         s1 = (self.n[i] - self.n[i-1] + d) *
>>>                 (self.q[i+1] - self.q[i]) /
>>>                 (self.n[i+1] - self.n[i])
>>>         s2 = (self.n[i+1] - self.n[i] - d) *
>>>                 (self.q[i] - self.q[i-1]) /
>>>                 (self.n[i] - self.n[i-1])
```

```
>>>         new_val = new_val + m1 * (s1 + s2)
>>>         return new_val
>>>
>>>     def _adjust_linear(self, i, d):
>>>         new_val = self.q[i]
>>>         new_val = new_val + d * (self.q[i + d] - self.q[i]) /
>>>                             (self.n[i+d] - self.n[i])
>>>         return new_val
```

Now that we have this time-oriented method, we should convince ourselves that it
works reasonably well with a toy example. We first try sampling data points from one
distribution and then abruptly changing to another. In each case, we are sampling for
the 40th percentile, although based on the configuration of 10 histogram points, we
are maintaining a histogram that indicates the 0th, 10th, 20th, …90th, 100th percen-
tiles. This is helpful because it means we can have a fairly detailed description of the
changing distribution. For this toy example, we focus on just the 40th percentile
(qt.q[4]), which results in the plot in Figure 15-5:

```
## python
>>> qt = PQuantile(10, 0.3)
>>> qt_ests = []
>>>
>>> for _ in range(100):
>>>     b.next_obs(uniform())
>>>     if len(b.q) > 10:
>>>         qt_ests.append(qt.q[4])
>>> for _ in range(100):
>>>     b.next_obs(uniform(low = 0.9))
>>>     qt_ests.append(qt.q[4])
>>>
>>> plt.plot(qt_ests)
```

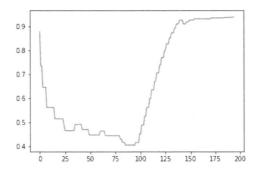

*Figure 15-5. When we discount older measurements heavily (multiplying by a smaller
discount factor), we more quickly see that the underlying distribution has changed.*

In contrast, we see a slower adoption to the changing quantile in the case of a larger
discounting factor (see Figure 15-6):

```
## python
>>> qt = PQuantile(10, 0.8)
>>> qt_ests = []
>>>
>>> for _ in range(100):
>>>     b.next_obs(uniform())
>>>     if len(b.q) > 10:
>>>         qt_ests.append(qt.q[4])
>>> for _ in range(100):
>>>     b.next_obs(uniform(low = 0.9))
>>>     qt_ests.append(qt.q[4])
>>>
>>> plt.plot(qt_ests)
```

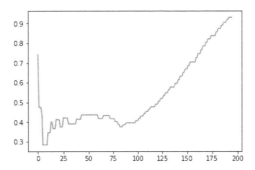

Figure 15-6. When we discount older measurements less (multiplying by a larger discount factor), our quantile estimation is slower to recognize that the underlying distribution has changed.

Now we apply this rolling quantile to a subset of our data (see Figure 15-7). We do not do the entire data set, not because of the computational challenge, but because it was overly taxing for my workaday laptop to graph all the recorded quantiles!

```
## python
>>> import numpy
>>> nrows = 1000000
>>> qt_est1 = np.zeros(nrows)
>>> qt_est2 = np.zeros(nrows)
>>> qt_est3 = np.zeros(nrows)
>>> qt_est4 = np.zeros(nrows)
>>> qt_est5 = np.zeros(nrows)
>>> qt_est6 = np.zeros(nrows)
>>> qt_est7 = np.zeros(nrows)
>>> qt_est8 = np.zeros(nrows)
>>> qt_est9 = np.zeros(nrows)
>>> for idx, val in enumerate(df.LagTime[:nrows]):
>>>     qt.next_obs(val)
>>>     if len(qt.q) > 10:
>>>         qt_est1[idx] = qt.q[1]
```

```
>>>        qt_est2[idx] = qt.q[2]
>>>        qt_est3[idx] = qt.q[3]
>>>        qt_est4[idx] = qt.q[4]
>>>        qt_est5[idx] = qt.q[5]
>>>        qt_est6[idx] = qt.q[6]
>>>        qt_est7[idx] = qt.q[7]
>>>        qt_est8[idx] = qt.q[8]
>>>        qt_est9[idx] = qt.q[9]
>>>
>>> plot(qt_est9, color = 'red')
>>> plt.plot(qt_est7, color = 'pink')
>>> plt.plot(qt_est5, color = 'blue')
>>> plt.plot(qt_est3, color = 'gray')
>>> plt.plot(qt_est2, color = 'orange'
```

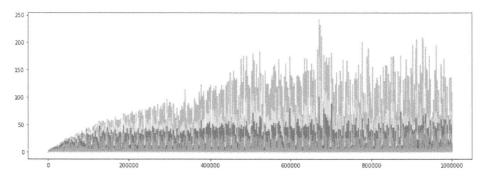

Figure 15-7. The 90th, 70th, 50th, 30th, and 20th percentile values over time for the first 100,000 rows in the data set when sorted by closed date.

Figure 15-7 shows the 90th, 70th 50th, 30th, and 20th percentile values plotted over time for the first 100,000 rows in the data set when sorted by closed date. Because they were sorted by closed date, it is likely that many quickly resolved 311 complaints were front-loaded, which explains the much smaller quantile estimations in the first portion of the data set.[9]

So did the distribution change? Visually it seems to, for a few reasons. One is left censoring, described shortly, which reflects how we sorted and selected data. The fact that we ordered the data by the ClosedData column, combined with the fact that this data set does not appear to have infinite lookback (that is, presumably 311 complaints made before a certain date did not make it into this system), makes the lag times we see at the beginning dates appear shorter. In other words, this apparent

---

[9] The 20th percentile is so small it can't really be seen when plotted on the scale of the other quantiles presented here. It appears as a solid horizontal line at the base of the other distributions, but you could see it better by changing to log-scaling or by plotting it on its own.

change over time is simply an artifact of our incomplete data (and an incomplete underlying data set) combined with our choice of sorting.

On the other hand, we can see features that nonetheless do suggest changes in distributions over time. There appear to be peaks and troughs in the quantile curve estimates, and we might even consider whether there may be periodic behavior in our curves, as there may be predictable times when the quantile values rise and fall due to exogenous organizational factors (perhaps a drive to close out complaints before the end of a month, or certain funding cycles that increase the number of workers available to close out complaints at certain times of the year).

Given that we now have preliminary results, the best option would be to establish some significant dates (i.e., where do we see spikes or periodic behavior?) and try to cross-reference them against any institutional facts we can establish about the rhythm of work. We should also run simulations to assess how we think the left censoring should affect early quantile estimations under different scenarios. In this way we could develop better qualitative and quantitative understanding of the unknown aspects of our system, and this information would be extremely helpful for making a final determination as to whether the distribution of resolution times is evolving over time and, if so, how.

Let's say we had taken these steps—then what? We would need to look for a methodology to compare distributions for similarity difference where only the quantiles of the distribution rather than all the sample points were available. One way we could do this would be to run simulations/bootstrap the entire process. This would lead to an answer where we could fully articulate and control the assumptions that went into the model by coding our simulations. In fact, many statistical approaches to doing such comparisons also focus on bootstrap methods.

Left Censoring, Right Censoring, and Interval Censoring

For time series analysis, and related pursuits such as survival analysis, we must be aware of how our data selection, sorting, or presentation may affect our analysis, as we determined was the case here in that the earlier time periods appear to have a much narrower distribution than the later time periods. This is likely due to the fact that we sorted the data by closed date, a form of *left censoring*, meaning that the "event of interest" has already occurred before the start of the "study." In this case, our event of interest is the start and end dates of the complaints, and we are limited by the start date of the data set. In contrast, the related concept of *right censoring* means the event of interest would occur after an event was over. Finally, *interval censoring* refers to situations in which we only have an inexact recording, such as an interval, for when an event occurred, which will also necessarily affect our analysis.

Remaining Questions

Our visualization suggests new queries. One relates to the possibility of cyclical or seasonal behavior. There seem to be periodic bumps in all the estimated quantiles. We might consider further investigating this, and we have a number of options for doing so:

- We could attempt to fit harmonics (sines and cosines) to these quantile curves and see whether a common periodicity emerged.

- We could model the quantile itself as an ARIMA or SARIMA process and look for evidence of seasonality. This would also entail preliminary steps such as exploring the ACF and PACF of the curves that we would model as time series.

- We could ask the agency running the 311 service for more information and see whether they recognize any periodic behavior induced by their organizational structure and operating procedures.

In addition to the periodic behavior, we can also see a jump in the estimated quantile values at just under the index = 70,000 location. Given that all the quantiles jumped, it seems unlikely that this is due to just a single or a handful of outliers. There are a few ways we might investigate this:

- Go back to the raw data for this time period and see what features may suggest an explanation. Was there a surge in 311 complaints? Or a surge in a particular kind of complaint that tends to take longer to resolve?

- Alternately, we could revisit the raw data to back out the approximate date of this jump in the quantiles and cross-reference this against local news, preferably with the help of someone who could point us in the right direction. Assistance from someone at the agency, or someone savvy about city government, might be most valuable. It's also possible, however, that this date could correspond to a massive event in NYC that would explain the jump, such as superstorm Sandy in 2012.

Further Improvements

We could make this an even more time-aware algorithm. Our modification to the P-square algorithm discounts prior observations, but it assumes that all observations are evenly spaced. This is manifested in the fact that the input of the next observation does not feature a timestamp, and the same discounting factor is always applied. We could craft a more flexible algorithm by using the change in timestamp to old information as compared to new information such that the discounting would depend on the change in time since the last update was measured. It would also be more accurate for our 311 data set. This is left as an exercise for the reader but only involves changing a few lines of code. Hint: the discounting factor should become a function of time.

We could also look into other ways of estimating quantiles over time, either with an online or window measurement. Because the importance of online data is growing—and especially for online big data—there is a variety of emerging research on this topic. Both statistical and machine learning approaches have dealt with this over the last years, and there are a good number of approachable academic papers that are accessible to practicing data scientists.

More Resources

Ted Dunning and Otmar Ertal, "Computing Extremely Accurate Quantiles Using t-Digests," research paper, 2019, https://perma.cc/Z2A6-H76H.
> The f-digest algorithm for extremely efficient and flexible computation of quantiles of online time series data is rapidly gaining traction as a leading technique for handling quantile estimation of online time series even for the case of nonstationary distributions. Implementations are available in a number of languages, including Python and high-performance C++ and Go variations. This approach is particularly useful because there is no need to decide in advance which quantiles interest you—rather, the entire distribution is modeled as a set of clusters from which you can infer any quantile you like.

Dana Draghicescu, Serge Guillas, and Wei Biao Wu, "Quantile Curve Estimation and Visualization for Nonstationary Time Series," Journal of Computational and Graphical Statistics 18, no. 1 (2009): 1–20, https://perma.cc/Z7T5-PSCB.
> This article illustrates several nonparametric methods to model nonstationary distributions in time series quantile estimation. It is helpful because it covers real-world data, simulation, and nonstandard distributions (such as non-Gaussian distributions). There is also sample code (unusual for a statistics academic journal article), albeit behind a paywall.

András A. Benczúr, Levente Kocsis, and Róbert Pálovics, "Online Machine Learning in Big Data Streams," research paper, 2018, https://perma.cc/9TTY-VQL3.
> This reference discusses technical approaches to a variety of common machine learning tasks related to time series. Particularly interesting is the discussion of online data processing for many kinds of machine learning and technical tips about parallelization of online tasks.

Sanjay Dasgupta, "Online and Streaming Algorithms for Clustering," lecture notes, Computer Science and Engineering, University of California San Diego, Spring 2008, https://perma.cc/V3XL-GPK2.
> While not specific to time series data, these lecture notes give a general overview of unsupervised clustering for online data. These notes are enough to get you started putting together potential solutions for a time series–specific application.

Ruofeng Wen et al., "A Multi-Horizon Quantile Recurrent Forecaster," research paper, November 2017, https://perma.cc/22AE-N7F3.

This research paper from Amazon provides an example of using quantile information from data as a way to train recurrent neural networks effectively for time series forecasting. The researchers demonstrate effective training of a neural network that can produce probabilistic assessments rather than point estimates. This paper is a good illustration of another potential use case for quantile information, an underused resource in time series analysis.

## CHAPTER 16

Time Series Packages

In the past several years, there have been a number of packages and papers released by large tech companies related to how they deal with the massive number of time series they collect as digital organizations with enormous customer bases, sophisticated logging, cutting-edge business analytics, and numerous forecasting and data processing needs. In this chapter we will discuss some of the main areas of research and development related to these ever-expanding time series data sets, specifically: forecasting at scale and anomaly detection.

Forecasting at Scale

For many large tech companies, dealing with time series is an increasingly important problem and one that arose naturally within their organizations. Over time, several of these companies responded by developing smart, automated time series packages specifically targeted to "forecasting at scale" because so many forecasts were needed in a wide variety of domains. Here's how two data scientists at Google who developed the company's automated forecasting package described the circumstances that motivated their product in a 2017 blog post (*https://perma.cc/6M7J-MWDY*) (emphasis added):

> The demand for time series forecasting at Google grew rapidly along with the company over its first decade. *Various business and engineering needs led to a multitude of forecasting approaches, most reliant on direct analyst support.* The volume and variety of the approaches, and in some cases their inconsistency, called out for an attempt to unify, automate, and extend forecasting methods, and to distribute the results via tools that could be deployed reliably across the company. That is, for an attempt to develop methods and tools that would facilitate accurate large-scale time series forecasting at Google.

There is so much relevant data and so much to forecast that it would be extremely expensive and organizationally challenging to onboard and employ enough analysts to generate every prediction of organizational interest. Instead, these packages move to a "good enough" philosophy; that is, a reasonably good forecast is far better than having no forecast while waiting for the perfect, artfully crafted one from a time series expert with domain knowledge. Next we discuss two automated forecasting frameworks in more detail, from Google and Facebook.

Google's Industrial In-house Forecasting

Google has released some information (*https://perma.cc/N3AU-5VWK*) about its in-house automated forecasting tool, which grew out of an effort led by several data scientists in the company's search infrastructure division. The task was to write a unified approach to making automated forecasts throughout the organization. Because the tasks were automated, this meant that the results had to be safe—they couldn't go too far off the rails and had to come with some estimation of uncertainty in the prediction. Moreover, because the team sought a widely applicable solution, the methodology had to address common problems in time series data sets related to humans, such as seasonality, missing data, holidays, and behaviors evolving over time.

The solution rolled out in Google comprises three interesting steps related to what we have covered in the previous chapters:

1. Automated and extensive data cleaning and smoothing

2. Temporal aggregation and geographic/conceptual disaggregation of the data

3. Combining of forecasts and simulation-based generation of uncertainty estimates

We will discuss each step in turn.

Automated and extensive data cleaning and smoothing

In the Unofficial Google Data Science Blog post mentioned earlier, two leaders of the in-house project indicated that the data cleaning and smoothing tackled a number of problems:

Imperfect data effects
- Missing data

- Outlier detection

- Level changes (such as those due to product launches or sudden but permanent changes in behavior)

- Transforming data

Imperfect data is a fact of life. Missing data can occur due to technical glitches. Outliers can have similar causes or be "real values" but are not worth including in a forecast if they are not likely to repeat themselves. Level changes (*regime changes*) can occur for a myriad of reasons, including that baseline behavior is changing drastically (evolving world), what is being measured is changing drastically (evolving product), or what is being recorded is changing drastically (evolving logging). Finally, data can come in distributions that are far from the normality or stationarity presumed in many time series models. The Google approach addresses each of these data imperfections in turn with automated methodologies to detect and "correct" these problems.

Calendar-related effects
- Yearly seasonality

- Weekly seasonality (effect of day of the week)

- Holidays

  The treatment of calendar-related effects was particularly tricky for an organization such as Google, with operations and users all over the world. The yearly seasonality would look very different in different parts of the world, particularly in opposite hemispheres with their own weather patterns and also in cultures with different underlying calendars. As the blog post pointed out, sometimes the same holiday could occur more than once in one (Gregorian) calendar year. Likewise, the "seasons" under which a particular group of people operate could shift within the Gregorian calendar year, which can often occur because the Islamic calendar has a different periodicity from the Gregorian calendar.

Temporal aggregation and geographic/conceptual disaggregation

The Google team found that weekly data worked well for most of their predictions of interest, so after they cleaned the data in the previous step, they aggregated it into weekly increments for purposes of making forecasts. In this sense, they performed temporal aggregation.

However, the team also found it was helpful to disaggregate the data, sometimes geographically, sometimes by category (such as device type), and sometimes by a combination of factors (such as geographic region by device type). The team found that in such cases, it was more effective to make forecasts for the disaggregated subseries and then reconcile these to produce a global forecast if both the subseries and global series were of interest. We discussed various ways to fit hierarchical time series in Chapter 6, and this method reflects the methodology of starting with lower-level forecasts and propagating them upwards.

Combining forecasts and simulation-based generation of uncertainty estimates

Google uses ensemble approaches, combining a number of different forecast model results to generate the final forecast. This is useful for a number of reasons. Google believes this generates a wisdom-of-experts style of forecast, drawing the benefits of many well-performing and well-justified forecast models (such as exponential smoothing, ARIMA, and others). Also, an ensemble of forecasts generates a distribution of forecasts, providing a basis to determine whether any of them are so different from the "crowd" as to be unreasonable. Finally, ensembling provides a way to quantify the uncertainty regarding the forecast, which Google also does by simulating forward propagation of the errors through time.

Ultimately, as the data science team concedes, Google's approach benefits from large-scale and highly parallel approaches appropriate to the enormous computing resources available at the company. It also benefits from establishing parallel tasks (such as simulating the error propagated forward in time many times) and automated processes with room for input from an analyst but no need for it. In this way reasonable forecasts can be generated at scale and for a variety of data sets.

While your work may not have the same advantages or need the high level of automation required at Google, there are many ideas you can incorporate into your workflow from their model even as a solo data scientist. These include:

- Building a framework or "pipeline" for cleaning your data as a baseline "good enough" version of how you would like to prepare every time series data set before modeling.
- Building an ensemble of respected forecasting models as your "go to" toolkit.

Facebook's Open Source Prophet Package

Facebook open-sourced its automated time series forecasting package, Prophet, at around the same time Google released information regarding its in-house package. In its own blog post (*https://perma.cc/V6NC-PZYJ*) about Prophet, Facebook highlighted some of the very same issues emphasized in Google's approach, particularly:

- "Human-scale" seasonalities and irregularly spaced holidays
- Level changes
- Missing data and outliers

Additionally, the Facebook package brought strength, such as:

- Capacity to handle data sets at various granularity levels, such as minute-by-minute or hourly data in addition to daily data

- Trends that indicate a nonlinear growth, such as reaching a saturation point

Facebook noted that the results with its package were often just as good as those produced by analysts. Like Google, Facebook indicated that it had found many instances of highly parallel tasks while developing its time series fitting pipeline. Facebook said that Prophet had developed into such a reliable forecaster that its forecasts were used not only internally but in outward-facing products. Also, Facebook said it had developed a work pattern of "analyst-in-the-loop" such that the automated process could be supervised and corrected where necessary, ultimately leading to a product that could assist or replace human analysis depending on the level of resources devoted to a task.

Facebook's approach is quite different from Google's; it includes three simple components:

- Yearly and weekly seasonal effects
- A custom lists of holidays
- A piecewise linear or logistic trend curve

These components are used to form an *additive regression model*. This is a nonparametric regression model, which means that no assumptions are made about the form of the underlying regression function and no linearity is imposed. The model is more flexible and interpretable than a linear regression but comes with a trade-off of higher variance (think of the bias-variance trade-off generally known in machine learning and statistics) and more problems in overfitting. This model makes sense for the general task Facebook seeks, in that complex nonlinear behaviors must be modeled in an automated way that avoids unduly taxing methodologies.

There are many advantages to Prophet, including:

- Simple API
- Open source and under active development
- Full and equal APIs in Python and R, which helps multilingual data science teams

Prophet is easy to use. We end this section with an example code snippet taken from the Quick Start guide (*https://perma.cc/9TLC-FFRM*) so you can see the minimal code needed to roll out an automated forecast. We pair our use of Prophet with use of `pageviews`, an R package that easily retrieves time series data related to Wikipedia page views. First we download some time series data:

```
## R
> library(pageviews)
> df_wiki = article_pageviews(project = "en.wikipedia",
```

```
>                                   article = "Facebook",
>                                   start = as.Date('2015-11-01'),
>                                   end = as.Date("2018-11-02"),
>                                   user_type = c("user"),
>                                   platform = c("mobile-web"))
> colnames(df_wiki)
[1] "project"   "language"  "article"   "access"    "agent"     "granularity"
[7] "date"      "views"
```

Now that we have some data at daily temporal resolution and over a few years, we can try to forecast this data with Prophet in just a few simple steps (see Figure 16-1):

```
## R
> ## we subset the data down to what we need
> ## and give it the expected column names
> df = df_wiki[, c("date", "views")]
> colnames(df) = c("ds", "y")

> ## we also log-transform the data because the data
> ## has such extreme changes in values that large
> ## values dwarf normal inter-day variation
> df$y = log(df$y)

> ## we make a 'future' data frame which
> ## includes the future dates we would like to predict
> ## we'll predict 365 days beyond our data
> m = prophet(df)
> future <- make_future_dataframe(m, periods = 365)
> tail(future)
             ds
1458 2019-10-28
1459 2019-10-29
1460 2019-10-30
1461 2019-10-31
1462 2019-11-01
1463 2019-11-02

> ## we generate the forecast for the dates of interest
> forecast <- predict(m, future)
> tail(forecast[c('ds', 'yhat', 'yhat_lower', 'yhat_upper')])
             ds     yhat yhat_lower yhat_upper
1458 2019-10-28 9.119005   8.318483   9.959014
1459 2019-10-29 9.090555   8.283542   9.982579
1460 2019-10-30 9.064916   8.251723   9.908362
1461 2019-10-31 9.066713   8.254401   9.923814
1462 2019-11-01 9.015019   8.166530   9.883218
1463 2019-11-02 9.008619   8.195123   9.862962
> ## now we have predictions for our value of interest

> ## finally we plot to see how the package did qualitatively
> plot(df$ds, df$y, col = 1, type = 'l', xlim = range(forecast$ds),
>       main = "Actual and predicted Wikipedia pageviews of 'Facebook'")
```

```
> points(forecast$ds, forecast$yhat, type = 'l', col = 2)
```

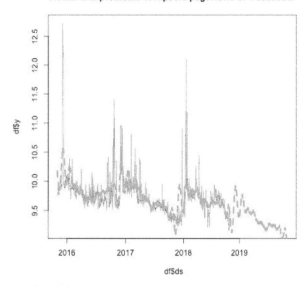

Actual and predicted Wikipedia pageviews of 'Facebook'

Figure 16-1. Plot of the Wikipedia page count data (thin solid line) and Prophet's predictions for that data (thick dashed line).

Prophet also offers the option to plot the components (trend and seasonal) that form the prediction (see Figure 16-2):

```
## R
> prophet_plot_components(m, forecast)
```

 Restrict your use of prophet to daily data. By its own description, the Prophet package was developed for, and works best for, daily data. Unfortunately, this narrow specialization means that the same techniques can be quite unreliable for data at different timescales. For this reason, you should approach this package and its associated techniques with caution when your data is not daily.

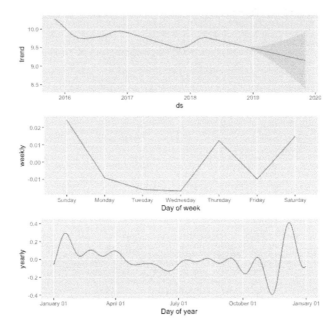

Figure 16-2. The prediction disaggregated into a trend, weekly, and yearly components. From this the prediction is formed by a sum of the components. Notice that different components are formed differently. The trend data has an underlying linear shape, while the yearly data is curvy due to its underlying Fourier series fit (read more in the Facebook blog post referenced earlier).

Over time, more automated time series open source packages and black box products are becoming available. These can be a good entry point for an organization that is new to forecasting and seeking to make a reasonable forecast. However, in the near future these packages likely will not make the best available forecast for every time series and every organization. When you can build domain knowledge and relevant organization constraints into your time series model, you will have a better result, and for now this remains the task of a human analyst until more general forecasting packages are built.

Anomaly Detection

Anomaly detection is another area where tech companies are making significant efforts and sharing them with the open source community. Anomaly detection is important in time series for a few reasons:

- It can be helpful to remove outliers when fitting models that are not sufficiently robust to such outliers.
- It can be helpful to identify outliers if we want to build a forecasting model specifically to predict the extent of such outlier events conditional on knowing they will happen.

Next we discuss the approach taken by Twitter in its open source work on anomaly detection.

Twitter's Open Source AnomalyDetection Package

Twitter open-sourced an outlier detection package, AnomalyDetection,[1] four years ago, and the package remains useful and well performing. This package implements Seasonal Hybrid ESD (Extreme Studentized Deviant), which builds a more elaborate model than Generalized ESD for identifying outliers. The Generalized ESD (*https://perma.cc/C7BV-4KGT*) test itself is built on another statistical test, the Grubbs test (*https://perma.cc/MKR5-UR3V*), which defines a statistic for testing the hypothesis that there is a single outlier in a data set. The Generalized ESD applies this test repeatedly, first to the most extreme outlier and then to successively smaller outliers, meanwhile adjusting the critical values on the test to account for multiple sequential tests. The Seasonal Hybrid ESD builds on Generalized ESD to account for seasonality in behavior via time series decomposition.

We can see the simple use of this package in the following R code. First we load some sample data provided by Twitter's package:

```
## R
> library(AnomalyDetection)
> data(raw_data)
> head(raw_data)
            timestamp   count
1 1980-09-25 14:01:00 182.478
2 1980-09-25 14:02:00 176.231
3 1980-09-25 14:03:00 183.917
4 1980-09-25 14:04:00 177.798
```

---

1 Read more at the project's GitHub repository (*https://perma.cc/RV8V-PZXU*) and on Twitter's blog (*https://perma.cc/6GPY-8VVT*)

```
5 1980-09-25 14:05:00 165.469
6 1980-09-25 14:06:00 181.878
```

Then we use Twitter's automated anomaly detection function with two sets of parameters:

1. We look for a large portion of anomalies in either the positive or negative direction.

2. We look for a small portion of anomalies in the positive range only.

These use cases are demonstrated here:

```
## R
> ## detect a high percentage of anomalies in either direction
> general_anoms = AnomalyDetectionTs(raw_data, max_anoms=0.05,
>                         direction='both')
>
> ## detect a lower percentage of anomalies only in pos direction
> high_anoms = AnomalyDetectionTs(raw_data, max_anoms=0.01,
>                         direction='pos')
```

We plot the results in both cases in Figure 16-3.

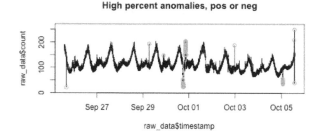

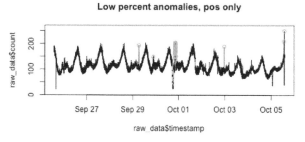

Figure 16-3. The anomalies reported back from Twitter's AnomalyDetectionTs() function both with very inclusive settings (top) and with more limited settings (bottom).

We see that a large number of the anomalies are reported all in one portion of the time series, so we also crop our plot down to this portion of the time series to get a better understanding of what is happening (Figure 16-4).

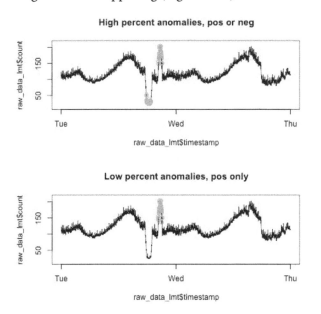

Figure 16-4. *The same identified anomalies, now focusing on the anomalies that occurred all in a cluster on the same day.*

We can better understand why these points are anomalies after taking this closer look. They deviate from the daily pattern that is otherwise present. As to why we might want to look only for positive deviations, imagine that we are building infrastructure for a high-traffic, data-intensive website. While downward/negative anomalies may be of interest, the make-or-break business anomalies for us will be the times when our infrastructure cannot handle a high-traffic opportunity. We are most interested in identifying spikes, for a number of reasons:

- If these numbers are fake, we'd like to weed them out so we can know the true ceiling of realistic usage. Anomalously high numbers will drive us to buy computing resources we don't need, which isn't true of anomalously low numbers. Using anomaly detection can help us clean our data as a preprocessing step.

- If computing equipment is cheap, we'd rather just buy more even to accommodate these anomalies. If we can label anomalies, that is the first step toward generating labeled data to attempt to predict these anomalies. However, by definition anomalies are difficult to predict, so you should not have high expectations of such efforts most of the time!

There are many parameters (*https://perma.cc/BR4K-R8GL*) you can use with Twitter's automated anomaly detection, and it's a good tool to have available when you're exploring a new data set both for cleaning and modeling data.[2]

Other Time Series Packages

In this chapter, we have largely focused on widely used packages developed by some of the largest tech companies in conjunction with the enormous data sets and related forecasts these companies produce as part of their core business operations. However, these companies are far from being the main or most senior providers of time series packages. There is an enormous ecosystem of time series packages dedicated to:

- Time series storage and infrastructure
- Time series data sets
- Breakpoint detection
- Forecasting
- Frequency domain analysis[3]
- Nonlinear time series
- Automated time series forecasting

This is not an exhaustive list. There are packages for everything, literally dozens or even hundreds of time series packages. The most extensive listing of open source packages (*https://perma.cc/HWY6-W2VU*) is maintained by Professor Rob Hyndman on R's official CRAN repository web page. It is worth taking a look at this listing, both to find specific packages that might fit your analytical needs for a particular project, but also more generally to educate yourself about the range of time series analysis methods available and being actively deployed in the community. Much like this book, that page offers an overview and delineation of the tasks associated with time series data. In Python, there is no similarly extensive and consolidated listing of time series modules, but data scientist Max Christ has compiled a very helpful listing (*https://perma.cc/GEQ3-Q54X*) of them.

---

2 It is worth noting that Twitter also released a level change detection package, BreakoutDetection, at the same time that it released the AnomalyDetection package. In the case of BreakoutDetection, the package serves to identify locations where there has been a level shift in a time series. This package is similarly accessible and easy to use, though it has not gained as much of a following and is not as much of a standout as the AnomalyDetection package. There are many alternative breakpoint detection packages that have been more extensively tested and deployed.

3 Frequency domain analysis is mentioned only briefly in this book, but it remains an important area of time series analysis and widely used in some disciplines, such as physics and climatology.

More Resources

StatsNewbie123, "Is it Possible to Automate Time Series Forecasting?" post on Cross Validated, December 6, 2019, https://perma.cc/E3C4-RL4L.

This recent StackExchange post asks whether it's possible to automate time series forecasting for any time series. There are two very useful and detailed responses with an overview of automated packages as well as a discussion of all the challenges inherent in such a task.

CausalImpact, Google's open source package for causal inference, https://perma.cc/Y72Z-2SFD.

This open source Google package is built on top of another Google release, `bsts`, the Bayesian Structural Time Series package, which we used in Chapter 7. The CausalImpact package uses `bsts` to fit models and then, once fit, to construct counterfactual control examples to assess causality and effect size in time series data. The package's GitHub repo contains links to relevant research papers and a helpful video overview by one of the package's creators. It's also worth checking out the associated research paper (*https://perma.cc/Q8K9-ZP7N*).

Murray Stokely, Farzan Rohani, and Eric Tassone, "Large-Scale Parallel Statistical Forecasting Computations in R," in JSM Proceedings, Section on Physical and Engineering Sciences (Alexandria, VA: American Statistical Association, 2011), https://perma.cc/25D2-RVVA.

This document offers a high-level, detailed explanation of how R packages were written for Google's in-house time series forecasting to do time series in a highly parallel and scalable way. This is good reading not just for the details related to forecasting, but also for better understanding how to build time series data pipelines for large organizations with large quantities and varying types of time series data.

Danilo Poccia, "Amazon Forecast: Time Series Forecasting Made Easy," AWS News Blog, November 28, 2018, https://perma.cc/Y2PE-EUDV.

One recent automated time series model we didn't cover is Amazon's new forecasting service, Amazon Forecast. It's not open source, but there are many promising reviews. It offers a way to use the models Amazon developed with its retail expertise to help companies make business forecasts. While it's a paid service, you can try the Free Tier, which offers fairly generous options. The service was designed to emphasize both accuracy and usability, and it is a good alternative for those organizations looking for a "good enough" model for high-volume forecasting situations. Amazon's package uses a mixed approach of deep learning and traditional statistical models, similar to how the simple LSTNET model from Chapter 10 combined a deep learning model with an autoregressive component. It's worth reading about Amazon's signature neural network architecture for forecasting, DeepAR (*https://perma.cc/DNF9-LJKC*).

## CHAPTER 17

Forecasts About Forecasting

There are many good quotes about the hopelessness of predicting the future, and yet I can't help wanting to conclude this book with some thoughts about what's coming.

Forecasting as a Service

Because time series forecasting has fewer expert practitioners than other areas of data science, there has been a drive to develop time series analysis and forecasting as a service that can be easily packaged and rolled out in an efficient way. For example, and as noted in Chapter 16, Amazon recently rolled out a time series prediction service, and it's not the only company to do so. The company's model seems deliberately general, and it frames forecasting as just one step in a data pipeline (see Figure 17-1).

These forecasting-as-a-service modeling endeavors aim for a good enough general model that can accommodate a variety of fields without making terribly inaccurate forecasts. Most of them describe their models as using a mix of deep learning and traditional statistical models. However, because the service is ultimately a black box, it will be difficult to understand what could make forecasts go wrong or even to retrospectively investigate how they can be improved. This means there is a reasonably high quality level for the forecasts but probably a performance ceiling as well.

This service can be valuable for companies that need many forecasts but do not have the personnel available to generate them individually. However, for companies that have substantial amounts of historical data where more general heuristics and "laws" of their data could be discovered, it's likely that an analyst could outperform these algorithms given familiarity with the domain.

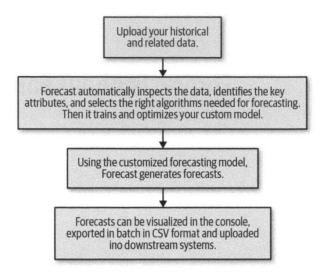

Figure 17-1. An example time series forecasting-as-a-service pipeline. Such services are being offered by a variety of small startups and tech behemoths, most notably Amazon, which has rolled out a suite of products tailored to forecasting time series data automatically and at scale via a suite of products driven by both deep learning and statistical models.

Notice that a fair amount of the product that is being sold in the area of forecasting-as-a-service also has to do with good visualizations of the forecasts and pipeline utilities to easily revise forecasts, change the forecasting frequency, and so on. Even if your organization will ultimately build out its own forecasting analyses, it can be helpful to see what is emerging as an industry standard.

Deep Learning Enhances Probabilistic Possibilities

In the past few years, many of the largest tech companies have made some information public about how they do their own forecasting for their most important services. In this case we are not talking about the need for many parallel forecasts of the large number of metrics that affect the company's business, but rather core concerns. In these cases, where the quality of the forecast is paramount, companies are often giving indications that they are using deep learning with a probabilistic component.

For example, Uber has blogged (*https://perma.cc/3W54-BK8C*) about forecasting demand for car rides, and Amazon has developed a well-regarded autoregressive recurrent neural network (*https://perma.cc/UL77-BY3T*), inspired by statistical thinking for making predictions about product demands. The more that researchers are able to integrate statistical methodologies, such as the injection of priors for domain knowledge and the quantification of uncertainty, the less reason there will be to seek out statistical models when a deep learning model can offer the strengths of both statistics and deep learning.

However, making reasonably interpretable deep learning models—so that we can know just how "wrong" or extreme a forecast can be—remains a difficult endeavor, so it's unlikely that traditional statistical models, with greater theoretical understanding and mechanistic clarity, will be discarded. For critical forecasts, where health and safety may be at risk, people reasonably may continue to rely on what has worked for decades until more transparent and inspectable methods can be developed for machine learning forecasts.

Increasing Importance of Machine Learning Rather Than Statistics

Empirically there seems to be less and less use of proper statistics in modeling data and generating predictions. Do not despair: the field of statistics continues to thrive and answer interesting questions that are related to statistics. And yet—particularly for low-stakes forecasts that merely need to be good enough—machine learning techniques and results-oriented statistical methods, rather than fancy theories and closed-form solutions or proofs of convergence, are winning out in actual deployment and real-world use cases.

From a practitioner's perspective, this is a good thing. If you happily left your problem sets behind long ago with no desire to prove things, you needn't fear a return of proper proofs and the like. On the other hand, this is a worrying trend as these technologies make their way into more and more fundamental aspects of life. I don't mind surfing a retailer's website that uses machine learning to guesstimate my likely future actions as a buyer. But I'd like to know that the time series predictions that modeled my health outcomes or my child's academic progressions were more thorough and were statistically validated, because a biased model could really hurt someone in these core areas.

For now, the leaders in time series thinking for industrial purposes are working in low-stakes areas. For problems such as predicting revenues from an advertising campaign or a social media product rollout, it's not important whether the forecasts are fully validated. As more fundamental aspects of the care and feeding of human beings come into the modeling domain, let's hope that statistics plays a more fundamental role in high-stakes forecasts.

Increasing Combination of Statistical and Machine Learning Methodologies

A number of indications point toward combining machine learning and statistical methodologies[1] rather than simply searching for the "best" method for forecasting. This is an extension of increasing acceptance and use of ensemble methods for forecasting, a phenomenon we have discussed throughout the book.

An example of an extraordinarily robust test with many real-world data sets is the recent M4 competition (*https://perma.cc/68AC-BKN7*), a time series competition measuring forecasting accuracy on 100,00 time series data sets (*https://perma.cc/76BQ-SZW9*), as mentioned briefly in Chapters 2 and 9. The winning entry to this competition combined elements of a statistical model and a neural network. Likewise, the runner-up incorporated both machine learning and statistics, in this case by using an ensemble of statistical models but then using a gradient boosted tree (XGBoost) to choose the relative weightings of each model in the ensemble. In this example we see two distinctive ways machine learning and statistical approaches can be combined: either as alternative models assembled together (as in the case of the winning entry) or with one method determining how to set the metaparameters of the other method (as in the case of the runner-up). A comprehensive and highly accessible summary of the competition results (*https://perma.cc/T8WW-6MDN*) was subsequently published in the *International Journal of Forecasting*.

As such combinations gain traction, we will likely see research develop in the area of determining what problems are most amenable to combining statistical and machine learning models, as well as best practices emerging for tuning the performance of these models and selecting architectures. We expect to see the same refinement as in other complex architectures, such as neural networks, whereby standard design paradigms emerge over time with known strengths, weaknesses, and training techniques.

More Forecasts for Everyday Life

More consumer-facing companies, such as mobile health and wellness applications, have rolled out (*https://perma.cc/QXT9-4B8T*) or been asked (*https://perma.cc/M8W7-EDCE*) to roll out, personalized predictions. As people grow more aware of just how much data their applications store about them and others, they are looking to take advantage by getting tailored forecasts for metrics such as health and fitness goals. Similarly, people often go looking for forecasts for everything from future real

---

[1] Special thanks to technical reviewer Rob Hyndman who suggested this topic (in addition to many other helpful suggestions throughout the book).

estate values (*https://perma.cc/R5WR-T7XP*) (difficult to predict) to the likely arrival dates (*https://perma.cc/5LTM-WRPB*) of migratory bird species.

More products will be explicitly driven by the demand to make forecasts, be it on esoteric subjects or for individual metrics. This means that more forecasting pipelines will be integrated into places where they have not been too likely before, such as mobile applications and websites for casual readers rather than industry specialists. The more common this becomes, the more likely that time series jargon will be part of everyday speech. Hopefully people will also acquire enough time series education to understand the limits and assumptions of forecasts so that they do not rely too much on these products."

Index

A

abstract data type, 131
accuracy, performance versus, 357
ACF (see autocorrelation function)
actuarial tables, 3
additive regression model, 453
agent-based modeling, 134
aggregation
    Google's in-house forecasting framework,
      451
    of government data, 431
AirPassengers data set, 82, 101
    2D visualizations, 105-113
    3D visualizations, 113-116
    smoothing data, 57
Akaike information criterion (AIC), 173
alpha parameter (smoothing factor), 58
AlphaVantage, 404
Amazon time series prediction service, 461, 463
AnomalyDetection (Twitter package), 457-460
AR(p) model, 167
    choosing parameters for, 171-176
    forecasting with, 176-180
    as moving window function, 177
ARCH (autoregressive conditional heteroske-
    dasticity), 201
ARIMA (see autoregressive integrated moving
    average)
Aristotle, 6
assumptions, in machine learning models, 291
astronomy, 9
attention mechanism, for feed forward neural
    network, 321-323
attention weights, 322

Augmented Dickey–Fuller (ADF) test, 84
autocorrelation, 91
autocorrelation function (ACF), 91-92
    AR(p) model and, 171
    of ARIMA residuals, 190-193
    MA/AR processes versus, 183
    of nonstationary data, 100-101
    PACF versus, 94-101
    of stationary data, 95
autoencoder, 334
automatic differentiation, 298
autoregressive (AR) model, 12, 166-181
    (see also vector autoregression (VAR))
    ACF/PACF versus, 183
    choosing parameters for AR(p) model,
      171-176
    forecasting many steps into the future,
      178-180
    forecasting one time step ahead, 176-178
    forecasting with an AR(p) process, 176-180
    MA process as infinite order AR process,
      181
    using algebra to understand constraints on
      AR processes, 166-171
autoregressive integrated moving average
    (ARIMA) model, 186-196
    about, 186-189
    ARMA model and, 187-189
    automated model fitting, 194
    flu prediction example, 371-376
    manual model fitting, 190-194
    parameter selection, 189-196
    SARIMA, 201

context vector, 322
continuous glucose monitor (CGM) dataset
    (see blood glucose level prediction)
convolution, defined, 324
convolutional neural networks (CNNs),
    324-329
    alternative models, 327-329
    causal convolutions, 327-328
    converting a time series into pictures, 329
    learning resources, 342
    simple model, 325-327
correlation
    cointegration, 103
    in a deterministic system, 92
    partial autocorrelation function, 94-101
    spurious, 102-103
CSV files, performance and, 361
cuDNN interface, 332
custom rolling functions, 91
CVAR (cointegrated vector autoregression)
    model, 201
cyclical time series, 63

D

data
    assembling a time series data collection
        from disparate sources, 27-32
    characteristics, 150
    cleaning of, 40-69
    discarding, 435
    found time series construction, 33-35
    government data sets, 23
    prepared data sets, 18-24
    retrofitting a time series data collection
        from a collection of tables, 26-35
    simulating (see simulating time series data)
    storing (see storing temporal data)
    symbols versus, 295
    timestamping difficulties, 35-39
    where to find, 18-26
data imputation (see missing data, methods for
    handling)
data sets (see specific data sets)
data storage formats, computational perfor-
    mance and, 361
    (see also storing temporal data)
data.table package, 42, 43
databases
    characteristics of time series data, 150

general-use NoSQL databases, 154-156
    InfluxDB, 153
    learning resources, 160
    popular time series database and file solu-
        tions, 152-156
    Prometheus, 154
    SQL versus NoSQL, 148-156
    temporal data storage with, 148-156
    time-series-specific databases and related
        monitoring tools, 152-154
daylight savings time, 66
decision trees, 264-272
    classification versus regression, 271
    code example, 268-271
    gradient boosted trees, 267
    random forests, 265
decomposition, 62
deep learning, 289-342
    assumptions, 291
    automatic differentiation, 298
    basics, 289-292
    CNNs, 324-329
    combination architectures, 335-339
    concepts, 292-294
    data, symbols, operations, layers, and
        graphs, 294-298
    enhancement of probabilistic possibilities,
        464
    feed forward neural network, 318-323
    learning resources, 340-342
    machine learning versus, 290
    multitask learning applied to time series,
        354
    popular frameworks, 295
    preprocessing financial data for, 410-416
    RNNs, 330-334
    softmax function, 322
    time series data simulations, 141
    for time series, 289-342
    training pipeline for, 298-318
demography, origins of, 3
Dennis, Richard, 9
depmixS4 package, 224-229
diabetes (see blood glucose level prediction)
Dickey–Fuller test, 84
differencing, 186
differentiation, automatic, 298
dilated causal convolution, 327
diluting the forecasting task, 414

imperative programming style, 294
imputation, 40
    (see also missing data, methods for han-
        dling)
    comparison of methodologies, 50
    health care data, 393
    moving average for, 46-48
    preparing a data set to test methodologies,
        41-44
    using data set's mean for, 48
influenza (see flu prediction)
InfluxDB, 153
informed prediction, 185
interpolation, 40, 49
interval censoring, 445
Ising model, 134, 136
iterators, for training pipeline, 308

J

Java, improving computational performance by
    using, 364

K

Kalman filter, 58, 210-217
    code for, 212-217
    learning resources, 235
    mathematics overview, 210-212
kernel density estimate, 262
Kwiatkowski–Phillips–Schmidt– Shin (KPSS)
    test, 85

L

lag (backshift) operator, 182
lag(), 81
layers, in deep learning, 296
left censoring, 445
legal issues, in data storage, 148
life (actuarial) tables, 3
LightGBM, 271
linear Gaussian state space models, 210
    (see also Kalman filter)
linear interpolation, 49
    moving average versus, 50
    when not to use, 50
    when to use, 50
linear regression, 163
Linux command-line sort tool, 432
Ljung–Box test, 176, 184

local time, UTC time versus, 37
LOESS (locally estimated scatter plot smooth-
    ing), 58, 62
Lomb–Scargle periodogram, 247
long short-term memory (LSTM), 331
longest common subsequence, 283
lookahead
    backward fill and, 46
    defined, 28
    model validation and, 355
    preventing, 67

M

M4 competition, 466
MA(q) process
    forecasting, 184-186
    parameter selection for, 182-184
machine learning, 290
    (see also deep learning)
    assumptions, 291
    clustering, 272-287
    deep learning versus, 290
    future combinations of statistical methodol-
        ogies with, 466
    learning resources for, 287
    statistics versus, 465
    time series analysis origins, 13
    time series classification, 260-272
    for time series, 259-288
    visualization and, 262
magnet, modeling of, 134-140
Markov Chain Monte Carlo (MCMC) simula-
    tion, 135, 232
Markov process, 218
    (see also Hidden Markov Models (HMMs))
matrix multiplication, 197
Mcomp, 24
mechanical trading, 9
medicine, 2
    (see also healthcare applications)
    medical instruments, 4-6
    population health, 2-4
    as time series problem, 2
memoization, 223
meteorology (see weather forecasting)
missing data, methods for handling, 40-52
    backward fill, 46
    data.table package in R, 42
    forward fill, 44-46

transactional data, 150
transformations
    assumptions with, 86
    making time series stationary with, 85
ts object, 89
tsfeatures (R package), 248
tsfresh (Python module), 244, 252-255
Twitter (AnomalyDetection package), 457-460

U

UCI Machine Learning Repository, 18-21
UEA and UCR Time Series Classification
    Repository, 21
unbiased estimators, 165
uncertainty, estimating, 350-353
unemployment data, 41-44
unit root, 83
univariate time series, 21
universal (UTC) time, 37
upsampling, 53
    for aligning inputs in model, 54
    converting irregularly sampled series to reg-
        ularly timed series, 54
    defined, 52
    of government data, 431
    interpolation based on variable's usual
        behavior, 54

V

vanishing gradients, 332
variables, exogenous versus endogenous, 197
vector autoregression (VAR), 196-201
visualizations

1D (Gantt charts), 104
2D, 105-113
3D, 113-116
converting a time series into pictures, 329
exploratory data analysis, 104-116
learning resources, 118
machine learning and, 262
Viterbi algorithm, 222
volatility, 410

W

weak stationarity, 170
weather forecasting, 6
window functions
    custom rolling functions, 91
    expanding windows, 89
    exploratory data analysis and, 86-91
    rolling windows, 86-88
Wold's theorem, 187

X

Xarray, 159
XGBoost, 267-272, 397, 466
xts objects, 89

Y

Yahoo Finance, 404
Yule, Udny, 12, 102

Z

zoo objects, 88

About the Author

Aileen Nielsen is an New York City–based software engineer and data analyst. She has worked on time series and other data in a variety of disciplines, from a healthcare startup to a political campaign, from a physics research lab to a financial trading firm. She currently develops neural networks for forecasting applications.

Colophon

The animal on the cover of *Practical Time Series Analysis* is a Bluefaced Leicester sheep (*Ovis aries*). The Bluefaced Leicester is a British breed also known as the *Dishley Leicester*; it was first bred in Dishley, Leicestershire, in the eighteenth century. Today, farmers throughout the UK and Canada breed this specialized strain of sheep.

Bluefaced Leicester sheep have white wool that makes a fine, shiny fleece and quality handspun yarn. Their brown eyes are prominent against their blue-gray faces. The sheep are hardy with long backs, strong shoulders and necks, broad muzzles, and dark hooves. They have no wool on their legs or faces.

When full-grown, a Bluefaced Leicester can weigh between 200 and 240 pounds. The average height from hoof to shoulder is 36 inches for rams and 33 inches for ewes. Like most sheep, the Bluefaced Leicester graze on grasses, weeds, and flowers. They graze in flocks of 20 ewes, on average.

Bluefaced Leicester sheep are highly sought after because the rams have a knack for producing what the industry call *mules*. Mules are ewes with high reproduction rates that breeders cross with other breeds to create lambs for the market.

Many of the animals on O'Reilly covers are endangered; all of them are important to the world.

The cover illustration is by Karen Montgomery, based on a black-and-white engraving from *Meyers Kleines Lexicon*. The cover fonts are Gilroy Semibold and Guardian Sans. The text font is Adobe Minion Pro; the heading font is Adobe Myriad Condensed; and the code font is Dalton Maag's Ubuntu Mono.

O'REILLY®

There's much more where this came from.

Experience books, videos, live online training courses, and more from O'Reilly and our 200+ partners—all in one place.

Learn more at oreilly.com/online-learning

Milton Keynes UK
Ingram Content Group UK Ltd.
UKHW010015200824
447150UK00002B/3